Textbook of Microbiology
for Paramedicals

Textbook of Microbiology for Paramedicals

Logeswari Selvaraj MD PhD
Professor and Head
Department of Microbiology
Theni Medical College
Theni, Tamil Nadu
India

JAYPEE BROTHERS MEDICAL PUBLISHERS (P) LTD

New Delhi • Ahmedabad • Bengaluru • Chennai • Hyderabad
Kochi • Kolkata • Lucknow • Mumbai • Nagpur

Published by

Jitendar P Vij

Jaypee Brothers Medical Publishers (P) Ltd

B-3 EMCA House, 23/23B Ansari Road, Daryaganj, **New Delhi** 110 002
INDIA   Phones: +91-11-23272143, +91-11-23272703, +91-11-23282021
+91-11-23245672, Rel: +91-11-32558559   Fax: +91-11-23276490
+91-11-23245683, e-mail: jaypee@jaypeebrothers.com
Visit our website: www.jaypeebrothers.com

Branches

❏ 2/B, Akruti Society, Jodhpur Gam Road Satellite,
    Ahmedabad 380 015, Phones: +91-79-26926233, Rel: +91-79-32988717
    Fax: +91-79-26927094 e-mail: ahmedabad@jaypeebrothers.com
❏ 202 Batavia Chambers, 8 Kumara Krupa Road, Kumara Park East
    Bengaluru 560 001   Phones: +91-80-22285971, +91-80-22382956
    +91-80-22372664, Rel: +91-80-32714073   Fax: +91-80-22281761
    e-mail: bangalore@jaypeebrothers.com
❏ 282 IIIrd Floor, Khaleel Shirazi Estate, Fountain Plaza, Pantheon Road
    Chennai 600 008, Phones: +91-44-28193265, +91-44-28194897
    Rel: +91-44-32972089   Fax: +91-44-28193231
    e-mail:chennai@jaypeebrothers.com
❏ 4-2-1067/1-3, 1st Floor, Balaji Building, Ramkote, Cross Road
    Hyderabad 500 095   Phones: +91-40-66610020, +91-40-24758498
    Rel: +91-40-32940929   Fax:+91-40-24758499
    e-mail: hyderabad@jaypeebrothers.com
❏ Kuruvi Building, 1st Floor, Plot/Door No. 41/3098, B & B1, St. Vincent Road
    Kochi 682 018 Kerala   Phones: +91-484-4036109, +91-484-2395739
    Fax: +91-484-2395740   e-mail: kochi@jaypeebrothers.com
❏ 1-A Indian Mirror Street, Wellington Square
    Kolkata 700 013, Phones: +91-33-22651926, +91-33-22276404
    +91-33-22276415, Rel: +91-33-32901926   Fax: +91-33-22656075
    e-mail: kolkata@jaypeebrothers.com
❏ Lekhraj Market III, B-2, Sector-4, Faizabad Road, Indira Nagar
    Lucknow 226 016   Phones: +91-522-3040553, +91-522-3040554
    e-mail: lucknow@jaypeebrothers.com
❏ 106 Amit Industrial Estate, 61 Dr SS Rao Road, Near MGM Hospital, Parel
    Mumbai 400 012   Phones: +91-22-24124863, +91-22-24104532
    Rel: +91-22-32926896,   Fax: +91-22-24160828
    e-mail: mumbai@jaypeebrothers.com
❏ "KAMALPUSHPA" 38, Reshimbag, Opp. Mohota Science College
    Umred Road, **Nagpur** 440 009 Phone: Rel: +91-712-3245220
    Fax: +91-712-2704275 e-mail: nagpur@jaypeebrothers.com

Textbook of Microbiology for Paramedicals

> This book has been published in good faith that the material provided by author is original. Every effort is made to ensure accuracy of material, but the publisher, printer and author will not be held responsible for any inadvertent error(s). In case of any dispute, all legal matters are to be settled under Delhi jurisdiction only.

First Edition: **2008**

ISBN 81-8448-199-3

Typeset at   JPBMP typesetting unit

Printed at   Gopsons Papers Ltd., Noida

Dedicated to

My Parents

Mrs Thulasi Selvaraj

and

Mr M Selvaraj (Advocate)

Preface

The positive approach of a teacher to teach a student is to standardise them right from the foundation so as to professionalise them appropriately.

All efforts are given to condense the recent developments on medical microbiology in recent years.

The portion of clinical microbiology is thoroughly updated with latest clinical results and it will be a good companion to all paramedicals and laboratory technologists.

The language of presentation of this book is easy, simple and understandable, which will be of great help to all students of microbiology.

Logeswari Selvaraj

Acknowledgements

The positive approach of a teacher to teach a student is to standardise them right from the foundation so as to professionalise them appropriately.

With such an aim, let this text be commonly accepted by my students with scientific assessment and knowledge.

My thanks goes to my school and college, namely Seventh-Day Adventist School and Chenglepet Medical College, Chenglepet.

My utmost thanks to Mr KC Balaraman, my philosopher and guide right from my childhood with whose love I still stay inspired to serve the society with good knowledge. I thank Dr Swamy Jyothi, Professor of Anatomy who has been and still is my godmother.

Dr P Arunachalam, Professor of Cardiology. MMC, Dr Manoharan, Professor of Cardiology MMC, Dr Manoharan, Professor of Cardiology CMC and last but not the least—Late Dr Narendra Babu Professor of General Medicine, whose recent demise has brought a vaccum in all our hearts–a person for perfection and simplicity.

My sincere thanks to M/s Jaypee Brothers Medical Publishers (P) Ltd, and to Mr R Jayanandan, Sr Author co-ordinator, Chennai branch who is always inspiring me to write more books.

My thanks also goes to my students especially, Dr Caleb and Dr Dhanasekar who were of great inspiration in all my career of teaching.

Wishing all my students
The very best in life

Contents

Section 4: Virology

Section 5: Parasitology

Section 6: Mycology

Section 7: Medical Entomology

Section 8: Laboratory Investigation of Microbial Infections

SECTION

1

GENERAL
MICROBIOLOGY

1

Introduction to Microbiology

Microbiology deals with the study of microscopic organisms. The diagnostic microbiology laboratory involves in the identification of infectious agents.

The procedures involved in the microbiological analysis varies from one branch to another. For example, examination of parasites are totally different from the routine diagnostic procedures employed in bacteriology and virology.

Identification of the infectious agent is the principle function of the diagnostic microbiology laboratory, moreover, the diagnostic laboratory also provides guidance in therapeutic management.

For example, in case of bacterial infections the laboratory provides information regarding the most effective antimicrobial agent and its dosage to be used for the particular patient.

HISTORY OF MICROBIOLOGY

As microbes are invisible to the unaided eye, definitive knowledge about them had to await the development of microscopes. The credit for having first observed and reported bacteria belongs to Antony van Leeuwenhoek, whose hobby was grinding lenses and observing diverse materials through them. Leeuwenhoek coined the word of 'little animalcules' as he called them, represented only curiosity of nature.

The development of microbiology as a scientific discipline dates back to Louis Pasteur (1822-95). Though trained as a chemist, his studies on fermentation led him to take an interest in microoganisms. The basic principles and techniques of microbiology were evolved by Pasteur during his enquiry into the origin of microbes, he introduced techniques of sterilisation and developed the steam steriliser, hot-air oven and autoclave. He also established the differing growth needs of different bacteria.

He attenuated cultures of the anthrax bacillus by incubation at high temperature (42-43°C) and proved that inoculation of such cultures in animals induced specific protection against anthrax. It was Pasteur who coined the term vaccine for prophylactic preparations to commemorate the first of such preparations, namely cowpox, employed by Jenner for protection against smallpox.

Pasteur and Koch had many disciples who discovered the causative agents of several bacterial infections. In 1874 Hansen described the leprosy bacillus; in 1879 Neisser described the *Gonococcus*; in 1881 Ogston discovered the *Staphylococcus*; in 1884 Loeffler observed the tetanus bacillus in soil; in 1886 Frenkel described the *Pneumococcus*; in 1887 Bruce identified the causative agent of Malta fever; in 1905 Schaudinn and Hoffmann discovered the spirochaete of syphilis.

Roux and Yersin (1888) identified a new mechanism of pathogenesis when they discovered the diphtheria toxin. Similar toxins were identified in tetanus and some other bacteria. The toxins were found to be specifically neutralised by these antitoxins.

The causative agents of various infectious disease were being reported by different investigators in such profusion that it was necessary to introduce criteria for proving. These criteria, first indicated by Henle, were enunciated by Koch and are known as "Koch's postulates".

1. The bacterium should be constantly associated with the lesions of the disease.
2. It should be possible to isolate the bacterium in pure culture from the lesions.
3. Inoculation of such pure culture into suitable laboratory animal should reproduce the lesions of the disease.
4. It should be possible to reisolate the bacterium in pure culture from the lesions produced in the experimental animals.
5. The specific antibodies to the bacterium should be demonstrable in the serum of patient suffering from the disease.

The next major discovery in immunity was Pasteur's development of vaccines for chicken-cholera, anthrax and rabies. Pfeiffer (1893) demonstrated bactericidal effect *in vivo* by injecting live cholera vibrios intraperitoneally in guinea pigs previously injected with killed vibrios. The vibrios were shown to undergo lysis.

Apart from the obvious benefits such as specific methods of diagnosis, prevention and control of infectious disease, medical microbiology has contributed to scientific knowledge and human welfare in many other ways.

Microscopy

The study of the morphology of very small organisms is of importance to the microbiologist that a microscope is essential. Microscopy consists of optical principle and construction of the microscope.

Microscopes designed by different manufacturers differ greatly in the details of their construction and mode of operation, so that the manufacturer's description and instructions for an instrument must be consulted and followed.

Monocular microscopes have a single eyepiece. They are convenient for use by beginners, who may have difficulty in focusing the images from a binocular microscope. A monocular microscope is required for photography, but a binocular instrument may be fitted temporarily with and interchangeable monocular head for this purpose.

Binocular microscopes have two eyepieces. They are to be recommended where much microscopic work has to be done, e.g. in routine examinations, by the use of both eyes, much eye strain and fatigue is avoided. Such an instrument is illustrated in Figure 2.1. The upright *stand* (5) rests on a heavy *foot* and bears at its upper end an inclined binocular head with two *eyepieces* (1), and a revolving. *nosepiece* (3) bearing several *objective.*

The middle of the stand is a horizontal platform, or *stage (6),* with a central hole over which the slide with specimen is held by clips. For focusing, the stage is racked upwards or downwards by turning the heads *of coarse* and *fine focusing adjustments* (14, 15). For searching different areas of the specimen, slide can be moved in two directions by turning *mechanical stage adjustments* (11). A built-in *lamp* (16) in the foot of the microscope passes a beam of light upwards through a *field diaphragm* (iris) (13). The beam is focused on to the specimen by a *substage condenser* (8). In the binocular head (2) the rays of light from the objective are divided equally. Each half of the rays is directed into its appropriate, eyepiece by means of prisms.

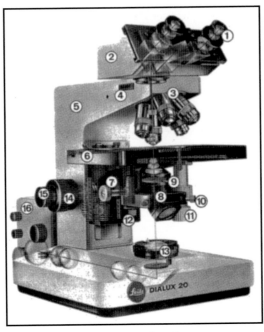

Fig. 2.1: Binocular microscope

MAGNIFICATION

The purpose of the microscope is to produce an enlarged, well defined image of objects too small to be observed with the naked eye. The degree of enlargement is the magnification.

It is perfectly possible to design an optical system that will give enormous magnifications, e.g. × 1, 000,000, but after a certain degree of magnification the sharpness of the image is lost, the larger images are increasingly blurred and no further detail is revealed. Excessive magnification of this type is known as empty masticator and is valueless to the microbiologist. The limit of useful magnification, up to which increasing detail is observed, is set by the resolving power, the limit of resolution of the microscope lenses, which itself is subject to a limit imposed by the wavelength of the light rays used. With the most powerful lenses, including the 2 mm oil-immersion objective, the limit of resolution is about 0.2 μm and the greatest useful magnification × 1000 or a little higher.

Magnification is effected in two stages, the first by the objective lenses and the second by the eyepiece lenses. To enable their use at different magnification, microscopes have revolving nosepiece bearing several objectives with different focal lengths and thus different magnifying powers.

The three objectives most commonly used in microbiology are (i) a low-power 'dry' objective with focal length 16 mm and magnification × 10, (ii) a high- power 'dry' objective focal length 4 mm and magnification × 40, and (iii) an oil immersion objective with focal length '2 mm' and magnification × 100. When these objectives are used with eyepieces magnifying x 10, the overall magnifications of the microscope are, respectively × 100, × 400 and × 1000.

Numerical Aperture

The numeral aperture (NA) may be defined simply as the ratio of the diameter of the lens to its focal length, but is expressed more precisely by the formula:

$$NA = N\ Sin\ U$$

Resolution

The limit of useful magnification of a microscope is set by its resolving power, i.e. its ability to reveal closely adjacent structural details as separate and distinct.

Definition

Spherical aberration is due to the rays passing through the edge of the lens not being brought to the same focus as those passing nearer the centre. *Chromatic aberration* occurs because white light traversing a lens is separated into its component colours of different wavelength which are refracted to different extents and not recombined at the same focus. The result may be a hazy image fringed with the colour of the spectrum.

Eyepieces

Functions of the eyepiece are to form a Lead and virtual image of the real image by the objective.

Condensers

The condenser has the function of focusing light on the object. It is mounted below the stage with a rack.

Centration of the Condensers

After the microscope and illuminant have been set up, close the condenser iris diaphragm to its limits.

METHOD OF USING LIGHT MICROSCOPE

1. Place the microscope at a convenient position on the bench. The eyepieces, i.e. microscope should be at the level with and close to observer's eye while

he/she is sitting in a upright position. The forearms or elbows be rested on the bench in position so that hands conveniently access the focusing adjustment.

2. Check to ensure that the objectives and eyepieces are free from dust and immersion oil. If they are not, clean them with a fresh lens tissue. Use only benzol or xylol to remove hardened oil.

3. Rack up fully the substage condenser. When the condenser adjustment is racked to its uppermost position, the top of the condenser is too high.

4. Fully open the substage and lamp irises. Switch on the lamp at low intensity and then increase the intensity.

5. First view the specimen with a *low-power* dry objective and use the *coarse* focusing adjustment to focus the specimen on the slide.

6. Adjust the distance between the eyepieces so that a single field is seen. Focus the microscope on the object.

7. Before using an oil-immersion (e.g. 2 mm) objective, rack down the stage to give adequate clearance between the slide and the objective, and place a moderately large drop of immersion oil on the middle of the specimen central to the light beam focused from the condenser. Take care not to spill oil on the microscope.

8. Apply the eyes to the eyepieces and, while watching slowly and focus the objective *away* from the slide by lowering the stage with the coarse focusing adjustment. Stop as soon as focus is reached and the specimen is seen. Then and only then, obtain and maintain exact focus by use of the fine adjustment.

9. After use wipe the oil from the objective with a clean tissue and clean off any oil that has been spilt on the stage or elsewhere on microscope. Turn down the light intensity control to low and then switch off the light.

DARKGROUND MICROSCOPE

Unstained living organisms may be observed in a wet film with the ordinary light microscope, they are seen only faintly, contrasting poorly with their background, and if scanty are difficult to detect. They may be seen much more clearly and with better resolution, with the darkground microscope. Optical systems that enhance the contrast of unstained bodies. The surfaces and denser internal bodies of bacterial and other cells are clearly read.

The darkground microscope has proved particularly useful for demonstrating the smaller spirochetes, such as those of syphilis, which are so thin they are practically invisible with the ordinary light microscope, whether in unstained wet films or in fixed films stained by ordinary methods.

The principle of darkground microscopy is that the specimen is illuminated only by rays of light so oblique that unless they are 'scattered' by objective, e.g. bacteria, of different refractive index from the suspending medium, they fail so that the scattered rays from the dense objectives enter the objective and reach the eye, so that these objects appear gleaming brightly against a dark background.

There are three requisites for adapting an ordinary microscope for darkground illumination (i) a darkground condenser, while focuses only oblique rays of light on the specimen, (ii) a suitable high-intensity lamp, and (iii) a funnel stop which reduces the numerical aperture of the objective to less than 1.0.

Darkground condenser: The special condenser incorporates concentric reflecting mirrors. A central one prevents light rays from passing directly up through the specimen in to the objective and reflects them outwards on to a peripheral mirror. The latter reflects the rays inwards on to the specimen at a very oblique angle.

FLUORESCENCE MICROSCOPY

Fluorescence occurs when something absorbs light of one wavelength and emits it at another in fluorescence microscopy, the specimen is stained with a fluorescent dye. In fluorescence microscopy, the field appears dark and the stained specimen appears brightly coloured.

The light source in a fluorescent microscope (Fig. 2.2) should be one that emits strongly at the wavelength absorbed by the dye. Some dyes require ultraviolet light. In all cases, filters are used to remove as much of the unwanted wavelengths as possible. The filter that controls the wavelength of light that illuminates the specimen is called the exciter filter. A second filter, called a barrier filter, is inserted in the light path above the specimen between the objective and ocular lenses. Its function is to remove all light except that wavelength emitted by the fluorescent dye.

When light passes through transparent or translucent objects, the characteristics of the waves are often changed. The light is said to have *changed* phase. A phase-contrast microscope is designed to detect light that has been forced to change phase by passing through the specimen.

The phase ring in the body of the microscope acts much like the barrier filter does in a fluorescence microscope. It blocks the light that is still in phase and allows only the changed light.

Phase-contrast microscopy uses a condenser that arranges the light rays so they are in phase. When the light passes through the specimen, some of it

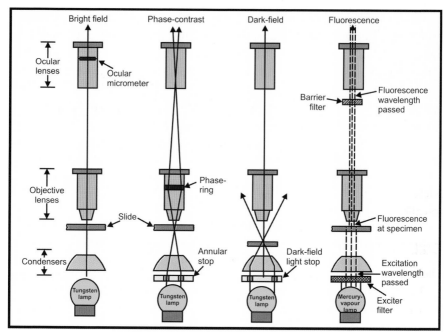

Fig. 2.2: The light source in a fluorescent microscope

is made to go out of phase. The phase plate, usually located in the specially designed phase-objective lenses enhances the apparent relative brightness of the out-of-phase light. This has the effect of causing a marked change in contrast between the specimen and the background medium. Phase-contrast microscopy is considerably more sensitive than dark-field microscopy, and almost always makes it possible to visualise internal cellular organelle.

Chapter

3

Anatomy and Physiology of Bacteria

Bacteria are unicellular and do not show true branching, except in the so-called 'higher bacteria' (Actinomycetales).

SIZE OF BACTERIA

The unit of measurement used in bacteriology is the micron (micrometre, μm)

1 micron (μ) or micrometre (μm) = one thousandth of a millimetre

1 millimicron (mμ) or nanometre (nm) = one thousandth of a micron or one millionth of a millimetre.

1 Angstrom unit (Å) = one tenth of a nanometre.

The limit of resolution with the unaided eye is about 200 microns. Bacteria, being much smaller, can be visualised only under magnification.

SHAPE OF BACTERIA

Depending on their shape, bacteria are classified into several varieties.

1. Cocci (from *kokkos* meaning berry) are spherical or oval cells.
2. Bacilli (from bacillus meaning rod) are rod shaped cells.
3. Vibrios are comma shaped, curved rods and derive the name from their characteristic vibratory motility.
4. Spirilla are rigid spiral forms.
5. Spirochaetes (from sphira meaning coil and chaite meaning hair) are flexuous sprial forms.
6. Actinomycetes are branching filamentous bacteria, so called because of a resemblance to the radiating rays of the sun.
7. Mycoplasmas are bacteria that are cell wall deficient and hence do not possess a stable morphology.

Bacteria sometimes show characteristic cellular arrangement or grouping. Thus, cocci may be arranged in pairs (diplococci), chains (streptococci), groups of four (tetrads) or eight (saricina), or as grape-like clusters (staphylococci).

Some bacilli too may be arranged in chains (streptobacilli) others are arranged at angles to each other, presenting a cuneiform or Chinese letter pattern (corynebacteria).

The outer layer or cell envelope consists of two components—a rigid cell wall and beneath it a cytoplasmic or plasma membrane. The cell envelope encloses the proteoplasm, comprising the cytoplasm, cytoplasmic inclusions such as ribosomes and mesosomes, granules, vacuoles and the nuclear body. The cell may be enclosed in a viscid layer, which may be a loose slime layer, or organised as a capsule. Some bacteria carry filamentous appendages protruding from the cell surface—the flagella which are organs of locomotion and the fimbriae which appear to be organs for adhesion.

Cell Wall

The cell wall accounts for the shape of the bacterial cell and confers on its rigidity. Chemically, the cell wall is composed of mucopeptide (peptidoglycan or murein) scaffolding formed by N-acetyl-glucosamine and N-acetyl muramic acid molecules alternating in chains, which are crosslinked by peptide chains. In general, the walls of the gram-positive bacteria have simpler chemical nature than those of gram-negative bacteria.

Cytoplasmic Membrane

The cytoplasmic (plasma) membrane is a thin (5-10 nm) layer lining the inner surface of the cell wall and separating it from the cytoplasm. It acts as a semipermeable membrane controlling the inflow and outflow of metabolites to and from the protoplasm. Chemically, the membrane consists of lipoprotein with small amounts of carbohydrate. Sterols are absent, except in Mycoplasma.

Cytoplasm

The bacterial cytoplasm is a colloidal system of a variety of organic and inorganic solutes in a viscous watery solution. The cytoplasm stains uniformly with basic dye. The cytoplasm contains ribosomes, mesosomes, inclusions and vacuoles.

Ribosomes

These are centres of protein synthesis. They are slightly smaller than the ribosomes of eukaryotic cells (sedimentation constant 70 S) and are seen integrated in linear strands of mRNA to form polysomes.

Mesosomes

Mesosomes (chondroids) are vesicular, convoluted or multilaminated structures formed as invaginations of the plasma membrane into the cytoplasm. They are more prominent in gram-positive bacteria. They are the principal sites of respiratory enzymes in bacteria.

Intracytoplasmic Inclusions

Volutin granules *(metachromatic* or *Babes-Ernst granules)* are highly refractive, strongly basophilic bodies consisting of polymetaphosphate. Polysaccharide granules may be demonstrated by staining with iodine, and lipid inclusions with fat soluble dyes such as Sudan black. They appear to be storage products. Vacuoles are fluid-containing cavities separated from the cytoplasm by a membrane.

Nucleus

Bacterial nuclei can be demonstrated by acid or ribonuclease hydrolysis and subsequent staining for nuclear material. They appear as oval or elongated bodies, generally one per cell. Some cells may possess two or more nuclear bodies. Bacterial nuclei have no nuclear membrane or nucleolus. Bacteria may possess extranuclear genetic elements consisting of DNA. These cytoplasmic carriers of genetic information are termed *plasmids or episomes.*

Slime Layer and Capsule

Many bacteria secrete a viscid material around the cell surface in the *Pneumococcus*, it is known as the capsule.

Flagella

Motile bacteria, except spirochaetes, possess one or more unbranched long, sinuous filaments called flagella, which are the organs of locomotion. Each flagellum consists of three distinct parts, the filament, the hook and the basal body. The hook-based body portion is embedded in the cell enveloped.

The flagella are 3-20 μm long and are of uniform diameter (0.01-0.013 μm) and terminate in a square tip. Flagella are made up of protein (flagellin) similar to keratin or myosin. Flagellar antigens induce specific antibodies in high titres. Flagellar antibodies are not protective but are useful in serodiagnosis.

The presence or absence of flagella and their number and arrangement are characteristic of different genera of bacteria. Flagella may be arranged all round the cell, **peritrichous** or situated at one or both ends of the cell-**polar**. Polar flagella may be single **monotrichous** in tufts, **lophotrichous** or with flagella at both poles, **amphitrichous.**

Fimbriae

Some gram-negative bacilli carry very fine, hair-like surface appendage called fimbriae or pili. Fimbriae functions as organs of adhesion.

GROWTH AND MULTIPLICATION OF BACTERIA

Bacterial growth may be considered at two levels, increase in the size of the individual cell and increase in the number of cells. Growth in number can be studied by bacterial counts. Two types of bacterial counts can be made—total count and viable count.

1. Direct counting under the microscope using, counting chambers.
2. Counting in an electronic device as in the Coulter counter.
3. Direct counting using stained smear.

The viable count measures the number of living cells, that is, cells capable of multiplication. Viable line will be obtained, counts are obtained by dilution or plating method.

BACTERIAL GROWTH CURVE

Lag Phase

Immediately following the seeding, of a culture medium, there is no appreciable increase in numbers, though there may be an increase in the size of the cells. This phase is called lag phase.

Lag (Logarithmic) or Exponential Phase

Following the lag phase, the cells start dividing and their numbers increase exponentially.

Stationary Phase

After a varying period of exponential growth, cell division stops due to depletion of nutrients and accumulation of toxic products.

Phase of Decline

This is the phase when the population decreases due to cell death. The principal constituent of bacterial cells is water, which represents about 80 per cent of the total weight. Proteins, polysaccharides, lipids, nucleic acids, mucopeptides and low molecular weight compounds make up the rest. For growth and multiplication of bacteria, the minimum nutritional requirements are water, a source of carbon, a source of nitrogen and some inorganic salt.

Bacteria which derive their energy from sunlight are called *phototrophs* and those that obtain energy from chemical reactions are called *chemotrophs*. Bacteria that can synthesise all their organic compounds are called *autotrophs*. Those that are unable to synthesise their own metabolites and depend on preformed organic compounds are called heterotrophs.

Bacteria require a supply of inorganic salts, particularly the anions phosphate and sulphate, and the cations sodium, potassium, magnesium, iron, manganese and calcium.

Some bacteria require certain organic compounds in minute quantities. These are known as growth factors or bacterial vitamins.

OXYGEN REQUIREMENT AND METABOLISM

Bacteria are divided into aerobes and anaerobes. Aerobic bacteria require oxygen for growth. They may be obligate aerobes like the cholera vibrio, which will grow only in the presence of oxygen, or facultative anaerobes which are ordinarily aerobic but can also grow in the absence of oxygen, though less abundantly. **Anaerobic bacteria**, such as clostridia, grow in the absence of oxygen and the obligate anaerobes may even die on exposure to oxygen. Microaerophilic bacteria are those that grow best in the presence of a low oxygen tension.

CARBON DIOXIDE

All bacteria require small amounts of carbon dioxide for growth.

TEMPERATURE

The temperature at which growth occurs best is known as the 'optimum temperature', which in the case of most pathogenic bacteria is 37°C. Bacteria which grow best at temperatures of 25 - 40°C are called *mesophilic. Psychrophilic* bacteria are those that grow best at temperatures below 20°C.

Another group of nonpathogenic bacteria, the thermophiles, grow best at high temperatures, 55-80°C.

MOISTURE AND DRYING

Water is an essential ingredient of bacterial protoplasm and hence drying is lethal to cells. Drying in vacuum in the cold (freeze drying or lyophilisation) is a method for the preservation of bacteria, viruses and many labile biological materials.

H⁻ ION CONCENTRATION

Bacteria are sensitive to variations in pH. The majority of pathogenic bacteria grow best at neutral or slightly alkaline reaction (pH 7.2-7.6). Some acidophilic bacteria such as lactobacilli grow under acidic conditions.

LIGHT

Bacteria grow well in the dark. Cultures die if exposed to sunlight. Exposure to light may influence pigment production.

4

Sterilisation and Disinfection

Sterilisation is defined as the process by which an article, surface or medium is freed of all living microorganisms either in the vegetative or spore state. Disinfection means the destruction or removal of all pathogenic organisms, or organisms capable of giving rise to infection. The term antisepsis is used to indicate the prevention of infections, usually by inhibiting the growth of bacteria in wounds or tissues.

The various agents used in sterilisation can be classified as follows.

PHYSICAL AGENTS

1. Sunlight
2. Drying
3. Dry heat: Flaming, Incineration, Hot air
4. Moist heat: Pasteurisation, Boiling, Steam under normal pressure, Steam under pressure.
5. Filtration: Candles, Asbestos pads, Membranes
6. Radiation.
7. Ultrasonic and sonic vibrations.

Sunlight

Sunlight possesses appreciable bactericidal activity and plays an important role in the spontaneous sterilisation that occurs under natural conditions.

Drying

Moisture is essential for the growth of bacteria. Four-fifths of the weight of the bacterial cell is due to water. Drying in air has therefore, a deleterious effect on many bacteria.

Heat

Heat is the most reliable method of sterilisation. Materials that may be damaged be heat can be sterilised at lower temperatures, for longer periods or by repeated cycles. The factors influencing sterilisation by heat are.

Dry Heat

Flaming: Inoculating loop or wire, the tip of forceps and spatulas are held in a Bunsen flame till they become red hot.

Incineration: This is an excellent method for safely destroying materials such as contaminated cloth, animal carcasses, and pathological materials,

Hot air oven: This is the most widely used method of sterilisation by dry heat. A holding period of 160°C for one hour is used to sterilise glassware, forceps, scissors, scalpels, all-glass syringes, swabs, some pharmaceutical products such as liquid paraffin, dusting powder, fasts and grease. The oven is usually heated by electricity, with heating elements in the wall of the chamber. It must be fitted with a fan to ensure even distribution of air and elimination of air pockets. It should not be overloaded. Glasswares should be perfectly dry before being placed in the oven. Test tubes and flasks should be wrapped in paper. Rubber materials, except silicon rubber, will not withstand the temperature. Cutting instruments such as those used in ophthalmic surgery, should ideally be sterilised for two hours at 150°C. The oven must be allowed to cool slowly for about two hours before the door is opened, since the glassware may crack due to sudden or uneven cooling.

Sterilisation Control

The spores of a nontoxigenic strain of *C. thermophilus* are used as a microbiological test of dry heat efficiency. Paper strips impregnated with 10^6 spores are placed in envelopes and inserted into suitable packs. After sterilisation, the strips are removed and inoculated into thioglycollated or cooked meat media and incubated for sterility test under strict anaerobic conditions for five days at 37°C.

A Browne's tube (green spot) is available for dry heat and is convenient for routine use. After proper sterilisation a green colour is produced.

Moist Heat

Temperature below 100°C

For pasteurisation of milk: The milk is heated at either 63°C for 30 minutes (the holder method) or 72°C for 15-20 seconds (the flash process) followed by cooling quickly to 13°C or lower. Serum or body fluids containing coagulable proteins can be sterilised by heating for one hour at 56°C in a water bath on several successive days. Media such as Lowenstein-Jensen and Loeffler's serum are rendered sterile by heating at 80 - 85°C for half an hour on three successive days in an inspissator.

Temperature at 100°C

Boiling: Vegetative bacteria are killed almost immediately at 90-100°C, but sporing bacteria require prolonged periods of boiling.

In cases where boiling is considered adequate, the material should be immersed in the water and boiled for 10-30 minutes. The lid of the steriliser should not be opened during the period.

Steam at Atmospheric Pressure (100°C)

The steamer consists of a tinned copper cabinet with the walls suitably tagged. The lid is conical, enabling drainage of condensed steam, and a perforated tray fitted above the water level ensures that the material placed on it is surrounded by steam. A single exposure of ninety minutes usually ensures sterilisation.

Steam under Pressure

Sterilisation by steam under pressure is carried out at temperature between 108°C and 147°C. By using the appropriate temperature and time, a variety of materials such as dressings, instruments, laboratory ware, media and pharmaceutical products can be sterilised.

Several types of steam sterilisers are in use:

1. Laboratory autoclaves
2. Hospital dressing sterilisers
3. Bowl and instrument sterilisers and
4. Rapid cooling sterilisers.

    Even the domestic pressure cooker can be used as a sterilisers.

The laboratory autoclave consists of a vertical or horizontal cylinder of gunmetal or stainless steel, in a supporting sheet iron case. The lid or door is fastened by screw clamps and made airtight by a suitable washer. The autoclave has on its lid upper side a discharge tap for air and steam, a pressure gauge and a safety valve that can be set to blow off at any desired pressure. Heating is by gas or electricity.

Sufficient water is put in the cylinder, the material to be sterilised is placed on the tray, and the autoclave is heated. The lid is screwed tight with the discharge tap open. The safety valve is adjusted to the required pressure.

The steam pressure rises inside and when it reaches the desired set level, the safety valve opens and the excess steam escapes. From this points, the holding period is calculated, the heater is turned off and autoclave allowed to cool till the pressure gauge indicates that the pressure inside is equal to the atmospheric pressure.

Sterilisation Control

Paper strips impregnated with 10^6 spores are dried at room temperature and placed in paper envelopes. These envelopes are inserted in different parts of the load. After sterilisation, the strips are inoculated into a suitable recovering medium and incubated for sterility test at 55°C for five days.

Filtration

Filtration helps to remove bacteria from heat labile liquids such as sera and solutions of sugars of antibiotics used for preparation of culture media. The following types of filters have been used.

Candle Filters

These are manufactured in different grades of porosity and have been used widely for purification of water for industrial and drinking purpose.

Asbestos Filters

They are disposable, single-use discs. They have high adsorbing capacity and tend to alkalinise filtered liquid.

Sintered Glass Filters

They are prepared by heat fusing finely powdered glass particulars of graded sizes.

Membrane Filters

Made of cellulose esters of other polymers, they have largely replaced other types of filters. They are routinely used in water purification and analysis, sterilisation and sterility testing, and for the preparation of solutions for parenteral use.

Radiation

Two types of radiations are used for sterilisation, non-ionising and ionising. Infrared and ultraviolet rays are of the non-ionising low energy type, while gamma rays and high energy electrons are the high energy ionising types.

Non-ionising Radiation

Here electromagnetic rays with wavelengths longer than those of visible light are used. Hence, infrared radiation can be considered as a form of hot air sterilisation.

Ionising Radiation

X- rays, gamma rays and cosmic rays are highly lethal to DNA and other vital constituents. They have very high penetrative power. Since there is no appreciable increase in temperature in this method, it is referred to as cold sterilisation.

CHEMICAL AGENTS

1. Alcohols: Ethyl, Isopropyl, Trichlorobutanol
2. Aldehydes: Formaldehyde, Glutaraldehyde
3. Dyes
4. Halogens
5. Phenols
6. Surface-active agents
7. Metallic salts
8. Gases: Ethylene oxide, formaldehyde, beta propiolactone.
   Several chemical agents are used as antiseptics and disinfectants.

Alcohols

Ethyl alcohol (ethanol) and isopropyl alcohol are the most frequently used. They are used mainly as skin antiseptic and act by denaturing bacterial proteins. They have no action on spores. Methyl alcohol is effective against fungal spores and is used for treating cabinets and incubators affected by them. A pad moistened with methanol and a dish of water are kept inside, and the incubator is left at working temperature for several hours. Methyl alcohol vapour is toxic and inflammable.

Aldehydes

Formaldehyde is active against the amino group in the protein molecule. In aqueous solutions, it is markedly bactericidal and sporicidal and also has a lethal effect on viruses. It is used to preserve anatomical specimens, and for destroying anthrax spores in hair and wool.

Formaldehyde gas is used for sterilising instruments and heat sensitive catheters and for fumigating wards, sick rooms and laboratories. Under properly controlled conditions, clothing, bedding, furniture and books can be satisfactorily disinfected.

Glutaraldehyde

This has an action similar to formaldehyde. It is specially effective against tubercle bacilli, fungi and viruses. It is less toxic and irritant to the eyes and skin than formaldehyde.

Dyes

Two groups of dyes, aniline dyes and acridine dyes are used extensively as skin and wound antiseptics. The aniline dyes in use are brilliant green, malachite green and crystal violet. They are more active against gram-positive organisms. The acridine dyes are more active against gram-positive organisms than against gram-negative but are not as selective as the aniline dyes.

Halogens

Iodine in aqueous and alcoholic solution has been used widely as a skin disinfectant. Chlorine and its compounds have been used as disinfectants for many years.

Phenols

Phenol (carbolic acid) is a powerful microbicidal substance. This and other phenolic disinfectants derived from coal tar are widely used as disinfectants for various purposes in hospitals. Lysol and cresol are active against a wide range of organisms.

Gases

Ethylene Oxide

This is a colourless liquid with a boiling point of 10.7°C, and at normal temperature and pressure is a highly penetrating gas with a sweet ethereal smell. It diffuses through many types or porous materials and readily penetrates some plastics. It is specially used for sterilising heart-lung machines, respirators, sutures, dental equipments, books and clothing. It is unsuitable for fumigating rooms because of its explosive property.

Formaldehyde Gas

This is widely employed for fumigation of operation theatres and other rooms. After sealing the windows and other outlets, formaldehyde gas is generated by adding 150 gm of $KMnO_4$ to 280 ml formalin for every 1000 cu.ft (28.3 cu.m) of room volume.

Beta-Propiolactone (BPL)

It has a rapid biocidal action but unfortunately has carcinogenic activity. For sterilisatin of biological products 0.2 per cent BPL is used.

TESTING OF DISINFECTANTS

In the Rideal Walker test, suspensions containing equal numbers of typhoid bacilli are submitted to the action of varying concentrations of phenol and of the disinfectant to be tested. In the Chick Martin test, the disinfectant acts in the presence of organic matter. Various other modifications have been introduced, but no test is entirely satisfactory.

5

Culture Media

Bacteria grow as bacterial population rather than single bacterial cells. Hence, they require separate culture so as to isolate pure culture.

Numerous culture media have been devised. The original media used by Louis pasteur were liquids bacteria growing in liquid media grow diffusely whereas when grown on solid media, bacteria have distinct colony morphology and exhibit many other characteristic features such as pigmentation or hemolysis.

Agar or agar-agar is universally used to solidify media. These are obtained from some types of seaweeds and the chief constituent is long chain polysaccharides.

Another almost universal ingredient of common media is peptone. It is a complex mixture of partially digested proteins. Its constituents are proteoses, polypeptides and amino acids, a variety of inorganic salts including phosphates, potassium and magnesium, potassium and magnesium, etc. Special brands of peptone such as neopeptone and proteose peptone are available for special uses. Commerically available peptones or digest broth can be used. Meat extract is also available and is commercially known as lab lemco.

Media have been classified in many ways.

1. Solid media, liquid media, semisolid media
2. Simple and complex media
3. Synthetic or defined media
4. Special media are further divided into enriched media, enrichment media, selective media, sugar media and transport media.
5. Aerobic media and anaerobic media.

TYPES

Media have been classified in many ways:

1. Solid media, liquid media, semisolid media.

2. Simple media, complex media, synthetic or defined media, semidefined media, special media. Special media are further divisible into enriched media, enrichment media, selective media, indicator or differential media, sugar media, and transport media.
3. Aerobic media, anaerobic media.

Simple Media

It consists of peptone, meat extract, sodium chloride and water. Nutrient agar, made by adding 2 per cent agar to nutrient broth is the simplest and most common medium in routine diagnostic laboratories. If the concentration of agar is reduced to 0.2-0.5 per cent, semisolid or sloppy agar is obtained which enables motile organisms to spread. Increasing the concentration of agar to 6 per cent prevents spreading or swarming by organisms such as *Proteus*.

Complex Media

These have added ingredients for special purposes or for bringing out certain characteristics or providing special nutrients required for the growth of the bacterium.

Synthetic or Defined Media

These media are prepared from pure chemical substances and the exact composition of the medium is known. These are used for various special studies such as metabolic requirements. Simple peptone water medium, 1 per cent peptone with 0.5 per cent NaCl in water, may be considered as a semi defined medium, since its composition is approximately known.

Enriched Media

In these media, substances such as blood, serum, or egg are added to a basal medium. They are used to grow bacteria which are more exacting in their nutritional needs.

Enrichment Media

Substances which have, a stimulating effect on the bacteria to be grown or an inhibitory effect on those to be suppressed are incorporated in the medium. If such substances are added to a liquid medium, result is an absolute increase in the numbers of the wanted bacterium relative to the other bacteria. Such media are called enrichment media, for example tetrathionate broth.

Selective Media

As in the above case, if the inhibiting substance is added to a solid medium, it enables a greater number of the required bacterium to form colonies than the other bacteria.

Indicator Media

These media contain an indicator which changes colour when a bacterium grows in them.

Differential Media

A medium which has substances incorporated in it, enabling it to bring out differing characteristics of bacteria and thus helping to distinguish between them, is called a differential medium.

Sugar Media

The term 'sugar' in microbiology denotes any fermentable substance. They may be:
1. Monosaccharides
    a. pentoses, e.g. arabinose, xylose,
    b. hexoses, e.g. dextrose, mannose
2. Disaccharides, e.g. saccharose, lactose
3. Polysaccharides, e.g. starch, inulin
4. Trisaccharides, e.g. raffinose
5. Alcohols, e.g. glycerol, sorbitol
6. Glucosides, e.g. salicin, aesculin
7. Noncarbohydrate substances, e.g. inositol.

Transport Media

Special media are devised for transporting. These are termed transport media, for example, Stuart's medium—a non-nutrient soft agar gel containing a reducing agent to prevent oxidation and charcoal to neutralise certain bacterial inhibitors—for gonococci, and buffered glycerol-saline for enteric bacilli.

Anaerobic Media

These media are used to grow anaerobic organisms, for example, Robertson's cooked meat medium.

PEPTONE WATER (PW)

Peptone	1.0 gm
Sodium chloride	0.5 gm
Distilled water	100 ml

Dissolve the ingredients and filter through filter paper. Adjust pH to 7.6. Distribute into small test tubes in 4 ml quantities and autoclave.

BBL Trypticase Peptone is superior to others when used for indole test.

Use

1. This is used as a base in sugar fermentation, tests.
2. Used to test for indole production by *Enterobacteriaceae*.

Alkaline Peptone Water (APW)

Prepare peptone water and adjust pH 8.4 +/– 0.2

Use

This medium is used as an enrichment broth for vibrios.

Alkaline Nutrient Agar

Peptone	1.0 gm
Sodium chloride	0.5 gm
Lab Lemco	0.3 gm
Agar	1.5 gm
Distilled water	100 ml

Mix the ingredients in distilled water. Adjust pH to 8.6 heat to dissolve and autoclave. Mix and pour plates.

Use

For selective growth of *Vibrio cholerae*

Nutrient Broth (NB)

Peptone	1.0 gm
Beef extract (Lab Lemco)	0.3 gm
Sodium chloride	0.5 gm
Distilled water	100 ml

Weigh out all the ingredients as above, peptone should be taken last, because it sticks to the paper on exposure. Mix the ingredients and dissolve them by heating. When cool, adjust the pH to 7.4-7.6. Distribute in tubes, bottles or flasks and sterilise by autoclaving.

Use

This is a basal medium and is also used to grow nonfastidious organisms for various purposes.

Nutrient Agar (NA)

Agar powder	1.5 to 1.8 gm
Nutrient broth	100 ml

Mix the agar powder, nutrient broth and heat to dissolve. When cool adjust the pH to 7.5-7.6. Sterilise by autoclaving. Pour as plates or slopes. To make deeps, reduce agar concentration to 05 per cent.

Use

This is used as a base for many media. Only nonfastidious organisms will grow on this.

MacConkey Agar (MA)

Peptone	2.0 gm
Sodium chloride	0.5 gm
Bile salt	0.5 gm
Lactose	1.0 gm
Agar	1.5 gm
Distilled water	100 ml

Dissolve the ingredients except lactose in distilled water by heating. Adjust pH to 7.6. Add 1 ml of 1 per cent neutral red solution to every 100 ml of medium with lactose. Sterilise by autoclaving at 121°C for 15 minutes.

Use

This is a partially selective and a differential medium, used for the differentiation of lactose fermenting and non-lactose fermenting enteric bacteria.

Blood Agar (BA)

Sterile defibrinated sheep blood	7 ml
Nutrient agar melted	100 ml

Pour about 7 ml of melted nutrient agar, as a base, into sterile Petri dishes and allow setting. This forms a thin base for pouring in the blood agar. Add sterile defibrinated sheep blood (5-7%) to nutrient agar, the latter should be cooled to about 45-50°C before blood is added. Mix well and pour about 15 ml of blood agar over the base in each Petri dish. **Human blood is not recommended for the, preparation 'of blood' agar.**

Alternately blood agar may be made with no agar base.

Use

It serves as an enriched medium and a differential medium for haemolytic organisms. Most common pathogens grow on it.

Blood Agar (4%)

Agar	4 gm
Nutrient broth	100 ml
Defibrinated sheep blood	7 ml

Follow instructions given for, blood agar...

This medium is used to suppress the swarming growth of *Clostridium sporogenes* and *Clostridium tetani*.

Chocolate Agar (CA)

Chocolate agar is heated blood agar.

Sterile defibrinated blood	10 ml
Nutrient agar (melted)	100 ml

Melt the nutrient agar. When the temperature is about 45 to 50°C add the blood and mix well. After the addition of blood, heat in a water bath slowly bringing up the temperature to 75°C with constant agitation. Special care should be taken to avoid fluctuation in the temperature. Heating is continued till the blood changes to chocolate colour. This colour is very critical. Remove from the water bath. Cool to about 50°C and pour about 20 ml into plates with sterile precautions. Special care must be taken to avoid air bubbles.

Use

This is an enriched medium used for the cultivation of pathogenic *Neisseriae* and *Haemophilus influenza*.

Chocolate Agar Modified

TSA	
Trypticase soy agar	20 gm
Distilled water	250 ml

Autoclave at 121°C for 15 minutes and cool to 50°C.

2% Haemoglobin solution	5 gm
Haemoglobin distilled water	250 ml

Mix the haemoglobin in 5-6 ml of distilled water to form a smooth paste. Continue mixing as the rest of the water is added. Autoclave and cool to 50°C.

Isovitalex   5 ml per 500 ml medium

Mix all the ingredients by gently swirling the flask and avoid forming bubbles. Pour into Petri dishes.

Use

This medium is used for isolating subculturing *Haemophilus influenzae.*

Chocolate Agar with VCN and Isovitalex (Thayer and Martin Medium)

Chocolate Agar	100 ml
VCN Inhibitor	1 ml
Isovitalex	2 ml

Cool chocolate agar to 50°C add VCN and Isovitalex mix thoroughly and pour about 20 ml into sterile plates.

Use

This is used for isolating *N. gonorrhoeae* from clinical specimens likely to be contaminated with other bacteria.

Deoxyribonuclease (DNase) Test Agar

DNA	0.2 gm
Phytone	0.5 gm
Sodium chloride	0.5 gm
Trypticase	1.5 gm
Agar	1.5 gm
Distilled water	100 ml

Mix all ingredients and adjust pH 7.3

Mix the ingredients well in the distilled water. Sterilise by autoclaving at 121°C for 15 minutes and pour into sterile Petri dishes.

Use

To test the production of the 'enzyme DNase' by certain organism.

Deoxycholate Citrate Agar (DCA)

Proteose peptone	10.0 gm
Lactose	10.0 gm
Sodium citrate	20.0 gm
Ferric ammonium citrate	2.0 gm
Sodium desoxycholate	5.0 gm
Bacto agar	15.0 gm
Neutral red (1% solution)	2.3 ml
Beef infusion broth	1000 ml

Add proteose peptone, sodium citrate, ferric ammonium citrate and bactor agar to the beef infusion broth, pH of which has already been adjusted to 7.4. Recheck pH. Keep it in a water bath for 30 minutes for dissolving. Then add lactose, sodium desoxycholate and neutral red solution mix and pour plates.

Use

A good selective medium used for the isolation of *Salmonella* and *Shigella*.

Mannitol Motility Medium

Agar	0.3 gm
Peptone	2.0 gm
Distilled water	100 ml

Dissolve and adjust pH to 7.6 then add the following:

Mannitol	0.2 gm
Potassium nitrate	0.1 gm
1% Phenol red solution	0.4 ml

Distribute in small test tubes in 3 to 4 ml quantities and autoclave at 110°C for 10 minutes.

Mannitol Salt Agar

Beef extract	0.1 gm
Proteose peptone no. 3	0.1 gm
Sodium chloride	7.5 gm
Mannitol	1.0 gm
Agar	1.5 gm
Phenol red	0.0025 gm
Distilled water	100 ml

Mix the ingredients well in distilled water. Sterilise by autoclaving at 121°C for 15 minutes. Pour into sterile Petri dishes.

Mueller-Hinton Agar (MHA)

Beef extract	2.0 gm
Acidicase peptone	7 5 gm
Starch	1.5 gm
Agar	17.0 gm
Distilled water	1000 ml

Dissolve the ingredients in one litre of distilled water. Mix thoroughly. Heat with frequent agitation and boil for one minute. Adjust pH to 7.4/–0.2 sterilise by autoclaving. Do not overheat.

Use

Standard medium for antimicrobial susceptibility testing.

Mueller-Hinton Haemoglobin Agar for Haemophilus

Ref. Manual for National Surveillance of anti-use microbial resistance WHO, CDC 1991.

MHA	3.8 gm
Distilled water	50 ml

Autoclave and cool to 50°C.

Haemoglobin	2 gm
Distilled water	50 ml

Autoclave and cool to 50°C.

Mix agar and haemoglobin solutions by gently swirling the flasks. Avoid forming bubbles. Add NAD solution and pour plates in 25 ml quantities.

Use

This medium is for testing the antibiotic susceptibility of *H. influenzae*.

Biphasic MacConkey's Medium (BPMM)

Agar

Peptone	20.0 gm
Lactose	10.0 gm
Nile salts	5.0 gm
Sodium chloride	5. 0 gm
Agar	30.0 gm
Neutral red (1%)	4.0 ml
Distilled water	1000 ml

Broth

Peptone	30.0 gm
Lactose	7.5 gm
Bile salts	7.5 gm
Neutral red (1%)	7.5 ml
Distilled water	1000 ml

Dissolve ingredients under 'agar' in distilled water by boiling, and adjust pH to 7.6. Filter through gauze and dispense in 25 ml amounts in screw-capped prescription bottles. Sterilise at 121°C for 15 minutes. Remove the bottles from autoclave while hot and place them in a horizontal position, so as to form slants on the broader side. Allow solidifying.

Dissolve ingredients under 'broth' in distilled water and sterilise by autoclaving at 121°C for 15 minutes. Add 30 ml to each bottle under strict aseptic conditions. Incubate all bottles at 37°C for 48 hours to check sterility.

Use

This medium is used for direct inoculation of blood for culture.

Bismuth Sulfite Agar (BSA) (Wilson and Blair Medium)

Polypeptone peptone	10.0 gm
Beef extract	5.0 gm
Dextrose	5.0 gm
Disodium phosphate	4.0 gm
Ferrous sulfate	0.3 gm
Bismuth sulfite indicator	8.0 gm
Brilliant green	0.025 gm
Agar	20.0 gm
Distilled water	1000 ml

pH 7.5 +/− 0.2

Mix the ingredients in distilled water. Keep for 5 minutes and mix again thoroughly. When uniform suspension has been obtained heat with frequent agitation and boil for 1 minute. Cool to about 50°C. Mix well and pour into sterile plates. The plated medium should be used on the same day it is prepared.

Use

This is a selective medium for hydrogen sulfide producing *Salmonella*.

Brain Heart Infusion Broth (BHIB)

Sodium citrate	1 gm
Sodium chloride	4 gm
Sodium phosphate	5 gm
Dextrose	10 gm
Peptone	10 gm
Brain heart infusion	
Brain infusion broth	250 ml
Heart infusion broth	750 ml
Sodium polyanethol sulphonate	0.25 gm.

Obtain ox brain and heart. Remove all fat from the heart. Cut into small pieces and grind. Add distilled water three times (v/w). Keep at 4°C overnight.

From the brain, remove meninges fully and then, weigh. Add distilled water, (3 times v/w) and mash by using hand. Keep in the cooler overnight.

Next morning boil the brain and heart separately, for 30 minutes. Then filter through cotton layer. Measure each broth separately.

Mix both infusions and the remaining ingredients. Dissolve well and adjust pH of the entire amount to 7.4 to 7. 6. Autoclave at 121°C for 15 minutes. Filter through filter paper and distribute in screw capped prescription bottles in 50 to 100 ml amounts. Autoclave once more at 115°C for 10 minutes.

Use

This is used for direct inoculation of whole blood, bone marrow and body fluids for culture.

Brain Heart Infusion Agar (BHIA)

Agar granules	15 gm
Brain Heart Infusion Broth (BHIB)	100 ml

Dissolve agar and autoclave it at 121°C (15 lbs. pressure) for 15 minutes. Cool to about 50°C and pour plates.

Use

This is used to culture dimorphic fungi and actinomycetes.

Bile Broth 20%

Ox gall	20 gm
Distilled water	100 ml

Mix and autoclave at 121°C for 15 minutes. Just before inoculating add 0.4 ml of this to 8 ml thioglycollate medium.

Castaneda Medium

Biphasic Trypticase Soy Agar (TSA) and Trypticase Soy Broth (TSB).

Triple Sugar Agar Slant TSB	32 gm
Triple Sugar Broth (powder)	12 gm
Sodium citrate	4 gm
Distilled water	800 ml

Adjust pH of both media to 7.4. Sterilise separately by autoclaving. Slant 20 ml of TSA each in sterile 4 oz prescription bottles. After the agar is set, add 20 ml of sterilised TSB with sterile precaution.

Use

Used for isolation of *Brucella* from blood, bone marrow, etc.

Carboxylase Test Medium

Basal medium	
Peptone	5.0 gm
Beef extract	5.0 gm
Bromcresol purple	0.01 gm
Cresol red (0.2%)	2.5 gm
Glucose	0.5 gm
Pyridoxal	5.0 mg
Distilled water	1000 ml

Mix all the ingredients and adjust pH to 6.0 amino acids.

L-lysine	Add to make a
L-arginine	final concentration
L-ornithine	of 1% of each

The basal medium is divided into four equal portions, one of which is distributed into tubes (1 ml amounts) without the addition of any amino acid. These tubes of basal medium are used as controls. To one of the remaining portions of basal medium add L-lysine, to the second portion add L-arginine and to the last add L-ornithine, get final concentrations of one per cent each. Distribute these also in one ml amounts, amino acids, used should be incorporated, since the microorganism apparently are only active against the L-forms. Sterilise by autoclaving at 115°C for 10 minutes.

Dysentery Enteric Cholera (DEC) Medium

Trypticase	5.0 gm
Lab lemco	5.0 gm
Bile salt	8.5 gm
Agar	15.0 g
Sodium thiosulphate	8.5 gm
Sodium citrate	8.0 g
Ferric citrate (ground)	7.5 gm
Lactose	12.5 g
Distilled water	1000 ml
Neutral red 1%	3.75 ml

Dissolve the first seven ingredients in distilled water. Heat and adjust pH to 7.6. Add ferric citrate, lactose and neutral red. Mix and pour plates.

Use

A highly selective medium for the isolation of *Salmonella, Shigella* and *Vibrios*.

Egg Yolk Agar (EYA) (Nagler's Medium)

Brain heart infusion agar	100 ml
Egg yolk from 1 egg	
Neomycin 1% solution	1 ml

Sterilise BHIA by autoclaving. Cool to 45-50°C. Add egg yolk and neomycin. Mix thoroughly and pour 25 ml per plate.

Use

This medium is used for the isolation and identification of *Clostridium lecithinase* activity of *C. perfringens* is particularly evident on this medium and hence used for specific identification by incorporating antitoxin to demonstrate Nageer's reaction and its inhibition. Also used for lipase reaction as shown by pearly layer formation.

EMJU Medium for Leptospira

Sodium phosphate dibasic	1 gm
Potassium phosphate monobasic	0.3 gm
Sodium chloride	1 gm
Ammonium chloride	0.25 gm
Thiamine	0.005 gm
Distilled water	900 ml

Dissolve the ingredients in distilled water. Sterilise by autoclaving. Cool to room temperature. Aseptically add 100 ml leptospira enrichment. Mix thoroughly and adjust pH 7.5 +1-0.2.

Distribute in 5 ml amounts in screw capped sterile test tubes.

Use

This medium is for cultivation of *Leptospira* from clinical specimens.

Esculin Broth

Peptone	5.0 gm
Dipotassium phosphate	1.0 gm
Andrades indicator	10.0 ml
Esculin	0.3 g
Ferric citrate	0.5 g
Distilled water	1000 ml

Heat to dissolve. Dispense in 3-4 ml amount sterilise at 121°C for 10 minutes.

Glucose Phosphate Broth or MR-VP Medium

Dipotassium phosphate	5.0 gm
Proteose peptone	5.0 gm
Glucose	5.0 gm
Distilled water	1000 ml

Suspend ingredients in distilled water and heat slightly to dissolve them. Tube and sterilise at 118°C for 15 minutes.

Kirchner's Medium

Disodium phosphate $(Na_2 HPO_4 \, 12H_2O)$	19.0 gm
Monopotassium phosphate anhydrous (KH_2PO_4)	2.5 gm
Magnesium sulfate	0.6 gm
Trisodium citrate	2.5 gm
Aspargine	5.0 gm
Glycerol	20.0 ml
Phenol red 0.4% aqueous	3.0 ml
Distilled water	1000 ml

Steam to dissolve and adjust pH to 7.4-7.6. Autoclave at 115°C for 10 minutes. Add 10 ml sterile horse or sheep serum to every 90 ml of medium and dispense in 5 ml quantities. For blood culture, it is dispensed in 50 ml amounts in prescription bottles.

Use

This is liquid medium used for cultivation of *M. tuberculosis* especially from blood and other sterile specimens.

Löeffler's Serum Medium

Sterile ox, sheep or horse serum	75.0 ml
Nutrient broth	25 0 ml
Glucose	0 25 gm

Dissolve the glucose in the broth and autoclave at 121°C for 10 minutes. Add the glucose broth to the serum with sterile precautions. Distribute in screw capped test tubes (7 ml per tube). Inspissate in a slanting position in an inspissator or a steam sterilizer. **Do not overheat.**

First day	80-85°C for 50 minutes
Second day	80-85°C for 30 minutes

Medium may be stored for long periods. Do not use if drying has occurred.

Use

This medium is used for the luxuriant growth of *C. diphtheriae* and to see the typical morphology of the organisms by special stains.

Löwenstein-Jensen Medium (LJ Medium)

(Modified from the International Union against Tuberculosis (IUT) recommendation).

Mineral Salt Solution.

Potassium dihydrogen phosphate (KH_2PO_4) (anhydrous)	2.4 gm
Magnesium sulphate	0.24 gm
Magnesium citrate $(Mg_3 C_6 H_{12} O_7)$	0.6 gm
Asparagine	3.6 gm
Glycerol	12 ml
Distilled water	600 ml
Malachite green solution 2%	20 ml
Egg solution	1000.0 ml

Dissolve the ingredients of mineral salt solution by heating. Autoclave at 121°C for 20 minutes to sterilise. This solution keeps indefinitely and may be stored in suitable amounts at 4°C.

Prepare a 2 per cent solution of malachite green in sterile water with sterile precautions by dissolving the dye in the incubator for 1-2 hours. This solution can be stored indefinitely and should be shaken before use.

Eggs must be fresh (not more than 4-day-old). Wash them carefully with soap warm water and brush rinse in running water for 30 minutes drain the water and place the eggs in a sterile tray, covet with a sterile paper and dry till next day. Alternatively, eggs may be dried by cleaning the shell with methylated spirit and burning it off.

Scrub hands with soap and water, dry with spirit and then crack the egg with a sterile knife and take the contents into a sterile beaker. Beat the egg with a sterile egg beater. There is no need to filter the beaten egg.

For complete medium mix as follows:

Mineral salt solution	600 ml
Malachite green solution	20 ml
Beaten egg	
(20-22 hen's eggs depending on size)	1000 ml

Distribute in 5 ml amounts in sterile screw capped tubes or McCartney bottles and screw the caps tightly on.

Lay the tubes/bottles on their sides and inspissate at 80 to 85°C for 50 minutes. Leave the medium in the inspissator (or steriliser) overnight and expose the media to 80-85°C for an additional 30 minutes.

Use

To grow mycobacteria other than *M. leprae.*

LÖWENSTEIN-JENSEN MEDIUM WITH ANTIBIOTICS

Antibiotic Stock Solution

Make stock solutions of following concentrations using pure antibiotic substance.

Streptomycin	200 000 µg/ml
INAH	10 000 µg/ml
Cycloserine	200 µg/ml 140 µg/ml
Ethambutol	140 µg/ml, 250 µg/ml
Thiacetazone	160 µg/ml
Ethionamide	570 µg/ml, 800 µg/ml

Rifampicin   10000 µg/ml
Pyrazinamide   2500 µg/ml

Accurately weigh the antibiotics and dissolve the drugs (except thiacetazone, ethionamide and rifampicin) in distilled water to give the desired concentration. Dissolve thiacetazone and ethionamide in triethylene glycol (Trigol). And rifampicin in alcohol The original rehydrated bottles are kept at –20°C and may be used for 3 months. The working stock solutions are preserved at –20°C and may be used for 1 month.

Further antibiotic dilutions in LJ medium.

Mix stock solution with LJ medium to give the desired final concentrations as shown below. Details of dilutions for each antibiotic is given separately. Mark all flasks and tubes before use, with the drug symbol and final concentration they contain.

Final concentrations of antibiotics in LJ medium.

Streptomycin 2, 4, 8, 16, 32 and 64 µg/ml

INAH (Isoniazid)

0.05,0.1,0.2,0.4,1 and 5 µg/ml

Thiacetazone

	0, 25, 0,5,1,2,4,8,16 and 32 µg/ml
Cycloserine	5, 10, 14, 20, 28 and 40 µg/ml
Ethimonamide	20,28,40,57,80,114 and 160 µg/ml
Rifampicin	4, 8, 16, 32, 64 and 128 µg/ml
Pyrazinamide	12.5. 25. 50 and 100 µg/ml

Streptomycin

1. Dissolve 1 gm powder in 4. 25 ml distilled water to get 200, 000 µg/ml.
2. Dilute 1 ml of stock solution with 19 ml distilled water to get 10,000 µg/ml.
3. Add 0.96 ml of the above to 29.04 ml of salt solution with malachite green (SSMG) to get 30 ml of 320 µg/ml.
4. Arrange a row of six sterile flask (125 ml).
5. To the first flask add 22.4 ml of SSMG and to the remaining 24 ml each.
6. Mix and transfer 24 ml to the next flask and so one to the sixth. Discard 24 ml from the sixth flask.
7. Add 40 ml of egg solution to all the flasks to get final concentrations.

SSMG (ml)	Antibiotic	Egg (ml)	Final conc. (µg/ml)
22.4	25.6 ↓	40	64
24	24 ↓	40	32
24	24 ↓	40	16
24	24 ↓	40	8
24	24 ↓	40	4
24	24	40	2

INAH

1. Dissolve 50 ml mg powder in 5 ml distilled water to get 10000 µg/ml
2. Make further dilutions in distilled water as shown below:

Antibiotic	Amount (ml)	Water (ml)	To get µg/ml (µg/ml)
10,000	0.10	4.90	200
10,000	0.04	9.96	40
40	1.00	4.00	8
40	0.50	4.50	4

Subsequent Dilutions in LJ Medium

Antibiotic (Conc. ml)	Final conc. (µg/ml)	LJ Medium (ml)	Final conc. (µg/ml)
200	1.75	68.25	5.0
40	1.75	68.25	1.0
40	0.6	59.4	0.4
8	1.5	58.5	0.4
4	1.5	58.5	0.2
4	0.75	59.25	0.05

THIACETAZONE

Prepare Stock of (160 µg/ml)

SSMG (ml)	Antibiotic (ml)	Egg (ml)	Final conc. (µg/ml)
22.4	25.6 (160 µg/ml) ↓	40	32
24	24 ↓	40	16
24	24 ↓	40	8
24	24 ↓	40	4
24	24 ↓	40	2
24	24 ↓	40	1
24	24 ↓	40	0.5
24	24 ↓	40	0.25
	24 discard		

ETHIONAMIDE

1. Dissolve 40 mg powder in 2. 8 ml of trigol by keeping in a waterbath at 60°C.
2. Mix with 47.2 ml of SSMG to get 800 µg/ml
3. Mix 22 ml of this with 9.2 ml of SSMG to get 570 µg/ml.

SSMG (ml)	Antibiotic (ml)	Egg (ml)	Final conc. (µg/ml)
22.4	25.6 (800 µg/ml) ↓	40	160
24	24 ↓	40	80
24	24 ↓	40	40
24	24	40	20
	24 discard		
22.4	24 (570 µg/ml) ↓	40	114
24	24 ↓	40	57
24	24 ↓	40	28.5
24	24 ↓		
	24 discard		

RIFAMPICIN

1. Dissolve 0.1 gm of power in 1 ml of alcohol.
2. Add 10 ml of distilled water to get 10000 µg/ml.

Rifampicin Concentration µg/m	Amount ml	Water ml	To get µg/ml (µg/ml)
10,000	0.4	1.6	2000
10,000	0.3	2.7	1000
1,000	2.0	2.0	500

These dilutions of antibiotics are mixed with LJ medium as shown below.

Antibiotic conc.	ml	Water ml	To get µg/ml (µg/ml)
10000	0.64	50	128
10000	0.32	50	64
2000	0.8	50	32
1000	0.8	50	16
500	0.8	50	8
500	0.4	50	4

CYCLOSERINE

1. Dissolve 10 mg powder in 50 ml SSMG to get 200 µg/ml.
2. In another flash mix 21 ml of this with 9 ml of SSMG to get 140 µg/ml.
3. To prepare medium follow steps given below.

SSMG ml	Antibiotic ml	Egg ml	Final conc. (µg/ml)
22.4	25.6 (200 µg/ ml) ↓	40	40
24	24 ↓	40	20
24	24 ↓	40	10
24	24 24 discard	40	5
22.4	24 (140 µg/ ml) ↓	40	28
24	24 ↓	40	14
24	24 ↓ 24 discard		

Imspissate the media with antibiotics, once only at 85°C for 50 minutes. Can to stored for one month at refrigerated temperature.

Use

To test the antibiotic susceptibility of *M. tuberculosis*.

Lysine Iron Agar (LIA)

Peptone	5.0 gm
Yeast extract	3.0 gm
Glucose	1.0 gm
L-lysine	10.0 gm
Ferric ammonium citrate	0.5 gm
Sodium thiosulfate	0.04 gm
Brom cresol purple	0.02 gm
Agar	15.2 gm
Distilled water	1000 ml

Adjust pH to 6.7. Dispense in 4 mi amounts in 13 × 100 mm tubes and sterilise at 121°C for 12 minutes. Slant tubes so as to obtain a deep but and a short slant.

Methyl Red Indicator for MR Test

Methyl red	0.1 gm
Ethyl alcohol (95%)	300 ml

Dissolve dye in the alcohol and then add sufficient distilled, water to make 500 ml.

Methylene Blue Indicator

Solution 1.

6% glucose in distilled water.

Solution 2.

6 ml of N/10 NaOH, diluted to 100 ml with distilled water

Solution 3.

3 ml of 0.5% aqueous methylene blue diluted to 100 ml with distilled water (0.015%).

Each time the indicator solution is required, equal parts of the three stock solutions are mixed together in a test tube. The mixture boiled until the methylene blue is reduced.

Use

This used as an indicator for anaerobiosis.

Oxidation Fermentation (OF) Medium

(Hugh Leifson Medium, 1953)

Peptone (Bacto tryptone)	2.0 gm
Sodium chloride	5.0 gm
Dipotassium phosphate	0.3 gm
Bacto brom thymol blue	0.08 gm
Bacto agar 2.0 gm	1000 ml

Add ingredients to one litre distilled water and heat to dissolve. Adjust pH to 7.1.

Add glucose (or any required carbohydrate) 1 gm to 100 ml of the medium. Dispense in 3 ml amounts into small test tubes. Pour and autoclave.

Thioglycollate Broth

Trypticase peptone	7.0 gm
Phytone peptone	3.0 gm
Dextrose	6.0 gm
Sodium chloride	2.5 gm

Sodium thioglycollate	0.5 gm
Agar	0.7 gm
L-cystine	0.25 gm
Sodium sulphite	0.10 gm
Resazurin	0.001 gm
Distilled water	1000 ml

Mix the ingredients in distilled water. Heat with frequent agitation and boil until solution is complete. Dispense in 15 × 150 mm test tubes to fill about 2/3 of it's length, i.e. approximately 10 ml. Sterilise by autoclaving at 118°C for 15 minutes. Store at room temperature. Do not use the medium if more than the upper third of the column is pink in colour.

Commercially available powder can be used. Follow manufacturer's instructions for preparation.

Use

For the cultivation of anaerobic and microaerophilic bacteria.

Heat in a boiling water bath and cool before use. Such restoration of anaerobic condition may be done once only.

Salmonella-Shigella Agar (SSA)

Beef extract	5.0 gm
Peptone (BEL polypeptone)	5.0 gm
Bile salts mixture	0.5 gm
Lactose	1.0 gm
Sodium citrate	0.85 gm
Sodium thiosulfate	0.85 gm
Ferric citrate	0.1 gm
Neutral red	2.5 mg
Brilliant green	0. 33 mg
Agar	1.5 gm
Distilled water	100 ml

Dissolve the ingredients except lactose in distilled water by heating. Adjust pH to 7.6. Add 1 ml of 1 per cent neutral red solution to every 100 ml of medium with lactose. Heat with frequent agitation and boil for 1 minute. Cool the medium to 45 to 50°C and pour into plates.

Use

This is highly selective and a differential medium, used for the isolation of *Salmonella* and *Shigella*. It has no added advantage over DCA or DEC.

Selenite-F-Enrichment Broth

Sodium acid selenite	4 gm
Thiotone peptone	5 gm
Lactose	4 gm
Sodium phosphate	10 gm
Distilled water	1000 ml

Mix the ingredients in one litre of distilled water. Adjust pH to 7.0 + 1-0.2. Dispense in 10 - 15 ml amounts in tubes. Sterilise by heat in flowing steam for 30 minutes. Do not autoclave. Excess heat is detrimental. The presence of a small amount of coloured precipitate will not interfere with performance.

Serum Telluite Agar (STA)

Basal medium	
Beef infusion broth	100 ml
Agar	1.5 gm
Peptone	1 gm
Sodium chloride	5 gm
10% Glucose solution (sterile)	2.0 ml
1% Potassium tellurite solution (Sterile)	2.0 ml
Ox serum (Sterile)	5.0 ml

Dissolve the agar, peptone and sodium chloride in beef infusion broth by gentle heating, adjust the pH to 7.4-7.6 and sterilise by autoclaving. Cool the medium to about 45 to 50°C and add the remaining ingredients. Mix well and pour plates.

Use

A selective medium for growing *C. diphtheriae* from clinical specimens. *C. diphtheriae* reduce tellurite to black tellurium intracellularly and so colonies appear black to greyish-black.

Simmon's Citrate Agar

Sodium chloride	5.0 gm
Magnesium sulphate	0.2 gm
Ammonium dihydrogen phosphate	1.0 gm
Potassium dihydrogen phosphate	1.0 gm
Sodium citrate	5.0 gm
Agar	20.2 gm
Bromthymol blue (1/500 aqueous solution)	40.0 ml
Distilled water	1000 ml

Mix the ingredients and adjust pH to 6.9. Sterilise by autoclaving and pour as slopes in enrichment medium for the isolation of small tubes.

Use

To test the ability of an organism to utilise citrate as the sole source of carbon.

Sodium Azide Broth *(Streptococcus Faecalis* (SF) Broth)

Tryptone	2.0 gm
Dextrose	0.5 gm
Dipotassium phosphate	0.4 gm
Monopotassium phosphate	0.15 gm
Sodium azide	0.05 gm
Brom cresol purple	0.0032 gm
Distilled water	100 ml

Dissolve the ingredients in distilled water by keeping in a boiling water bath. Adjust pH to 6.0.

Distribute in 1 ml amounts and autoclave.

Todd-Hewitt Broth (THB)

Dehydrated beef heart infusion	3 01 gm
Neopeptone	20.0 gm
Dextrose	2.0 gm
Sodium chloride	2.0 gm
Disodium phosphate	0 4 gm
Sodium carbonate	2.5 gm
Distilled water	1000 ml

Dissolve the ingredients in distilled water and adjust pH to 7.8 +/– 0.2 and sterilise by autoclaving.

Use

For the cultivation of streptococci including pneumococci particularly good for growing streptococci for grouping as well as M-and T-typing.

Triple Sugar Iron (TSI) Agar

Sodium chloride	0.5 gm
Yeast extract	0.5 gm
Peptone	2.0 gm
Agar	1.5 gm
Distilled water	100 ml

Dissolve by keeping in a boiling water bath and add the following:

Lactose	1.0 gm
Sucrose	1.0 gm
Dextrose	0.1 gm
Sodium thiosulphate	0.03 gm
Ferrous sulphate	0.02 gm

Adjust the pH to 7.6. Then add, phenol red 0.024 gm (Add 2.4 ml of 1 per centsolution). Distribute into small test tubes in 4 ml quantities and autoclave. Keep the tubes in a slanting position immediately after that, so as to get a deep butt and a short slant.

Use

This is an important screening medium used for the identification of GNB, by detecting their ability to ferment glucose, sucrose and or lactose. This medium also detects H_2S production.

Trypticase Soy Broth (TSB)

Trypticase	17 0 gm
Phytone	3.0 gm
Sodium chloride	5.0 gm
Dipotassium phosphate	2.5 gm
Dextrose	2.5 gm
Distilled water	1000 ml

Mix the ingredients in a litre of distilled water warm gently to dissolve. Dispense and autoclave.

Use

An all purpose medium. Is also suitable for blood cultures in place of BHI.

Urea Agar, Christensen's

Urea Solution

Sodium chloride	5.0 gm
Dextrose	1.0 gm
Trypticase	1.0 gm
Monopotassium phosphate	2.0 gm
Urea	20.0 gm
Distilled water	100 ml
Phenol red 1% solution (in alcohol)	1.2 ml

Urea Broth

Urea	2.0 gm
Monopotassium phosphate	0.91 gm
Disodium phosphate	0.05 gm
Yeast extract	0.01 gm
Phenol red	0.001 gm
Distilled water	100 ml

Dissolve the ingredients thoroughly in distilled water and adjust pH to 6.8. Sterilise by filtration. Alternatively the basic medium may be autoclaved at 115°C for 10 minutes and the filtered urea solution added. Distribute the broth in 1 ml amount in small sterile test tube.

Urea Agar Base

Agar	1.5 gm
Distilled water	90 ml

Dissolve the ingredients for the solution in distilled water and adjust pH to 6.8. Then add phenol red. Sterilise by filtration. Keep this stock solution in the cooler.

Dissolve the agar in distilled water and sterilize by autoclaving. Cool to 45°C and add 10 ml of urea solution. Dispense in 3-4 ml quantities in 12 × 100 mm test tubes. Allow it to solidify to form a small but and a long slant.

Use

To test the ability of an organism to hydrolise urea.

Urea Broth for Mycobacteria

Peptone	1.0 gm
Dextrose	1.0 gm
Sodium chloride	5.0 gm
Potassium phosphate	
Monobasic	0.4 gm
Urea	20 gm
Phenol red 1% solution	1 ml
Tween 80	0.1 ml
Distilled water	1000 ml

Dissolve ingredients in distilled water. Adjust the pH to 5.8 +/- 0.1. Sterilise by filtration. Distribute in 1.5 ml amounts in screw capped tubes. Store at 4°C up to two months.

Inoculate heavily from a solid medium. Incubate at 37°C for 1-7 days.

Xylose Lysine Desoxycholate Agar (XLD Agar)

Xylose	3.5 gm
Lysine	5.0 gm
Lactose	7.5 gm
Sucrose	7.5 gm
Sodium chloride	5.0 gm
Yeast extract	3.0 gm
Phenol red	0.08 gm
Agar	13.5 gm
Sodium desoxycholate	2.5 gm
Sodium thiosulfate	6.8 gm
Ferric ammonium citrate	0.8 gm
Distilled water	1000 ml

Dissolve all ingredients in distilled water by heating. Adjust pH to 7.4. Do not autoclave but boil. Pour into Petri dishes.

Use

This is a differential as well as a partially selective medium for *Salmonella and Shigella*.

Willis and Hobbs Medium (WHM)

Nutrient agar	100 ml
Neomycin solution 1%	1.0 ml
Neutral red solution 1%	0.3 ml

Dissolve the agar, lactose and neutral red by heating. Sterilise by autoclaving. Cool to 45-50°C. Add the sterilised skimmed milk, egg yolk suspension and neomycin solution. Mix thoroughly. Pour into sterile Petri dishes, 25 ml per plate.

Used for the isolation and identification of clostridia; especially for the specific identification of *C. perfringens*. Ideal medium for studying saccharolytic, proteolytic and lipolytic activities of clostridia.

Robertson's Cooked Meat Medium (RCM)

Beef infusion broth
Minced and dried meat

Use lean beef. Remove fat and connective tissue before grinding. Mix meat (one part) and water (two parts). Cool, refrigerate overnight and skim off any remaining fat. Boil for 30 minutes. Filter the mixture through two layers of gauze and spread the meat particles out to dry. Adjust the pH of filtrate to 7.4-7.6.

The dried meat particles are distributed in 15 × 150 mm or 12 × 100 mm test tubes to a height of 1.5 to 2.5 cm (1 part). The pH adjusted filtrate is then added to get 3 to 4 parts (v/v) liquid per tube (preferably screw cap tubes) The tubes are plugged and sterilised by autoclaving.

This is suitable for growing anaerobes preferably in the 15 × 150 mm tubes For the preservation of stock cultures use the 12 × 100 mm tubes with paraffin coated cork.

Sabouraud Dextrose Broth (SDB)

		Modified
Dextrose	40 gm	20 gm
Peptone	10 gm	10 gm
Distilled water	1000 ml	1000 ml
Adjust final pH	5.6	6.8 to 7.0

Dissolve the ingredients in water by heating. Adjust pH. Autoclave at 121°C for 10 minutes.

Use

A liquid medium for fungi, particularly to grow fungi from blood and tissue aspirates.

Sabouraud Dextrose Agar (SDA)

	Original	Emmon's
Dextrose	40 gm	20 gm
Neopeptone	10 gm	10 gm
Agar	20 gm	20 gm
Distilled water	1000 ml	1000 ml
Adjust final pH	5.6	6.8 to 7.0

Mix the ingredients in water by heating adjust the pH sterilise in the autoclave at 121°C for 10 minutes.

Use

For primary isolation of most fungi.

Sabouraud Dextrose Agar with Antibiotics

Sabouraud dextrose agar	100 ml
Chloramphenicol	5 mg
Gentamicin	0.5 mg

Cool 100 ml of sterile SDA to 50°C. Add 5 mg of chloramphenicol, dissolved 1 ml of 95 per cent alcohol and gentamicin. Mix and pour plates.

Use

This medium is used as a primary isolation medium for materials heavily contaminated with bacteria. Chloramphenicol partially inhibits actinomycetes.

SDA Diluted

Agar	15 gm
Glucose	2 gm
Peptone	1 gm
Potassium dihydrogen Phosphate	1 gm
Distilled water	1000 ml

Dissolve, autoclave and pour into Petri dishes.

Use

This medium induces sporulation.

Carbohydrate Fermentation Media

Prepare sugar solutions as described below for different groups of organisms and dispense in 3-4 ml quantities into test tubes (12 × 100 mm). Introduce Durham's tubes into glucose broth for the detection of gas production. Autoclave at 115°C for 10 minutes. Disaccharides like lactose and sucrose are better filtered and added to sterile basal medium.

For Enterobacteriaceae

(Glucose, lactose, sucrose, maltose, mannose, arabinose, trehalose, rhamnose, mannitol, dulcitol, adonitol, inositol, sorbitol and salicin).

Sugar	0.5 gm
Nutrient broth base	100 ml
Brom thymol blue indicator (0.2% alcoholic)	1.2 ml

Note:

Inulin — Prepare 1.25 gm in 100 ml.

Xylose— Prepare 1 gm in 100 ml and use.

Andrade's indicator.

For Corynebacteria

(Glucose and Sucrose)

Sugar	1.0 gm
Peptone water	100 ml
Andrade's indicator	1 ml

Andrade's Indicator

Acid fuchsin	0.5 ml
Sodium hydroxide (IN)	16 ml
Distilled water	100 ml

Starch, dextrin and maltose—prepare 0.5 gm in 100 ml and use bromothymol blue indicator.

For Anaerobes

Thioglycollate medium without dextrose and indicators 100 ml

Carbohydrates 1 gm

Add the indicatore (1 drop/tube) only after anaerobic incubation is over, as it is reduced in the anaerobic atmosphere.

For Candida

(Glucose, Sucrose, Maltose, Lactose and Galactose).

Sugar	1.0 gm
Nutrient borth	100 ml
Bromothymol blue (0.2% alcoholic)	12 ml

Distribute as 5 ml per tube and introduce Durham's tube inside all broth. Autoclave at 115°C for 10 minutes.

Carbohydrate media for assimilation tests of *Candida*.

Yeast Nitrogen Base

Yeast nitrogen base	6.7 gm
Distilled water	100 ml

Mix and sterilise by filtration

Agar Medium

Agar	2.0 gm
Distilled water	100 ml

Dissolve agar by heating and distributes as 13. 5 ml. Volumes into test tubes. Autoclave.

Carbohydrate Solution (20%)

(Dextrose, Galactose, Lactose, Maltose, Raffinose, Sucrose and Cellobiose)

Sugar	20.0 gm
Distilled water	100 ml

Dissolve and sterilise by filtration and keep as stock solution.

Crystal Violet Blood Agar (CVBA)

Trypticase soy agar medium	100.0 ml
Crystal Violet 0.1% aqueous solution	0.2 ml
Sheep blood	5-7 ml

Mix TSA and crystal violet. After autoclaving, cool to 45 to 50°C. Add sheep blood., mix and pour plates.

Use

To isolate streptococci from skin and throat. This medium inhibits staphylococci and *Bacillus subtilis.*

Gelatin Medium

Heart infusion broth	2.5 gm
Gelation	12.0 gm
Distilled water	1000 ml
Adjust pH 7.2-7.4	

Melt to dissolve the ingredients. Dispense 5 ml per tube. Autoclave.

Use

To test proteolytic activity of organisms.

Cornmeal Agar

Yellow cornmeal	40 gm
Distilled water	1000 ml
Agar	20 gm

Mix cornmeal in distilled water. Heat to 60°C for one hour. Filter through gauze and make up volume to 1000 ml. Add 20 grams of agar and steam for one hour. Filter through two layers of cotton and gauze. Distribute into tubes and sterilise again in autoclave at 121°C for 15 minutes.

Use

This is a good sporulation medium for dermatophytes.

Cornmeal Agar with 1% Tween 80

Tween 80	1 ml
Cornmeal agar	100 ml

Melt cornmeal agar and add the tween 80. Mix and autoclave at 115°C for 15 minutes.

Use

This is an excellent chlamydospore forming medium for *C. albicans*.

Cornmeal Agar with 1% Dextrose

Dextrose	1 gm
Cornmeal agar	100 ml

Add the dextrose to the melted cornmeal agar and autoclave at 115°C for 15 minutes.

Use

Since *T. rubrum* produce pigments on this medium, it can be used to differentiate *T. rubrum* and *T. mentagrophytes*.

Loeffler's Serum Medium

Sterile ox, sheep or horse serum	75.0 ml
Nutrient broth	25.0 ml
Glucose	0.25 gm

Dissolve the glucose in the broth and autoclave at 121°C for 10 minutes. Add the glucose broth to the serum with sterile precautions. Distribute in screw capped test tubes. (7 ml per tube). Inspissate in a slanting position in as inspissator or a steam steriliser. **Do not overheat.**

First day 80-85°C for 50 minutes

Second day 80-85°C for 30 minutes

Medium may be stored for long periods. Do not use if drying has occurred.

Use

This medium is used for the luxuriant growth of C. *diphtheriae* and to see the typical morphology of the organisms by special stains.

Monsur's Tellurite Taurocholate Gelatin Agar (TTGA) Medium

Sodium chloride	1 gm
Trypticase	1 gm
Gelatin	0.3 gm
Sodium taurocholate	0.5 gm
Agar	1.5 gm
Distilled water	100 ml

Mix the ingredients and heat to dissolve. Adjust pH to 8.6. Autoclave at 121°C for 15 minutes. Pour as plates.

Use

A highly selective medium for the isolation of V. *cholerae.*

Novy-Macneal-Nicolle (NMN) Medium

Agar	14 gm
Sodium chloride	6 gm
Distilled water	900 ml
Rabbit blood defibrinated	10 ml

Add salt and agar into boiling water. After the agar is dissolved, allow the mixture to cool slightly and dispense in screw capped tubes (one third full).

Plug the tubes and sterilise by autoclaving. Remove and store in a refrigerator.

For use place the tubes in hot water to melt the agar and allow it to cool to 48 to 50°C. Add one-third the original volume of sterile defibrinated blood and mix the contents thoroughly. Allow the tubes to cool in a slanted position.

Use

Used for the cultivation of haemoflagellates. Use freshly prepared only.

Potato Dextrose Agar

Use commercially available potato dextrose agar powder and follow instructions given by manufacturer. Medium contains potato infusion, dextrose and agar.

Rice Grain Medium

Rice grain	8 gm
Water	25 ml

Use

This medium is used to differentiate the *Microsporum* species.

BASIC TECHNIQUES IN THE PREPARATION OF CULTURE MEDIA

Some basic tecnhiques need to be mastered before you are ready to prepare the culture media. The microbes may fail to grow if the media are not properly made and this leads to false-negative results.

pH Adjustment of Culture Media

Most media reqpuire pH adjustment, which should be done prior to sterilization and in the liquid phase. Use 1 N Sodium hydroxide solution (40 gr of NaOH pellets dissolved in distilled water and made up to 1000 ml in a volumetric flask).

Determination of pH

Determine the pH of the medium by pH paper or pH indicator solutions. Indicator solutions with varying pH may be prepared in the laboratory by combining 5 ml of buffer with different pH values (6.0 , 6.5, 7.0, 7.5, 8.0) and 0.5 ml of 0.2 per cent phenol red indicator.

Titration

Determine the amount of alkali needed to increase the pH to the desired level by the following steps:
a. Add 1 ml of the medium (at room temperature) in a conical flask (100 ml).
b. Add 1 ml of phenol red indicator (0.02%) w/v, aqueous solution.
c. Titrate the medium with 0.1 N sodium hydroxide (dilute 1 N 1:10) taken from a burette. Always add the titrant one drop at a time during titration, thoroughly mix, observe the colour change, and then add the next drop.
d. Stop adding the titrant (0.1 N NaOH) when the colour of the titrated fluid attains the desired pH, as indicated by matching the colour with the buffer standards.
e. Note the amount of 0.1 N NaOH required for bringing 10 ml of the medium to the desired pH.

Addition of Alkali

This is done by adding the requisite amount of 1 N sodium hydroxide solution to the medium. The amount of alkali needed for the pH adjustment is determined by the following calculation:

Required amount of 1 N NaOH = $(V_t \times V_m)/10 \times 0.1$

V_t = Volume of titrant required for the adjustment for pH adjustment of pH with 10 ml aliquot

V_m = Total quantity of culture medium for pH adjustment (100 ml)

1.0 = Aliquot (10 ml) of the medium taken for titration

0.1 = Alkali to be used for adjusting pH (1N) is 100 times more concentrated than the titrant (0.1N)

Making of Cotton Plugs

Cotton plugs are used for preventing contamination of culture media placed in tubes or Erlenmeyer flasks. Although they are cheap, the cotton plugs have to be replaced after repeated usage. The cotton plugs allow exchange of gases between the inner chamber of the culture tube and the outside atmosphere without allowing the organisms to through. Thus they are ideal for aerobic cultures.

Plugs are made by stuffing an appropriate amount of cotton into each tube or flask with a hard rod (e.g. handle of an inoculating loop).

1. Tear a strip of cotton 2 cm wide and 9 cm long for each tube. This size is good for culture tubes but as bigger size will be required for Erlenmeyer flasks.
2. Centre the cotton strip over the mouth of the tube or flask. Push the cotton plug into the tube with the handle of the inoculating loop.
3. Half of the cotton should be inside the mouth of the tube and half of it protruding from the tube.

Preparation of Culture Media

Procedure

Weight the medium (or ingredients) carefully and accurately. Measure the requisite amount of water and pour it in a heat—resistant beaker. The capacity of the container should be larger than the amount of medium prepared.

Place the weighted dry medium in the beaker with the water, and bring the water gradually to full boiling on a heat source like a Bunsen burner or a hot plate. While boilding the medium, stir it constantly with a glass stirring rod to prevent burning or boiling over. Once the medium comes to the boil, turn off the hot plate immediately. Take a small amount out, cool and check the pH. If necessary, adjust the pH. The hot agar solution is then transferred to test tubes (15 ml). After filling the requisite number of test tubes, the rest of the medium can be poured flasks or other containers. Sterilise the tubes and bottles in an autoclave.

Preparation of Agar Slant

Agar slants or slopes are merely test tubes containing an agar medium that was placed at an angle during cooling. The sloped surface of the agar provides more surface area, which is easily inoculated with a loop or needle. For stab cultures, a straight needle is used to streak the inoculum on the surface, and to stab down to the bottom of the agar medium called butt. The butt has lower oxygen tension than the surface. Hence, the biochemical reactions on the slope and the butt can be helpful in the identification of the pathogen on TSI (triple sugar iron slopes). In addition, when motility medium is used, stabbing helps to identify motility of an organism.

Place the medium (with 1.5 agar) in test tubes, and sterilise by auto-claclaving. If the tubes contain sterilised medium, but cooled in an upright position the agar surface does not have a slope. First melt the agar in a boiling water both and cool the medium in an inclined position. Increased inclination will allow more surface, but if a butt is needed, make the slope at 20° angle or less.

Preparation of Culture Plates

For aerobic culture of an organism, the use of Petri dishes is convenient. In addition, the petri dishes allow the organism to be isolated in pure colonies from the commensals. A culture plate of standard size contains 20 to 50 ml of sterile medium.

Procedure

1. Sterilise the medium by autoclaving.
2. Cool the agar to 55°C.
3. Pour the cooled melted agar into previously sterilised Petri dishes under aseptic conditions.
4. Allow the medium to solidify in a flat position and leave the Petri dishes overnight at room temperature in an inverted position; the lid of the Petri dish, which is bigger in size, is placed at the bottom, keeping of Petri dishes in the inverted position during storage and incubation prevents condensation on the lid.
5. Store the plates in the refrigerator in a cool place in inverted position.
6. Biphasic medium for blood culture is prepared by sterilising the medium in bottles with screw-caps and solidifying the agar in a slanted position. To this sterilised nutrient broth is added under aseptic conditions.

Chapter

6

Culture Methods

Various culture methods are performed for the following purposes:
1. Isolate bacteria in pure culture
2. Demonstrate their properties
3. Obtain sufficient growth for preparation of antigens and for others tests;
4. Type isolates by methods such as bacteriophage and bacteriocin susceptibility
5. Determine sensitivity to antibiotics
6. Estimate viable counts; and
7. Maintain stock cultures.

The methods of culture used ordinarily in the laboratory are the streak, lawn, stroke, stab, pour plate and liquid cultures. Special methods are employed for culturing anaerobic bacteria.

The streak culture (surface plating) method (Fig. 6.1) is routinely employed for the isolation of bacteria in pure culture from clinical specimens.

One loopful of the specimen is transferred onto the surface of a well dried plate, on which it is spread over a small area at the periphery. The inoculum is then distributed thinly over the plate by streaking it with the loop in a series of parallel lines, in different segments of the plate. The loop should be flamed and cooled between the different sets of streaks. On incubations, growth may be confluent at the site of original inoculation, but becomes progressively thinner, and well separated colonies are obtained over the final series of streaks.

The lawn or carpet culture provides a uniform surface growth of the bacterium and is useful for bacteriophage typing and antibiotic sensitivity testing (disc method). Lawn cultures are prepared by flooding the surface of the plate with a liquid culture or suspension of the bacterium, pipetting off the excess inoculum and incubating the plate. Alternatively, the surface of the plate may be inoculated by applying a swab soaked in the bacterial culture or suspension.

The stroke culture is made in tubes containing agar slope (slant) and is employed for providing a pure growth of the bacterium for slide agglutination and other diagnostic tests.

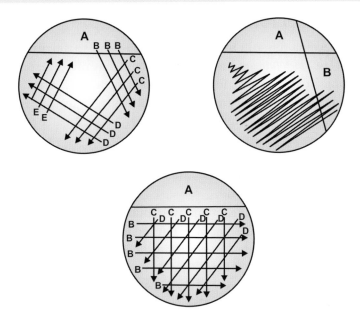

Fig. 6.1: Culture methods

Stab cultures are prepared by puncturing a suitable medium such as nutrient gelatin or glucose agar with a long, straight, charged wire. Stab cultures are employed mainly for demonstration of gelatin liquefaction and oxygen requirement of the bacterium under study. They are also used in the maintenance of stock cultures.

For preparing pour plate culture, tubes containing 15 ml of the agar medium are melted and left to cool in a water bath at 45-50°C. Appropriate dilutions of the inoculum (of 1 ml) are added to the molden agar, mixed well and the contents of the tubes poured into sterile Petri dishes and allowed to set. After incubation, colonies will be seen well distributed throughout the depth of the medium and can be enumerated using colony counters.

In the sweep plate method, the edges of the Petri dishes containing the culture medium are rubbed over the fabric, with the medium facing in. The dust particles stirred up from the cloth, settle on the culture medium, and colonies develop on incubation.

Liquid cultures in tubes, bottles or flasks may be inoculated by touching with a charged loop or by adding the inoculum with pipettes or syringes. Large inocula can be employed in liquid cultures and; hence, this method is adopted for blood cultures and for sterility tests.

ANAEROBIC CULTURE METHODS

Anaerobic bacteria differ in their requirement of and sensitivity to oxygen. Some, such as *C. histolyticum,* are aerotolerant and may produce some growth on the surface of aerobic plates, while others such as C. *tetani,* are strict anaerobes and form surface growth only if the oxygen tension is less than 2 mmHg. A number of methods have been described for achieving anaerobiosis.

1. Cultivation in vacuum was attempted by incubating cultures in a vacuum desiccators.

2. Displacement of oxygen with gases such as hydrogen, nitrogen helium or carbon dioxide is sometimes employed, but this method rarely produces complete anaerobiosis. A popular, but ineffective method is the candle jar. Here inoculated plates are placed inside a large airtight container and a lighted candle kept in it before the lid is sealed. The burning candle is expected to use up all the oxygen inside before it is extinguished, but some oxygen is always left behind. The candle jar provides a concentration of carbon dioxide stimulates the growth of most bacteria.

3. In the chemical or biological method, alkaline pyrogallol absorbs oxygen. This method, first introduced by Buchner (1888), has been employed with different modifications for providing anaerobiosis.

The most reliable and widely used anaerobic method is the McIntosh- Fildes anaerobic jar. It consists of a stout glass or metal jar with a metal lid which can be clamped air tight with a screw. The lid has two tubes with taps, one acting as the gas inlet and the other as the outlet. The lid also has two terminals which can be connected to as electrical supply. Leading from the terminals and suspended by stout wires on the underside of the lie is a small grooved porcelain spoon around which is wrapped a layer of palladinised asbestos.

Inoculated culture plates are placed inside the jar, with the medium in the bottom half of the plates, and the lid clamped tight. The outlet tube is connected to a vacuum pump and the air inside is evacuated. The outlet tap is then closed and the 'inlet tube connected to a hydrogen supply'. After the jar is filled with hydrogen, the electric terminals are connected to a current supply so that the palladinised asbestos is heated. This acts as a catalyst for the combination of hydrogen with the residual oxygen present in the jar.

The *Gas-pak* is now the method of choice for preparing anaerobic jars. The Gaspak is commercially available as a disposable envelope, containing chemicals which generate hydrogen and carbon dioxide on the addition of water. After the inoculated plates are kept in the jar, the Gas-pak envelope, with water added, is placed inside and the lid screwed tight.

The Gas-pak is simple and effective, eliminating the need for drawing a vacuum and adding hydrogen. An indicator should be employed for verifying the anaerobic condition in the jars. Reduced methylene blue is generally used for this purpose. It remains colourless anaerobically but turns blue on exposure to oxygen.

Robertson's cooked meat medium is probably the most widely used fluid medium for the culture of anaerobes. It consists of fat-free minced cooked meat in broth, with a layer of sterile vaseline over it. It permits the growth of even strict anaerobes and indicates their saccharolytic or proteolytic activities, the meat being turned red or black, respectively.

For fastidious anaerobes, particularly for quantitative cultures, pre-reduced media and an anaerobic chamber ('glove box') may be used. The anaerobic chamber is an airtight, glass-fronted cabinet filled with inert gas, with an entry lock for the introduction and removal of materials, and gloves for the hand.

MAKING OF WIRE LOOPS FOR INOCULATION

Loops are usually wire rings of approximately 1.5 mm to 3 mm in diameter. The thickness of the wire as measured by the standard wire gauge (SWG) is 26 or 27.
1. Wind the wire once round a metal rod of appropriate diameter.
2. With a pair of old scissors, cut one arm of the wire at the junction.
3. Bend back the loop to centre it.
4. Insert into a metal wire holder (commercially available).

Culture Methods

Microorganisms are commonly grown in Petri dishes or test tubes. While transferring the inoculum on the surface of agar in a Petri dish the following method is recommended:
1. Sterilise the loop (or needle).
2. Cool the loop in the air so that the hot needle does not cremate the living bacterial cells.
3. Pick up the specimen (or colony) to be plated with the sterilised loop.
4. Open the lid halfway and streak the plate.

Procedure of Streaking on Primary Medium

Specimens contain the infectious agent along with the members of normal flora (commensals) and contaminants. Pure culture of the infectious agent is necessary for the correct identification of the pathogen. Streaking is one of the common methods of obtaining pure cultures on the primary plate. In this process small amount of the specimen (called inoculum) is spread out (called streaking) on a solid medium which leads to the development of mixed discrete colonies of inhabiting organisms in the specimen. An isolated colony can then be transferred to a secondary medium to obtain a pure culture. The following two methods are more common in the laboratories of developing countries.

Discontinuous Streaking Method

1. Place a small amount of inoculum (a loopful of a broth culture or use a swab for the specimen) near the periphery of the plate.
2. Spread the inoculum over zone by a to-and-fro movement of the loop or swab.
3. Turn the plate 45°, overlap the previous streak at several points and streak on the second zone area. Finally stab the loop into the agar several times so that the organism growing below the agar surface will have reduced oxygen tension during the growth of the organism.
4. Turn the plate again to 45°; flame the loop, allow the loop to cool by touching the hot loop on the agar surface at the periphery. Start streaking and overlop the previous streak at several points; stab the loop as before to complete the streaking.
5. Finally, lift the loop and streak the centre of the plate with zig-zag motion.

While streaking, place the cover of the Petri dish on the bench face down, and complete the streaking fast.

Discrete colonies should be found in the central portion. If a blood agar plate is used, additional information of the haemolytic activity may be gained from the stabbed areas.

Continuous Streaking Method

This is recommended for broth culture.

1. Place a loopful of the inoculum near the periphery of the Petri dish and cover a zone with close parallel streaks.
2. Turn the plate at right angles, and streak approximately one half of the remaining portion without overlapping previous streaks.
3. Turn the plate 180° and streak the remainder of the plate, avoiding previous streaked areas. Do not sterilise the loop before competing step (3).

Inoculation of test tubes: While transferring the inoculum into a test tube, the following method is recommended. In general, the cotton plug or cap is removed with, and held in, the outer three fingers of the hand holding the inoculating needle. The lip of the tube is flamed before and after entry to prevent contamination of the culture. The transfer is performed with the sterilised inoculating loop.

1. Assemble the materials needed on the table.
2. Hold the tube in your left hand.
3. Hold the inoculating loop in your right hand (for right hander) and sterilise by flaming. Cool the loop.
4. Pick up the inoculum with the sterilised loop.
5. Remove the capall or other closure by wrapping the little finger of your right hand around the capall of the tube nearest to your right hand. Grab only the uppermost portion of the capall. The tube should not be held vertically once the capalls are removed. Do not allow the loop with the inoculum to touch any surface other than the medium.
6. Flame the neck of the tube immediately by passing it back and forth through the flame twice.
7. Transfer the inoculum (the small portion of the specimen held on the loop) to the surface of the tubed-medium.
8. Re-flame the neck of the tube.
9. Re-cap the tubes, replacing the capall, which is held by the little finger.
10. Re-flame the tubes, killing all the bacteria on it. Do not ever put down the loop without flaming.
11. Return the tubes and the loop to the test tube rack.

ANAEROBIC

Candle Jar Technique

1. Take a glass container that holds several Petri dishes and can be tightly sealed. If you are using a screw-capped jar, make sure that there is rubber seal. If glass jars are used, put grease on the rim so that the glass plate fits air-tight.
2. Place the inoculated plates in the jar in an inverted position (lid facing down).
3. Now affix a candle to the inside of a Petri dish lid, and set it in the jar on top of the inoculated plates.
4. Light the candle and seal the jar tightly. The oxygen inside the jar will be used up by the burning candle ad replaced by carbon dioxide from combustion. When the oxygen has been replaced by carbon dioxide, the candle will go out, and a 3 to 5 per cent carbon dioxide atmosphere will then be created. This suits the needs of microaerophiles.

THIOGLYCOLLATE BROTH METHOD

This is one of the most common methods of producing an anaerobic environment for microbes by introducing a reducing agent (sodium

Thioglycollate) into a liquid medium. The reducing agent removes oxygen from the liquid medium through a chemical reaction. Methylene blue or reazurin dye may be added to the medium as indicators. These dyes are colourless in the reduced state (when free oxygen is not present). When the dye molecules become oxidized, their respective colours reappear-blue for methylene blue and pink for resazurin. Resazurin is less inhibitory to delicate organisms than methylene blue and hence used more often. When allowed to stand, the upper levels of the thioglycollate broth, where oxygen had diffused into the medium, will have a pink tinge if the broth contains resazurin, or blue if the indicator is methylene blue. The surface of the medium allows the growth of aerobes. Facultative anaerobes thrive throughout the medium. Microaerophiles are localized under the surface while the obligate anaerobes (e.g. Clostridium, Actinomyces) tend to grow throughout the depth of the broth but are more concentrated at the bottom.

A small amount agar (0.5%) is also added to the medium which tends to localize bacterial growth and encourages anaerobiosis. Store the thioglycollate broth at room temperature and if the surface of the broth indicates oxygenation by the change of the colour of the indicator, use the broth after boiling and cooling. This will drive oxygen out of the medium. Repeated boiling and cooling are not recommended.

Alkaline-pyrogallol Method

In this method, the reducing agent is placed in the aerial space of the culture tube that allow the anaerobe to grow on an agar slant. This procedure helps in the isolation of the organism in pure from which is not possible with the broth culture. The method is tricky and hence not too popular. It is adopted only when it is absolutely necessary.

An anaerobic condition is established by activating pyrogallol with sodium carbonate, which then acts as a strong reducing agent and removes the free oxygen from a tightly stoppered tube.

Procedure

1. Use nutrient agar slant with a cotton plug and inoculate it with the anaerobic microorganism to be tested.
2. After completing the inoculation, return the cotton plug to the tube, and cut off the protruding portion of the cotton plug. Use the handle of your inoculating loop to push the remaining half of the plug down into the tube to within 1 cm from the tip of the agar slant. The cotton plug must fit tightly so that the chemicals placed on it will not seep into the agar until

you can invert the tube. At the same time, it must allow free passage of gases between the two compartments created by the cotton plug.

3. With a clean weighing spoon, now add pyrogallol crystals to within about 2 cm of the neck of the tube.

4. Add one dropper full of 10 per cent sodium carbonate (w/v aqueous solution) to the dry pyrogallol. Work as rapidly as possible from this point on to avoid seepage of these chemicals through the cotton onto the agar. Furthermore, the small amount of sodium carbonate and pyrogallol used can only reduce a limited amount of oxygen. It is a good idea to practice these steps with an empty tube.

5. Insert a rubber stopper into the neck of the tube and simultaneously invert the tube and twist the rubber stopper into place until an a airtight seal is formed.

6. To ensure a tight seal, apply a strip of masking tape over the top of the rubber stopper and wrap a second strip around the rubber stopper juncture several times.

7. Once the tube is inverted, keep it in the inverted position in a wire basket.

8. Incubate at 35°C for 24 to 48 hours (or longer).

9. Note the colour change in the medium which may help in the identification of the anaerobe.

Caution

Use extreme care when removing masking tape and rubber stopper (do this very slowly). Sometimes the organisms produce so much gas that the stopper and contents are forcibly expelled.

Brewer Anaerobic Jar Method and Gas-pak System for Anaerobic Culture

The Brewer anaerobic jar allows the culture of anaerobes on a solid medium placed in Petri dishes. This is convenient for the isolation of colonies which is not possible with thioglycollate broth culture and inconvenient by the pyrogallol method. The anaerobic of the Brewer jar is created by physically evacuating the air out with the help of a vacuum pump and the evacuated space is filled with an inert gas like hydrogen, nitrogen or carbon dioxide.

In the Gas-pak system (BBL, Becton Dickinson, USA), oxygen is chemically removed from the chamber. In setting up the chamber, water is introduced into the foil-packaged hydrogen generator which is then promptly placed inside the chamber along with the package of methylene blue indicator that shows the anaerobic condition. The colour of the dye changes if there is air leakage. The hydrogen released from the generator-package combines with oxygen present in the chamber with the mediation of the catalyst palladium. The latter is held by the lid towards the inner side of the chamber (the chamber must always be closed with the lid so that the palladium stays active). The methylene blue oxygen indicator must stay colourless in order to assure the anaerobic condition of the chamber. The chemical reaction is shown below:

Sodium borohydride + $H_2O \rightarrow 2H_2O$

$2H_2 + O_3 \xrightarrow{\text{Palladium}} H_2O$

Following the addition of water into the hydrogen generator package the chamber is tightly sealed. Droplets of moisture will begin to condense on the inside of the jar within 30 minutes, indicating that hydrogen has been generated which combined with the atmospheric oxygen. The entire Gas-pak jar, along with the culture plate, is kept in the incubator for the specified period (24-48 hours) for the laboratory culture of anaerobic organisms. If the methylene blue develops blue colour during the incubation period, leakage of air is suspected.

Chapter

7

Identification of Bacteria by Biochemical Methods

This chapter deals with the preparation of the biochemicals and the methodology and interpretation of biochemical reactions.

CARBOHYDRATE FERMENTATION

Test Procedure for Aerobes

Inoculate lightly from a young culture from agar slant or broth. Incubate at 37°C for one to five days.

Reading

Positive test is shown by acid production (yellow if bromthymol blue indicator is used and pink if Andrade's indicator is used) and or gas inside the Durham tube.

For example, *E. coli* produces acid and gas in glucose; *S. typhi* acid only.

Carbohydrate Fermentation for Anaerobes

Add 2 drops of a young culture from RCM into each tube. Incubate for 3-5 days in an anaerobic jar. Remove, add the indicator and read.

Reading and Interpretation

A positive test is indicated by the appropriate colour change of the indicator added.

Carbohydrate Fermentation of *Candida*

Culture the test organism in a tube of Nutrient agar for three generations. Add 2 ml of saline to the 3rd generation of yeast, suspend and pipette into a test tube.

Add 0.2 ml of the suspension into each of the sugars. Overlay the tubes with the liquid paraffin and incubate at 37°C for 10 days.

Change in colour indicates a positive tests.

Carbohydrate Assimilation Tests of *Candida*

1. Make saline suspension of test *Candida* as described above.
2. To this suspension add 1.5 ml of yeast nitrogen base.
3. Pour this into 13.5 ml of melted and cooled plain agar
4. Swirl and pour into Petri dish and let it solidify.
5. After the medium is set, place sugar soaked filter paper discs in the designated areas by using 2 mm diameter loop.
6. Incubate at 250°C. Observe for the growth. Those that can assimilate a particular carbohydrate, grow well around the disc pattern of assimilation can thus be noted.

OXIDASE TEST

Reagent

Tetramethyl-p-phenylene diaminé dihydro-chloride 1 per cent aqueous solution.

Prepare the reagent fresh before use.

Procedure

Soak a portion of a filter paper in a Petri dish with few drops of oxidase reagent.

Take a part of the colony to be tested from NA, BA or CA. Do not use colonies from MA or MHA. The colony is taken using a platinum loop or a glass rod. Rub on the soaked filter paper. Use *Ps. aeruginosa* as positive control.

Reading and Interpretation

A positive oxidase test is indicated by the development of a purple colour.

For example, *Neisseria* spp, *Pseudomonas* spp.
1. A rusted loop can give a false-positive reaction.
2. Colonies from MA will show a false-positive pink colour.
3. Colonies from MHA will show a false-negative reaction.

Oxidase test is usefully employed to pick out pathogenic *Neisseria* spp from a mixed culture of urethral pus or throat.

Organisms belonging to Enterobacteriaceae are oxidase negative.

CLUMPING FACTOR TEST

Requirement

Human or rabbit sterile plasma.

Place two drops of saline on separate areas on a clean.

COAGULASE TUBE TEST

Requirements

1. Young culture of staphylococci in 0.5 ml NB
2. Human plasma sterilised by filtration, pretested and found satisfactory at 1/4 dilution or undiluted.

Procedure

1. Add 0.5 ml of the plasma to 0.5 ml of the broth culture
2. Incubate at 37°C waterbath. Examine the tubes every 15 minutes by gently inverting the tube. Do not shake.

Reading and Interpretation

A positive test is shown by a clot or coagulum formation, e.g. *S. aureus.*
Absence of clotting shows a negative test.
Usually 80-85% of coagulase positive strains give positive clumping factor.

Citrate Utilisation Test

Inoculate lightly from a young culture over the entire surface of the slant of Simmon's citrate agar. Incubate at 37°C for 2 to 7 days.

Reading

Blue medium with a streak of growth is positive, e.g. *Klebsiella* spp.
Original green colour and no growth indicate a negative reaction.
For example, *E coli.*

Bile Solubility Test

Inoculate the colonies to be tested into two tubes of 1.0 ml NB containing 1- 2 drops of sterile serum. Incubate overnight at 37°C in a CO_2 atmosphere.

Adjust the pH to neutrality (7.0-7.2) with NaOH and phenol red (1 drop) as the indicator. Otherwise the broth solidifies on the addition of desoxycholate.

Add a few drops 10 per cent sodium desoxycholate to one tube. Add a comparable volume of sterile normal saline to the second tube.

A positive reactions is indicated by the clearing of the broth containing bile salt within 15 minutes in contract to the tube with normal saline.

Methyl Red (MR) Test

Inoculate glucose phosphate broth lightly from a young culture.

Incubate at 37°C for 48 hours.

Add 5 to 6 drops of MR reagent to the culture.

Note: Tests should not be carried out with cultures incubated less than 48 hours.

Reading and Interpretation

Results are read immediately. A red colour indicates a positive relation. Negative tests are yellow in colour.

A positive reaction indicates ability of organism to produce and maintain an acid pH, for example, *E. coli.*

Mannitol Motility Test

Stab once down the centre of the medium reaching the bottom but not touching the sides.

Incubate overnight and examine.

Reading and Interpretation

1. Mannitol fermentation is indicated by change of colour of the medium to yellow nitrogen if produced, is trapped inside as bubbles.
2. Motile organisms grow out from the stabbed line throughout the medium. For example, *E. coli.*

   Nonmotile bacteria grow only along the stab line.

   For example, *Klebsiella* spp.
3. This medium can also be used to test nitrate reduction by adding the mixture of reagents A and B as given in nitrate reduction test.

Nitrate Reduction Test

Reagents

Solution A

Dissolve 8.0 gm of sulphanilic acid in 1 litre of 5 N-acetic acid.

Solution B.

Dissolve 5.0 gm of alpha-naphthylamine in 1 litre of 5 N-acetic acid.

Procedure

Inoculate the organism into potassium nitrate medium.

Incubate at 37°C for 24 to 96 hours.

Immediately before use, make the test reagent by mixing equal volumes of solutions A and B.

Add 0.1 ml of the test reagent to the test culture.

Reading and Interpretation

A red colour developing within a few minutes indicates the presence of nitrite and hence the ability of the organism to reduce nitrate to nitrite.

For example, *Enterobacteriaceae.*

If red colour does not develop, a pinch of zinc powder is added and shaken. If a red colour develops on addition of zinc, it confirms that the test was negative.

For example, *Erwinia* spp

If no colour develops following addition of zinc, it confirms that nitrite was further reduced to nitrogen.

For example, *Ps. aeruginosa*

Oxidation Fermentation (OF) Test

Stab 2 tubes of OF medium (Hugh-Leifsons) with a young culture of the test organism. After inoculation one tube is over laid with sterile liquid paraffin to form a 5 mm layer.

Incubate both tubes at 37°C for one to four days.

Reading and Interpretation

If both tubes turn yellow because of acid production, it indicates fermentation of the sugars.

For example, *Enterobacteriaceae.*

Acid formation in the tube with out paraffin overlay only indicates oxidative utilization.

For example, *Ps. aeruginosa (Oxidises glucose)*

Lack of acid production in either tube indicates that the test organism does not utilize the sugar by oxidation or fermentation.

For example, *Acinetobacter*

Phenyl Pyruvic Acid (PPA) Test

N/10 hydrochloric acid.

Ferric chloride 10% (w/v).

Procedure

Inoculate PPA broth with the organism.

Incubate overnight.

After recording the result of the malonate test as described before, acidify the culture with NI 10 HCl drop by drop until the colour of the medium changes to a distinct yellow (acid reaction). Then add 4-5 drops of 10 per cent (w/v) ferric chloride shake gently and note the colour.

Reading and Interpretation

Record immediately, as the green colour indicating a positive reaction fades rapidly.

Strong positive - Dark green to bluish-green

For example, *Proteus* spp.; *Providencia* spp.

Weak positive - Light green.

Negative yellow to buff

Triple sugar iron agar test

Stab the centre of the butt of TSI and on the slope with a needle charged with a single colony of the test organism. Incubate at 37°C for 24 to 48 hours.

Reading and Interpretation

Glucose fermentation is shown by a yellow butt and red slant.

Glucose, sucrose and/or lactose fermentation is shown by a yellow butt and slant.

Gas produced if any, is seen trapped inside the medium.

For details of reactions of different organisms on TSI refer *Faeces* Section.

Note: Retraction of medium by drying at the juction of butt and slope should not be mistaken for gas production.

Urease Test by Christensen's Urea Agar

Inoculate test organism heavily over the entire slope surface and incubate at 37°C in a water bath for 1 hour to 4 hours. If negative at this time, keep for overnight incubation. Hold those with negative reaction for up to 4 days.

Reading and Interpretation

A positive reaction is indicated by a pink or red colour of the medium. Decomposition of urea by urease enzyme results in production of ammonia and CO_2. The alkaline pH produces changes the colour of the medium to pink or red.

Example: Most strains of *Proteus* spp give a positive reaction within 1 to 3 hours. A few give marked reaction by 24 hours. Some *Klebsiella* spp and *Citrobacter* spp give a weak positive reaction after 24 hours.

Salmonella spp and *Shigella* spp are urease negative.

Voges-Proskauer (VP) Test

Inoculate glucose phosphate broth lightly from a young culture.
Incubate at 37°C for 48 hours.

O'Meara's Method

Reagent

Potassium hydroxide	40 gm
Creatinine	0.3 gm
Distilled water	100 ml

Add 0.5 ml of this reagent to the above culture. Place tubes in a 37°C water bath for 4 hours. Aerate by shaking at intervals.

Reading and Interpretation

A positive reaction is denoted by the development of an eosin-pink colour, usually in 2-5 minutes, and indicate that acetyl methyl carbinol has been produced.

Barritt's Method

Reagent

40% Potassium hydroxide	1 ml
5% Alpha-naphthal	3 ml
in absolute ethanol	

Add this reagent to the culture and shake the tube vigorously to ensure maximum aeration.

Reading and Interpretation

A positive reaction is indicated by the development of a pink colour in 2-5 minutes becoming crimson in 30 minutes.

Example: *Klebsiella* spp

Streptococcus Faecalis Broth Test Procedure

Growth on NA or BA is emulsified with a cool loop in a drop of aqueous 0.5 per cent sodium desoxycholate solution placed on a slide.

Reading and Interpretation

In a positive reaction, a mucous like "string" can be observed as the loop is lifted away from the slide.

For example, *Vibrio* spp.

Aeromonas spp is negative.

String Test

Procedure

Growth on NA or BA is emulsified with a cool loop in a drop of aqueous 0.5 per cent sodium desoxycholate solution placed on a slide.

Reading and Interpretation

In a position rection, a mucous like "String" can be observed as the loop is lifted away from the slide.

For example, *Vibrio* spp.

Aeromonas spp is negative.

Indole Test

Ehrlich's Reagent

Paradimethylaminobenzaldehyde	4 gm
Ethyl alcohol (95%)	380 ml
Hydrochloric acid (conc.)	80 ml

Add benzaldehyde to alcohol. Then slowly add acid and mix. The reagent should be prepared in small quantities and refrigerated when not in use.

Alternately **Kovac's reagent** may be used.

Paradimethylaminobenzaldehyde	10 gm
Isoamyl alcohol	150 ml
Hydrochloric acid	80 ml

Procedure

Inoculate the peptone water lightly from a young culture.

Incubate at 37°C for 24 to 48 hours.

Add about 0.5 ml of the reagent along the side of the test tube to form a layer at the top.

Reading and Interpretation

A positive reaction is indicted by the formation of a pink ring at the junction.

For example, *E. coil.* Negative reaction is shown by *K. pneumoniae.*

Note: Indole may be extracted by shaking the 48 hours culture with 1 ml of ether or xylol and then tested with the indole reagent.

Hydrogen Sulfide Production

Saturate filter paper strips (5 × 50 mm) with lead acetate solution. Dry in Hot air oven at 70°C.

Procedure

Inoculate the culture into a tube of peptone water. Introduce one lead acetate paper strip and suspend it inside the tube and hold it tightly the cotton plug. The paper strip should not touch the culture.

Incubate at 37°C overnight.

Reading and Interpretation

H_2S production is seen by the blackening of the tip of the lead acetate paper.

For example, *Proteus* spp, most *Salmonella* spp.

Gelatinase Test

Inoculate the test organism as a stab into nutrient gelation.
Incubate at 37°C for 48 hours to a few days.

Reading

Keep the gelatin cultures in a breaker of ice or in the refrigerator before taking reading.

Liquefaction of gelation, i.e. absence of setting of gelation at low temperature indicates gelatinase activity.

For example, *S. aureus, Pseudomonas* spp.

Note: Gelatin normally melts at about 27°C and sets below 20°C.

Esculing Hydrolysis

Add 2 drops of young culture from NB or RCM into esculing broth. Incubate at 37°C aerobically or in the anaerobic jar for 3.5 days.

Reading and Interpretation

A positive reaction is indicated by the appearance of brownish-black precipitate.

For example, *B. fragilis.*

Decarboxylase Test

Inoculate lightly from a young culture into the three tubes with amino acid media, L-lysine, L-arginine and L-ornithine and also into the control tube Overlay with sterile mineral oil to a depth of 5 mm after inoculation.

Incubate at 37°C. Examine daily for 4 days.

Positive reactions are indicated by alkalinization of the media and a consequent change in the colour of the indicator system from yellow to violet or reddish-violet. Some organisms may give delayed reaction after 3 to 4 days as in the case of arginine decarboxylation for *Salmonella*.

8

Stains Used in Microbiology

Simple stains are used to detect the presence of organisms and their morphology in exudates. This gives a wide knowledge about the basis of pathogens and pathogenicity of the aetiological agents.

Thus, this provides us a provisional diagnosis as well as to proceed with further laboratory analysis that guide to prevention and cure.

The commonly used stains are as follows.

Loeffler's Methylene Blue

Methylene blue	0.2 gm
Absolute alcohol or	
rectified spirit distilled water	10.0 ml
Distilled water	90.0 ml

Dissolve the dye in alcohol and then add water. Filter through a filter paper.

Use

Used to make out clearly the morphology of the organism, e.g. *H. influenzae* in CSF, gonococci in urethral pus.

Polychrome Methylene Blue

This is made by allowing Loeffler's methylene blue to ripen slowly. The stain is kept in bottles which are half filled and shaken at intervals to aerate the contents. The slow oxidation of the methylene blue forms a violet compound that gives the stain its polychrome properties. The ripening takes 12 months, or more to complete or it may be ripened quickly by the addition of 1 per cent potassium carbonate to the stain.

Use

This stain is used to demonstrate McFadyean reaction of *B. anthracis*. The blue bacilli are surrounded by irregular purple capsular material.

Dilute Carbol Fuchsin

Add distilled water to concentrated Carbol Fuchsin (Ref: Ziehl-Neelsen stain) to get a 1110 to 1/15 dilution.

Uses

1. To stain throat swab in suspected Vincents. This is better than Gram's stain for *Borrelia.*
2. As counterstain in Gram's stain.
3. To demonstrate the morphology of *V. cholerae.*

DIFFERENTIAL STAINS

Gram's Stain

The Gram's stain is the most commonly used differential stain and is used to divide bacteria into two major groups. Those, which retain crystal violet after treatment with iodine and alcohol, appear purple or bluish-purple; these are said to be gram-positive. Those, which lose the crystal violet, show the colour of the counterstain employed, red, if safranine has been used; these are said to be gram-negative.

Crystal Violet

Crystal violet	1.0 gm
5% Sodium bicarbonate	1.0 ml
Distilled water	99 ml

    Add 1 gm of crystal violet and the sodium bicarbonate into a mortar and using a pestle, grind until you get a good paste. Then add water and mix well. Filter through a filter paper.

Gram's Iodine

Iodine crystals	2.0 gm
Sodium hydroxide	10.0 ml
Distilled water	90.0 ml

    Add NaOH to the iodine crystals kept in a mortar. Grind with pestle to get a good paste. Add distilled water and mix well. Filter through a filter paper.

Acetone 100%

Safranine

Safranine	0.34 gm
Absolute alcohol or rectified spirit	10.0 gm
Distilled water	90.0 gm

Grind the dye in alcohol and then add water. Filter through filter paper.

Use

Widely used in diagnostic bacteriology, to differentiate gram-positive and gram-negative bacteria and yeast and yeast like organisms.

Acid-Fast Stains

Acid fast bacilli are more difficult to stain than other bacteria, and simple, or Gram's stains do not usually give satisfactory results. Either very long exposure or application of heat is required for penetration of even intense stains. Once stained, the organisms do not give up the stain readily, when decolourised with acid. Hence, they are called acid fast organisms.

Auramine Stain

Auramine powder	0.3 gm
Phenol (Crystalline)	3.0 gm
Distilled water	97.0 gm

Dissolve the phenol in water with gentle heat. Add auramine gradually and shake vigorously, until dissolved. Filter and store in a dark bottle with a stopper.

1% acid alcohol decolouriser:

Conc. HCl	0.5 ml
Sodium chloride	0.5 gm
70% alcohol	100 ml

Pour 70 per cent alcohol into a large flask and place the flask in cold water in a container. Add conc. HCl and cover the top of the flask to prevent escape of fumes leave for 10 minutes decant into a labelled bottle.

Potassium permanganate counter stain:

Potassium permanganate	10 gm
Distilled water	1000 ml

Add potassium permanganate to water and shake to dissolve (0.1% solution)

Ziehl-Neelsen Stain

Carbol fuchsin

Basic fuchsin (powder)	1.6 gm
Phenol liquefied	22.5 ml

Alcohol (95% or absolute)	50 ml
Distilled water	422.5 ml

Dissolve the fuchsin in alcohol using a mortar and pestle. Mix phenol with water and add to the dissolved dye. Filter the mixture before use. Warm phenol at 45°C in a water bath and measure with a warm pipette.

3% Hydrochloric acid alcohol

Concentrated hydrochloric acid	3 ml
Absolute alcohol	97 ml

Sulphuric acid
20-25% H_2SO_4 in water (v/v) (Acid must be added to the water).

Loeffler's methylene blue
See under simple stains.

Acid-fast organisms like *Mycobacterium tuberculosis* will appear pink and other material on the slide (bacteria or cells) will be stained blue.

Gabbet's Stain

Carbol fuchsin—as in Ziehl-Neelsen stain

Gabbet's methylene blue

Methylene blue	2 gm
25% H_2SO_4	100 ml

Dissolve the methylene blue in sulphuric acid. Do not filter, as the acid will dissolve the filter paper.

Use

To demonstrate organisms which are acid fast, but not alcohol fast, e.g. smegma bacilli. This stain combines the decolourisation and counter-staining steps. Some prefer this stain for staining of urine smears.

Kinyoun Acid-fast Staining
Kinyoun's Carbol Fuchsin

Basic fuchsin	4.0 gm
Phenol liquefied	8.0 ml
Alcohol (95%)	20.0 ml
Distilled water	100.0 ml

Grind the basic fuchsin in alcohol and add water slowly while mixing. Add 8 ml of melted phenol to the stain. Filter the mixture before use.

1% Sulphuric Acid (Aqueous solution)
Conc. sulphuric acid   1 ml
Distilled water   99 ml
    Add 1 ml conc. sulphuric acid to 99 ml of distilled water.

Methylene Blue
Methylene Blue   2.5 gm
Alcohol (absolute)   100 gm
    This stain is used to differentiate *Nocardia* from *Streptomyces*. *Nocardia* species, particularly *N. asteroides* are partially acid-fast by this technique. The conidia of *Streptomyces* species are sometimes acid-fast.

Modified Ziehl-Neelsen Stain for *M. leprae* by Cold Staining

Carbol fuchsin—as for Ziehl-Neelsen stain
    Sulphuric acid, 5% solution

Use

This method is known as cold staining, since no heating is done. It is commonly employed to demonstrate and enumerate *M. leprae*. Organisms appear pink in a bluish background.

Modified Ziehl-Neelsen Stain for Spores

Carbol Fuchsin — as for Ziehl-Neelsen
    Acetone 100% or alcohol 95%
    Loeffler's methylene blue-as in simple stain
    Spores will be stained red, vegetative forms blue.

Modified Acid-Fast Stained for Cryptosporidia (Faeces Sample)

Carbol fuchsin—as for Ziehl-Neelsen stain
    1% HCl in alcohol.
    0.4% malachite green or Loeffler's methylene blue.
    Oocysts are characteristically round or slightly ovoid, measuring 4.5-5 µ. They are usually acid-fast, but staining reaction is variable.

Metachromatic Granules

Albert's Stain

Albert's Stain I
Toluidine blue   0.15 gm
Malachite green   0.20 gm

Glacial acetic acid	1.0 ml
Alcohol (95%)	2.0 ml
Distilled water	100 ml

Grind and dissolve the dyes in alcohol; add water and then finally acetic acid. Let stand for 24 hours and filter through a filter paper.

Albert's Stain II (Gram's Iodine)

Iodine	2.0 gm
Potassium iodine	3.0 gm
Distilled water	300 ml

Dissolve iodine and potassium iodine in water by grinding in a mortar with a pestle. Filter through a filter paper.

Use

To demonstrate metachromatic granules in C. *diphteriae.* They appear bluish-black whereas bacillary body appears green.

Ponder's Stain

Toluidine blue	0.02 gm
Glacial acetic acid	1.0 ml
Alcohol (95%)	2.0 ml
Distilled water	100 ml

Grind toluidine blue in alcohol, mix with water and then add glacial acetic acid. Filter through a filter paper.

Use

To stain metachromatic granules which stain purple and the body light blue.

BIPOLAR STAINING

Solution A

Basic fuchsin	0.20 gm
Methylene blue	0.75 gm
Ethyl alcohol	20.0 ml

Solution B

| Phenol 5% | 200 ml |

Pour A to B slowly and filter. Stain smears for 10-20 seconds. Wash with water and blot dry.

Use

The stain is for demonstrating bipolar staining.

CAPSULE STAIN

India Ink

Commercially available India ink is used undiluted.

Use

To demonstrate capsule which is seen as unstained halo around the organisms distributed in a black background, e.g. *Cryptococcus neoformans.*

FOR SPIROCHAETES

Fontana's Stain

Fixative (Ruge's Fluid)

Acetic acid	1 ml
Formalin	20 ml
Distilled water to	100 ml

Mordant

Phenol	1 ml
Tannic acid	5 ml
Distilled water to	100 ml

Stain

Silver nitrate	0.5 gm
Distilled water to	100 ml

Use

This stain is used to stain smears of spirochaetes.

STAIN FOR PNEUMOCYSTIS CARINII

Toluidine O stain	
Toluidine blue	0.3 gm
Concentrated HCl	2.0 ml
Absolute alcohol	140.0 ml
Distilled water	60.0 ml

Dissolve the stain in absolute alcohol. Add 60 ml water and finally add 2 ml of conc. HCl.

Sulfation Reagent

Glacial acetic acid	4.5 ml
Conc. sulfuric acid	1.5 ml

Keep the tube with acetic acid in cold water. Add sulfuric acid drop by drop stirring continuously.

Make fresh reagent for each day's use.

Use

For demonstrating cyst forms of *P. carinii*.

MOUNTING FLUIDS AND STAINS USED IN MYCOLOGY

1. 10% potassium hydroxide

Potassium hydroxide	10 gm
Distilled water	100 ml

Dissolve 10 grams of potassium hydroxide in 100 ml of distilled water.

Use

This technique is used to clear the keratin and debris from skin, hair and tissues so that the fungal filaments and spores are easily seen.

2. **20% potassium hydroxide** is prepared by dissolving 20 grams of potassium hydroxide in 100 ml distilled water.

Use

Hard materials like nail and bone are treated with this for better digestion of tissue.

Lactophenol Cotton Blue Mounting Medium

Phenol (liquefied)	20 ml
Lactic acid	20 gm
Glycerin	40 gm
Cotton blue	0.05 gm
Distilled water	20 ml

Melt the phenol crystals in a water bath, measure and mix with water. Grind 0.05 gm of cotton blue in a mortar with a pestle. Add the diluted phenol little by little and make into a nice paste. Add the remaining amount of diluted phenol and mix well. Then add lactic acid and glycerin and mix well. Filter through a filter paper.

Use

1. For preparing mounts of fungus cultures, particularly from slide cultures for microscopic examination. The filaments and spores are stained light blue.
2. For demonstrating *Acanthamoeba* cysts in corneal scrapings.

Ascospore Stain

Malachite Green 1%

Malachite green	1 gm
Phenol (liquefied)	1 ml
Distilled water	100 ml

Mix phenol with water. Grind malachite green and add diluted phenol. Filter through filter paper.

Safranine Solution

Safranine	0.5 gm
Distilled water	100 ml

Grind safranine in a mortar and add 100 ml distilled water. Filter through filter paper.

Use

To detect the presence of ascospores in *Saccharomyces cerevisea* and other Ascomycetes.

STAIN FOR PARASITES

Schaudin's Fixative

Stock solution

Mercuric chloride as saturated	
Aqueous solution	600 ml
Ethyl alcohol 95%	300 ml

Immediately before use, add glacial acetic acid 5 ml per 100 ml of stock solution.

PVC Fixative

Polyvinyl alcohol (PVA)	10 gm
Ethyl alcohol	62.5 ml
Mercuric chloride saturated aqueous	125 ml
Acetic acid glacial	10 ml
Glycerine	3 ml

Mix the liquid ingredients in a 500 ml beaker. Add PVA powder. Do not stir. Cover the beaker with a Petri dish or wax paper or foil. Allow PVA to soak overnight. Heat the solution slowly to 75°C. When this temperature is reached, remove the beaker and swirl the mixture, until a homogeneous slightly milky solution is obtained (30 seconds).

Trichrome Stain

D'Antoni's Iodine

Distilled water	100 ml
Potassium iodide	1 gm
Powered iodine crystals	1.5 ml

Trichrome Stain

Chromotroe 2 R	0.6 gm
Light green SF	0.3 gm
Phosphotungstic acid	0.7 gm
Acetic acid glacial	1 ml
Distilled water	100 ml

Add 1 ml of glacial acetic acid to the dry components. Allow the mixture to stand for 15-30 minutes, to 'ripen'. Add 100 ml of distilled water. A good stain is purple in colour.

Use

This stain is used to stain protozoa in stool samples. Microsporidia and blastocystis will also be stained.

Lugol's Iodine

Potassium iodide	10 gm
Iodine crystals	5 gm
Distilled water	100 ml

Dissolve potassium iodide in water and then add iodine crystals, enough to saturate and leave a few crystals un dissolved. Store in brown glass stoppered bottles. When the solution lightens, it is discarded.

This is used to stain cysts of protozoa found in the intestine.

GIEMSA STAIN

Giemsa powder	0. 3 gm
Glycerine	25. 0 ml
Acetone free methanol	25.0 ml

The stock solution is diluted before use by adding 1 ml of stain to 10 ml of distilled water.

Use

This stain differentiates various parts of a cell and so can be used to identify mammalian cells and also for demonstrating parasites in tissues and blood.

9 *Direct Microscopic Examination and Staining Procedures*

Direct examination of the specimen is the first step in the laboratory diagnosis of pathogenic organisms, and at times, it may be the only choice left when the organism can not be cultured in the particular laboratory. The presence of other cells like neutrophils and epithelial cells are also to be observed.

Direct microscopic examination of the specimen may help in the early identification of an infectious agent and can guide in the laboratory culture as the confirmatory report takes a longer time at least 18 hr and sometimes two to three weeks. This reveals the morphological features as well as the characteristic staining reactions of the microbe.

WET MOUNT TECHNIQUE

A wet mount is done without fixing the specimen. It helps in the identification of motile organisms (e.g. *Vibrio cholerae*, protozoa in faecal specimens). Examination of fixed stained smears help in establishing the characteristic staining reactions. Gram's staining reaction and acid-fast reaction are the commonly used stains.

PROCEDURE

Place a drop (2-3 mm diameter) of a wet specimen or microbial suspension at the centre of a clean glass slide using an inoculating loop. If the specimen is dry (e.g. organism growing on agar medium), put a drop of sterile saline on a microscopic slide and mix the specimen or the organism. Carefully place a coverslip over the drop.

An ideal wet mount should not have any fluid oozing out of the coverslip and no air bubbles should be traped under the coverslip. Examine the slide first under the low power objective (100 × magnification). As the bacteria are colourless, transparent, homogeneous structure, the low power objective helps to screen the entire slide quickly as it covers more area of the optical field. For further magnification, switch to the high power objective (high dry, 400 × magnification), and increase the light intensity as needed. Use the oil-immersion objective (10 × magnification) is not recommended with wet mount preparations, as the oil tends to stick the floating coverslip. If oil-immersion objective is needed, pour melted vaseline—wax mixture on the edge of the covership.

PROCEDURE OF HANGING DROP

Hanging drop is a special wet mount preparation to observe motility. These preparations are most frequently used in examination of faeces of patients suspected to have cholera (*Vibrio cholerae*). Use of a concave slide (or cavity slide) is recommended and is described here.

1. Place a small sheet of white paper on the table and place a square size coverslip on it. Then put a very small drop of mineral oil or vaseline on each corner.
2. Put a loopful of liquid specimen (faecal suspension or broth culture) in the centre of the coverslip and make sure that the specimen does not touch the oil at the corners.
3. Invert a concave slide over the coverslip and press gently to allow the oil to spread and form a complete seal. This will prevent the specimen for drying.
4. Turn the slide so that the coverslip is on top, and inspect the drop to be sure it is hanging freely.
5. Place the slide on the stage of the microscope and adjust the light, using the low power objective focused on the edge of the drop. Then switch to high power objective and observed closely, looking for any motility of the microbe.

DRY MOUNT TECHNIQUE

This is the examination of stained smear. To stain bacteria, the specimen is thinly spread on a microscope slide, allowed to air dry and then fixed by heat. The fixing process kills and hardens the bacteria to prevent further changes and also to make them adhere to the glass slide. Fixation is achieved by simply passing the back of the microscope slide though Bunsen burner flame two or three times. The slide with the bacterial film should not be heated.

SIMPLE STAINING

Methylene blue stain is most commonly used and safranin can be substituted for methylene blue. It has limited use for the study or gross morphology. The stain accentuates the otherwise colourless bacterial cell.

PROCEDURES

Spread a thin film of material on a clean microscope slide, dry by waving in the air, and then heat fix by passing the slide 2 to 3 times through the frame with the clean surface of the slide facing the flame.

Place the slide on the staining race. The staining rack is made by placing two glass rods across the sink or a staining tray. Flood the surface of the slide with methylene blue staining solution. Allow the stain to remain on the slide for 1 to 2 minutes. Wash the slide gently with running water to remove excess stain. Allow the slide to air dry or blot dry.

Examine the smear first with the low power objective. Switch to high power (dry) objective and then finally to the oil-immersion objective to facilitate the detailed study of the morphology. Note and record shape, size, arrangement and uniformity of staining of the microorganisms present. The method is used mostly for reporting the presence of *Corynebacterium diphtheriae* in a throat swab, and for mycotic agents.

The method is useful in the identification of *Haemophilus influenzae* in spinal fluid or *Neisseria gonorrhoeae* in urethral pus. Methylene blue can be replaced with safranin.

GRAM STAINING

This is a useful differential staining procedure used in bacteriology, which is in addition to determining gross morphology serves to differentiate bacteria into two distinctly separate groups gram-positive and gram-negative.

SPECIMEN

All types of specimen and laboratory cultures can be subjected to gram-stain.

PROCEDURE

Preparation of Smear

Take a clean glass microscope slide and label the slide with a grease pencil, sterilised distilled water or saline on the slide; resterilise the loop.

Transfer a small portion of the specimen to be examined into the drop of saline on the slide to prepare a light suspension of the material, specimen taken on cotton swabs (throat, nose, ear, ulcer, pustule, urethral discharge or cervix, etc.) can be direct used. Rub the swab lightly and and gently over the slide.

If the suspension is a liquid (pus, exudates, body fluids, urine, broth culture and others), apply directly, a loopful of the specimen on the slide. In case of laboratory culture, transfer a small portion of a colony, growing on the medium, into the distilled water placed on the slide. Spread out the suspension of the slide by holding the loop in a flat position and with an oval spiral movement of the loop, move it outwards from the centre. Leave sufficient space from the edges of the sliders. Resterilise the loop before setting the loop aside.

Drying

Allow the smear to dry in the air. The smear must be throughly dry before it is fixed.

Fixing

Fix the air dried smear by passing the slide 3 to 4 times through the flame very quickly with the smear side facing up. Heat only the bottom of the slide. It is important in keeping the bacterial cells from washing off the slide during the staining process.

Staining

Lay the fixed slide on a staining rack. Cover the smear completely with the crystal violet stain. Leave the stain on the slide for one minute.

Pour off the stain, and flood the slide with Gram iodine solution and wait for one minute. Gently drain off the iodine solution and rinse with running tap water; shake off the excess.

Decolourise quickly with alcohol–acetone (1:1) solution or with 95 per cent alcohol (rectified spirit), continue the decolourisation process until the purple stain just stops coming off the slide. This usually takes about 5 seconds gently wash the slide under running tap water and drain. Do not delay as this will lead to over decolourisation.

Flood the decolourised slide with the counter stain safranin and wash briefly with tap water. Drain and allow to dry in the air or dry with blotting paper.

Microscopic Examination

When the slide is completely dry, first examine the slide under the low power objective; then switch to the high dry and finally to the oil-immersion objective for higher magnification. Gram-positive bacteria will be purple and the gram-negative ones are pink.

ACID-FAST STAIN

The acid-fast stain is another differential stain used mainly to detect mycobacteria that cause tuberculosis (*Mycobacterium tuberculosis*) and leprosy

(*Mycobacterium leprae*). These organisms are extremely difficult to stain by ordinary method because of the lipid-containing cell walls. They have a unique characteristic, in binding carbol-fuchsin stain so tightly that they resist destaining with strong declourising agents such as alcohols and strong acids. Other bacteria (Acid-fast negative) readily lose the stain when treated with acid-alcohol solution.

Three methods of acid-fast staining will be described here-Ziehl-Neelsen "hot stain". Modified Ziehl-Neelsen : "Gold Stain" and Kinyoun "Cold stain". These are respectively recommended for *M. tuberculosis, M. leprae* and *Nocardia asteroids* (weekly acid-fast). Kinyoun cold staining is also applied for *M. leprae*. All organisms pick up the primary stain and appear red-coloured. During the subsequent acid-alcohol treatment, only the acid-fast organisms retain the colour of the dye while other (acid-fast negative) lose the colour. Following the counterstaining with methylene blue, the decolourised acid-fast negative organisms and other cells take the blue colour and stand in contrast with the red-coloured acid-fast organisms.

Ziehl-Neelsen Method ("Hot Stain")

This is recommended for the preliminary diagnosis of *Mycobacterium tuberculosis* infection by direct method.

Procedure

Preparation of Smear

Prepare a film of the specimen and heat fix it as described under Gram staining.

Special handling sputum specimen: Place a purulent portion of the sputum on the slide. Make the smear as thin as possible.

Staining: Place the heat fixed slide on the staining rack, flood it with freshly filtered carbol-fuchsin. Heat gently until steam rises. Continue to heat for 5 minutes so that steam is seen, but without boiling. Do not allow the stain to boil or dry on the slide. If the stain starts to dry during heating, add more carbol-fuchsin.

Rinsing: Cool and wash the stain off the slide with water. Continue rinsing until the water that runs off is colourless.

Decolourisation: Cover the slide with acid-alcohol solution for 3 minutes. Wash with running tap water and drain.

Counterstain: Cover the slide with methylene blue stain. Leave the stain on the slide for 30 seconds to 1 minute. Wash with tap water, allow the water to drain or, blot carefully.

Examination of the slide: Examine the slide thoroughly under the low power objective, and then examine closely under oil-immersion objective. AF (acid-fast) organisms will appear bright red on a blue background while other organisms (AF negative) are dark blue in colour.

Reporting of results: All AF organisms are not pathogenic and the report of preliminary staining must be confirmed by culture.

Number of AF organisms found in 10 fields	Report
None	Negative
0-1	Positive (1+)
1-10	2+
10-100	3+
> 100	4+

Modified Ziehl-Neelsen Stain ("Gold Staining")

The leprosy bacilli (*Mycobacterium leprae*) are weakly reactive with the carbol-fuchsin stain as compared to the tubercle bacill.

Reagents

1. Carbol-fuchsin (Same as used earlier for Ziehl-Neelsen staining).
2. Sulphuric acid or hydrochloric acid (1% v/v, aqueous solution).
3. Methylene blue 0.3 per cent w/v, aqueous solution.

Procedure

1. Prepare the smear as described earlier and flood the heat-fixed film with carbol-fuchsin. Wait for 12 to 15 minutes without heating.
2. Wash with water.
3. Decolourise with 5 per cent sulphuric acid.
4. Counterstain with methylene blue.
5. Wash with water, drain and blot dry. This is followed by gently heating of the slide.

Kinyoun Method ('Cold Stain')

This method is useful in the identification of *Nocardia asteroids* which is "weakly acid-fast".

Reagent

Kinyoun's carbol-fuchsin stain

Basic fuchsin	4 g
Phenol crystals, melted	8 ml
Alcohol, 95%	20 ml
Distilled water	100 ml

Dissolve by grinding in a mortar 4 gm of basic fuchsin in 20 ml of 95 per cent alcohol. Add 100 ml of distilled water slowly while shaking the preparation. Melt phenol crystals (carbolic acid) in a 56°C water bath. Add 8 ml of melted phenol to the stain, using a pipette with mechanical suction (do not pipette by mouth).

Acid-alcohol reagent: 1% v/v of concentrated hydrochloric acid (concentrated) in 95 per cent alcohol 1% v/v of concentrated sulphuric acid is also used.

Methylene blue counterstain: 0.3% w/v aqueous solution.

Procedure

1. Prepare the smear as described earlier. Flood the heat-fixed smear with Kinyoun's carbol-fuchsin stain. Do not heat.
2. Rinse with distilled water.
3. Decolourise by adding acid-alcohol reagent drop by drop with continual agitation until carbol-fuchsin no longer washes off. This requires approximately 2 minutes for a smear of average thickness.
4. Wash with tap water.
5. Counterstain with methylene blue for 20 to 30 seconds.
6. Wash quickly with water and blot dry, follow with gentle heating.

Special Staining

Ablert's Stain

This granule stain is specifically used for the identification of *Corynebacterium diphtheriae*. It demonstrates the presence of metachromatic granules found in *C. diphtheriae* which provides provisional diagnosis for diphtheria. Most frequently smear of throat and nose secretions are subjected to Albert's stain.

Procedure

1. Cover the heat fixed smear with Albert's stain (solution A) for 5 minutes.
2. Drain; do not wash.
3. Apply solution B and let stand for 1 to 2 minutes.
4. Wash briefly with tap water, drain, blot dry and examine under oil-immersion objective. Metachromatic granules stain-bluish-black whereas the bacillary body appears green or bluish-green.

Chapter

10

Antimicrobial Susceptibility Testing

SPECIMEN

Usually subculture of individual isolate from clinical specimen, though specimens such as CSF, PUS, urine may be used directly when smear examinations (for example, Direct Grams staining) reveal the presence of only one type of organism.

METHOD

The method used commonly for determination of susceptibility is the disc diffusion test. Discs impregnated with known concentrations of antimicrobials, are placed on an agar plate that has been inoculated with a culture of the bacterium to be tested. The plate is incubated 35-37°C for 16-18 hours.

If the organism is susceptible to a particular agent, there will be a zone of inhibition of growth around the disc.

If the organism is resistant, the organism will grow up to the edge of the disc.

Selection of Media

- Mueller-Hinton Agar—for gram-negative Bacilli, *Staphylococcus* spp and *Enterococcus* spp.
- Mueller-Hinton Blood Agar-*streptococcus pneumoniae* and other streptococci, Mueller-Hinton chocolate Agar (or) Haemophilus.
- Sensitivity test medium (HTM) for *Haemophilus* spp.
- Mueller-Hinton chocolate agar–*Neisseria* spp.
- Wellcotest agar–Any organism to be tested with sulphonamides and co-trimoxazole.
- The WHO recommends the Kirby-Bauer method of disc diffusion testing of Antibiotic susceptibility test.
- The Kirby-Bauer method.

Preparation of Plates

The medium is prepared and sterilised. Defibrinated blood may be necessary for test on fastidious organisms. Human blood is not recommended as it may contain antimicrobial substances. The medium should be poured into Petri dishes on a flat horizontal surface to a depth of 4 mm. Pure plates are stored at + 4°C and used within one week of preparation. These plates should be dried before inoculation, the pH of the medium should be checked and maintained 7.2 and 7.4.

Preparation of Inoculum

At least four morphologically similar colonies from an agar medium are touched with a wire loop and the growth is transferred to a test tube containing 1.5 ml of sterile suitable broth. The tubes are incubated for 2 hours at 35 to 37°C to produce a bacterial suspension of moderate turbidity.

Inoculation

Plates are inoculated within 15 minutes of preparation of the suspension so that the density does not change. A sterile cotton-wool swab is dipped into the suspension and excess removed by rotation of the swab against the side of the tube above the fluid level. The medium is inoculated by even streaking of the swab over the entire surface of the plate in three directions.

Antibiotic Disc

After the inoculum is dried, single disc is applied with forceps, a sharp needle or a dispenser and pressed gently to ensure even contact with the medium. Not more than six discs can be accommodated on an 85 mm circular plate and not more than twelve discs are accommodated on a 135 mm circular plate.

Discs should be stored at + 4 °C in sealed containers with a desiccant and should be allowed to come to room temperature before the containers are opened.

If the antimicrobial solution prepared in the laboratory are being used proceed as follows:

1. Pick up a 2 mm loopful of the standard antibiotic solution and lower carefully onto a paper disc which, when moistened will adhere to the loop.
2. Place the moistened disc on the surface of the inoculated plate in the appropriately labeled segment.
   Take care to avoid "contamination"
3. Repeat for each antimicrobial agent to be used placing the impregnated discs in their respectively labelled segments.

Incubation

Plates are incubated for 16 to 18 hours at 35 to 37°C aerobically or in CO_2 atmosphere for fastidious organisms.

Interpretation

The diameters of a zones are measured to the nearest millimetre with Vernier calipers or a thin transparent millimeter scale. The point of complete inhibition of growth, is considered as the zone edge.

Each zone is interpreted according to the organism in comparison to the reference table (Commercially available charts) known as zone diameter interpretive standards for each antibiotics.

Drug susceptibility testing for mycobacteria is a difficult procedure to standardise. Variables include stability of drugs, alteration of antimicrobial activity when drugs are incorporated in different media, etc.

There are two methods, namely direct and indirect test. The Direct susceptibility is done when AFB is seen in the smear. The indirect test is done after isolation of mycobacteria on primary culture.

PROCEDURE: RESISTANCE RATIO METHOD

1. Antibiotic containing medium (Antituberculous drugs like INH, ethambutol, etc are used)
2. Inoculum

Remove approximately 4 mg moist weight of bacterial culture form a LJ slope with a 3 mm external diameter (24-gauge) loop. Transfer to a 7 ml. Bijou bottle antoclaved with 12 small glass beads and 0.2 ml sterile distilled water. Mix thoroughly. Shake the bottle mechanically for one minute to produce a uniform suspension, then add 0.8 ml sterile distiled water and shake the bottle with hand.

Inoculation and Incubation

Using a 3 mm external diameter (26-gauge wire) loop inoculate one loopful of inoculum on each slope. Also inoculate as control on a drug free slope.

Include the standard H3RV stain as control with each batch. Incubate the tubes 37°C for 4 weeks.

Reading of Results

Tubes are examined for growth at the end of the 4th week. "Growth" is defined as the presence of 20 or more colonies. The control tube should show 100 or more colonies for the test to be satisfactory. Record the minimum (lowest) concentration of drug inhibiting the growth (MIC).

Report

The ratio of the MIC of the rest strain to the MIC of the test strain to the MIC of standard H 37 RV gives the resistant ratio (RR).

Interpret the RR as follows:

RR of 2 or less   ....SUSCEPTIBLE

RR of 8 or more   ....RESISTANT

RR of 4   ....DOUBTFUL

Repeat a doubtful test from the control. If a resistance ratio of 4 or more is still obtained, finally classify the strain as resistant. If the ratio in the respective test is 2 or less classify as sensitive.

SECTION

2

IMMUNOLOGY/
SEROLOGY

11

Basics of Immunology

Immunology is defined as the study antigen and antibody and their various reactions in the host resulting in harmful and harmless effects.

Antigen has been defined as any substance, which stimulates the production of antibody with which it reacts specifically and in an observable manner.

An Antigen introduced into the body reacts only with those particular immunocytes namely B-lymphocytes or T-lymphocytes which carry specific marker for the antigen and produces an antibody or cells respectively.

Antibody is also termed as "immunoglobulin" which includes glycoproteins which are of five types IgG, IgM, IgA, IgD and IgE. The structure of antibodies includes smaller chains known as light (L) chains and the larger ones as heavy (H) chains. Distinguishing properties of various immuno-globulin.

Some properties of immunoglobulin classes					
	IgG	*IgA**	*IgM*	*IgD*	*IgE*
Molecular weight	150,000	160,000	900,000	180,000	190,000
Serum concentration (mg/ml)	12	2	1.2	0.03	0.00004
Half-life (days)	23	6	5	2-8	1-5
Daily production (mg/kg)	34	24	3.3	0.4	0.0023
Intravascular distribution (%)	45	42	80	75	50
Carbohydrate (%)	3	8	12	13	12
Complement fixation					
Classical	++	–	+++	–	–
Alternative	–	+	–	–	–
Placental transport	+	–	–	–	–
Present in milk	+	+	–	–	–
Selective secretion by Seromucous glands	–	+	–	–	–

* IgA may occur in 7S, 9S and 11S forms

ANTIGEN-ANTIBODY REACTIONS

Antigen and antibodies combine with each other specifically. These reaction serve several purposes.

1. Antibody mediated immunity in infectious diseases.
2. Diagnosis of infections in the laboratory.
3. Identification of infectious agents and non-infectious agents like enzymes.
   Antigen-antibody reactions *in vitro* are known as serological reactions.
   The various serological reactions are:
   a. Precipitation reaction (Flocculation)
   b. Precipitation in gel (Immunodiffusion)
   c. Agglutination (Coombs' test)
   d. Complement fixation test
   e. Neutralisation tests (Toxin neutralisation)
   f. Immunofluorescence
   g. Radioimmunoassay
   h. Enzyme immunoassays (EIA)
   i. ELISA (Enzyme-linked immunosorbent assay)
   j. Immunoelectroblot techniques.

IMMUNODEFICIENCY DISEASE

These are conditions where the defence mechanisms of the body are impaired. This leads to repeated microbial infections of varying severity and malignancies.

Immunodeficiency may be classified as primary or secondary. Primary immunodeficiencies result from abnormalities in the development of the immune mechanism and secondary immunodeficiencies are consequences of disease, drugs, nutritional inadequacies.

HYPERSENSITIVITY REACTIONS

Hypersensitivity refers to the injurious consequences in the sensitised host.

Hypersensitivity reactions have been classified into 'immediate' and 'delayed' types.

The immediate and delayed reactions are clinically distinguished as follows.

I. Immediate hypersensitivity (B-cell or antibody mediated)
   a. Anaphylaxis/Atopy
   b. Antibody mediated cell damage
   c. Arthus phenomenon
   d. Serum sickness.
II. Delayed hypersensitivity (T-cell mediated)
   a. Infection (tuberculin) type
   b. Contact dermatitis type

AUTOIMMUNITY

Autoimmunity is a condition in which structural or functional damage is produced by antibodies against the normal cells of the body.

This may be due to

1. Elevated level of immunoglobulins.
2. Demonstrable autoantibodies.
3. Genetic predisposition, etc.

IMMUNOHEMATOLOGY

ABO Blood Group System

This contains four blood groups and is determined by the presence or absence of two distinct antigens A and B on the surface of red blood cells.

Red cells of group A carry antigen A, cell of group B carry antigen B and cells of group AB have both A and B antigens, while group O cells have neither A nor B antigen.

Rh Blood Group System

Rh positive or negative depends on the presence or absence of antigen D (Rho) on red cells and can be tested with anti-D (anti-Rh) serum.

Chapter

12

Introduction to Serology and Sero-diagnostic Procedures

A vaccination or immunisation is an artificial infection given to an individual by introducing a non-virulent pathogen; the body produces the desired antibody to combat future infection.

The immunologic or antigen-antibody reaction can be used as a diagnostic tool for the recognition of a specific antibody present in the patient's serum. Once this specific antibody is recognized, the identification of the offending pathogen (antigen) is diagnosed.

Antibodies are produced by lymphocytes and plasma cells. These special immune reacting leucocytes (also called immunocytes) are stimulated by foreign substances. The sensitized immunocytes evolve the nature of the antigen and produce a specific antibody through their protein production by humans. In order to produce an antibody by this active process, the individual should be immunologically competent. In other words, the immunocytes must be healthy and active. Most bacteria, viruses, and many fungi, elicit antibody production. Bacterial infections, however, stimulate the production of granulocytes as a result of a primary immunologic reaction while viral infections, in most cases, lead to lymphocytosis.

The principle of serodiagnosis is that the unknown aetiologic agent (antigen) is identified by the laboratory through a suitable detectable immunologic reaction, incorporating a specific reaction between the host's pathogen. Immunologic techniques are useful in identifying organisms which are the causal agent syphilis) or whose identification is difficult by routine laboratory techniques (e.g. *Salmonella*)

Immunity is imparted to an individual at two levels:

1. At the fluid level (humoral immunity)
2. At the cellular level (cell-mediated immunity).

The antibody that circulated through the fluid component of the blood (serum) protects the body from the invading pathogen, while cell-mediated immunity is responsible for graft rejection and immunity to certain organisms, e.g. fungi, intracellular pathogens. Antibodies are produced by lymphocytes originating from the bone marrow (B-lymphocytes), and are responsible for humoral immunity; (Fig. 12.1) the lymphocytes originating from the thymus related tissues (T-lymphocytes) are responsible for cell-mediated immunity.

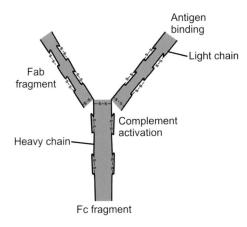

Fig. 12.1: Chemical structure of antibody

COMPLEMENT

Complement is a protein (globulin) present in normal serum. Complement takes part in the primary immune reaction (immediate, spontaneous, and non-specific) through various pathways, and assists in rendering the pathogen inactive. There are at least nine component system (C1-C9) that combine with the antigen-antibody complex to effect lysis. Once activated the components are involved in a great number of immune defense mechanisms including phagocytosis, leucocyte chemotaxis and others.

In routine serological investigation, other than the complement fixation test, complement is inactivated by heating the serum at 56°C for a period of 20 minutes so that it does not interfere in serological testing. If the inactivated serum is not used immediately, complement becomes active and needs to be re-inactivated (56°C for 20 minutes).

SERODIAGNOSIS

Serological tests to detect antibodies against the infecting microorganisms provide a useful means of indirect diagnosis. These tests are of special value for those organisms which cannot be isolated and cultured in the laboratory such as viral diseases, rickettsial diseases and the causal agent of syphilis (*Treponema pallidum*).

PRINCIPLES OF SERODIAGNOSTIC TESTS

The various methods applied in serodiagnosis attempts to detect the antigen-antibody reaction in the patient's body and also quantitate the amount of antibody formed. The basic principle behind the serodiagnostic technique is to identify the specific antibody in circulation and suggest the offending antigen. In some cases, however, the antigen is sought (e.g. in viral hepatitis) and the reagent used for the testing is the known antibody. The following principles are routinely applied in various serological tests (Fig. 12.2).

Precipitation

In the precipitation test, the invisible soluble antigen (called precipitinogen) and antibody (precipitin) combine together as a result of immunologic reaction and precipitate out of solution. The concept of the precipitation reaction, however, has been utilised in many modern immunologic techniques like immunodiffusion and nephelometry.

Flocculation

Flocculation is the aggregation of inert particles (latex, charcoal, cholesterol emulsion), coated with an antigen, as a result of immunologic reaction. The flocculation techniques is best applied in the VDRL test for the diagnosis of syphilis. Flocculates usually float in the routinely performed diagnostic tests.

Agglutination

This is one of the most common techniques utilised in serologic testing, and is essentially the same as the precipitation reaction, except that the size of the reacting particles are larger. Unlike precipitation, in case of agglutination, the antigen particles (red cells, bacterial cells, latex particles) are visible before the immune reaction occurs. When an antibody is added to a suspension of such particles, it combines with the surface antigens and links them together to form clearly visible aggregates or agglutinates. The agglutination reaction can be seen on a slide without the aid of a microscope or it can be done in a tube test. Most laboratories use the slide test for screening (qualitative) and the tube test for the determination of a titre (quantitative).

The haemagglutination reaction utilises group "O" blood in order to avoid complications due to the presence of anti-A or anti-B in the patient's serum. The red cells are treated with tannic acid and then coated with appropriate antigen. The agglutination of the red cells is seen by the deposition of cells at the bottom of the tube or in the well.

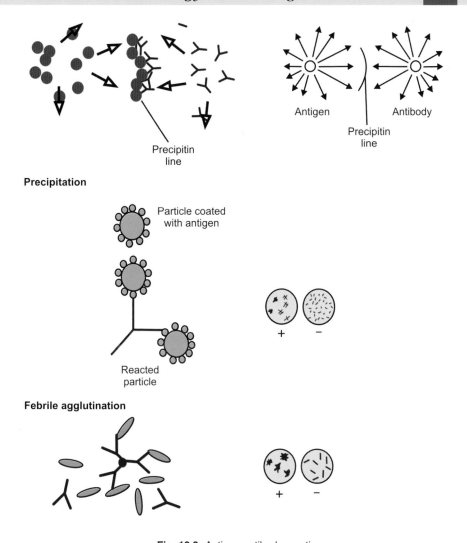

Precipitation

Particle coated with antigen

Reacted particle

Febrile agglutination

Fig. 12.2: Antigen-antibody reactions

Haemagglutination-inhibition Test

This is also referred to as the negative agglutination test and is the reverse of the direct agglutination technique. Unlike the agglutination reaction, a positive reaction (agglutination) here indicates a negative or non-reactive specimen. A specimen (serum or urine) carrying the antigen (e.g. human chorionic gonadotrophin or hCG, in case of pregnancy) is first reacted with its homologous antibody (e.g. anti-hCG). If the antigen is present in the specimen, it will neutralise the antibody (Fig. 12.3). In the subsequent step when the antigen-coated particles are mixed with the "treated specimen", the neutralised

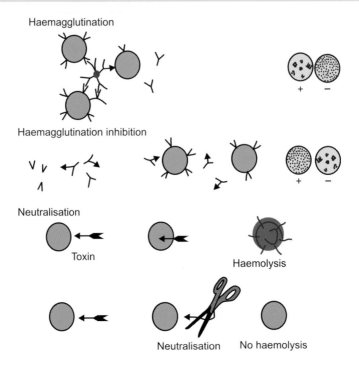

Fig. 12.3: Haemagglutination test and haemagglutination inhibition test

antibody will not bring about agglutination. Thus in the agglutination inhibition test, a positive agglutination indicates a non-reactive specimen (or no antigen present), and negative agglutination is indicative of a reactive specimen (or antigen present).

Complement Fixation

Some immunologic reactions are evidenced by the lysis of haemolysin-coated red cells (used as indicator cells) with the mediation of complement (Fig. 12.4). If the complement is fixed, the indicator cells do not lyse. Positive test (infected

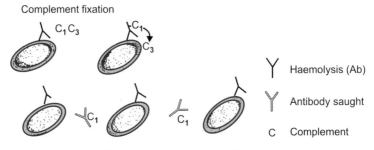

Fig. 12.4: Complement fixation text

condition) is indicated by lack of haemolysis while normal patient's serum will cause haemolysis of indicator cells.

Fluorescent Antibody Technique

The fluorescent antibody technique demonstrates antigen-antibody reaction at a microscopic level. This method is highly specific and is the confirmatory test for syphilis and other infections.

The fluorescent antibody technique (FAT) involves labelling an antibody with a fluorescent dye (Fig. 12.5) (e.g. acridine orange or fluorescein isothiocyanate) which can be excited by ultraviolet light rays. The excited molecules of the fluorescent dye then emit light in the visible range. The phenomenon is known as fluorescence. If the fluorescent dye is chemically conjugated to a protein (antibody), the location of the tagged protein can be recognised when viewed under the fluorescent microscope.

There are two main methods of immunofluorescent marking:

The Direct Test

This is applied to detect the antigen (e.g. pathogen). The known antibody is attached to the fluorescent dye and then applied directly to the tissue or to the suspected organism.

The Indirect Test

This is applied to detect the antibody. Here, the unlabelled patient's antibody attaches to the antigen of the tissue or bacterium, and then fluorescein-tagged anti-immunoglobulin (antihuman globulin) is applied.

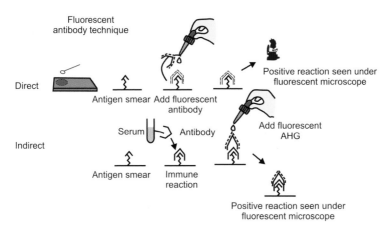

Fig. 12.5: Fluorescent antibody technique

MODERN IMMUNOLOGIC TECHNIQUES

Radial Immunodiffusion

In the case of immunodiffusion technique, solutions of antigen and antibody are allowed to react in gels, when they diffuse toward one another, and at the point at which they meet in optimal proportions they will form a visible precipitate, called the precipitin line. The precipitin lines can be seen in the single radial immunodiffusion method or the double immunodiffusion in two dimensions (Fig. 12.6). The single radial immunodiffusion technique is applied for the quantitative analysis of various antibodies (e.g. IgG, IgA, IgM) commercially available plates of semisolid media (e.g. cellulose acetate, agarose, agar, polyacrylamide gel, etc.), containing corresponding antibodies are used. Known quantities of patient's serum are put in the wells on the medium and the test system is incubated for 48 hours at 37°C. The antigen present in the serum diffuse into the gel, and reacts with the antibody present in the medium. As a result, a ring is formed whose diameter is proportional to the amount of antigen present in the serum.

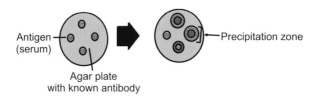

Fig. 12.6: Radial immunodiffusion

Immunoelectrophoresis

The immunoelectrophoresis technique is a further refinement of the precipitin reaction in gels. Here, a serum sample is initially electrophoresed through an agarose matrix which separates the various protein components according to their electrophoretic mobility. When this is achieved, antisera containing specific antibodies are applied parallel to the direction of migration and allowed to diffuse toward the separated protein antigens. Here the precipitin lines are curved (Fig. 12.7). The technique is useful in the investigation of immunoglobulin disorder.

Counterimmunoelectrophoresis

This technique is extensively used for detecting the presence of hepatitis associated antigen (HAA) in the donor's blood collected for transfusion. In principle it is the same as the immunodiffusion technique but here the rates of

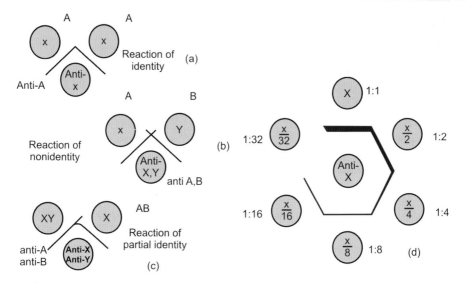

Fig. 12.7: Immunoelectrophoresis

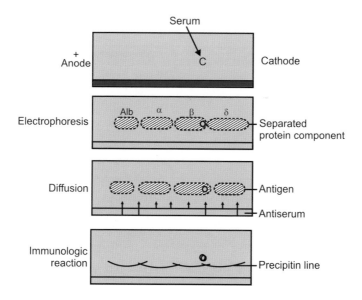

Fig. 12.8: Counterimmunoelectrophoresis

migration of the antigen and the antibody are expedited by placing the antigen and the antibody opposite the poles towards which they are attracted (antigen migrates towards anode and antibody towards cathode) (Fig. 12.8).

Chapter

13 *Laboratory Procedures in Serology*

Laboratory procedures in serology consist of collection and preparation of serum specimens and various immunologic test procedures for specific serodiagnosis. These include: Serological tests for syphilis (STS) which include screening (non-treponemal) and confirmatory (treponemal) tests: (a) Non-treponemal serological test for syphilis: VDRL (b) Treponemal serological test for syphilis: FTA—ABS. Agglutination tests: Widal, Weil-Felix, staphylococcal agglutination, haemagglutination and latex agglutination, C-reactive protein test (CRP), rheumatoid arthritis test (RA), antistreptolysin O-test (ASO), Pregnancy test.

DIAGNOSIS OF INFECTIONS

	Name of test	Name of disease
1.	Widal test	Enteric fever
2.	Weil-Felix test	Rickettsial diseases
3.	Standard agglutination test for brucellosis	Brucellosis
4.	Cold agglutination test	*M. pneumoniae* infections
5.	Paul-Bunnell test	Epstein-Barr virus infection
6.	ASO, ADNAse B,	Streptococcal infection
7.	Microagglutination test	Leptospirosis
8.	VDRL and TPHA (TPPA.)	Syphilis
9.	*Toxoplasma* antibodies	Toxoplasmosis

DIAGNOSIS OF AUTOIMMUNE DISORDERS

	Autoantibodies	Test used
1.	Rheumatoid factor	Latex agglutination
2.	Antinuclear antibody	Immunofluorescene
3.	Anti DS DNA antibodies	Enzyme immunoassay
4.	Anti thyroglobulin and antimicrosomal antibodies	Particle agglutination

Contd...

Contd...

5.	Anticardiolipin antibodies	Enzyme immunoassay
6.	Antibodies to extractable nuclear antigens (U,RNP, Jo-1, Cetromere Sc1-70, Ro, La	Enzyme immunoassay

DETECTION OF PARAMETERS OF INFLAMMATION

1. CRP   Latex agglutination
2. Total complement   Haemolytic assay
3. C3 and C4   Radial immunodiffusion.

COLLECTION AND PREPARATION OF SPECIMEN

Fresh serum is routinely used for all serological tests. Blood is collected in the morning, usually by venipuncture. If the specimen cannot be tested within 24 hours, it should be stored frozen.

Collection of Serum Specimen

Perform venipuncture. Use only a dry sterile syringe and needle. Collect 10 ml of blood. Remove the needle from the syringe and expel the blood into a centrifuge tube without any anticoagulant. Stopper the tube of blood immediately. Leave the blood to clot at room temperature for 30 minutes. Do not disturb the clotting blood during this period. After 30 minutes detach the clot from the wall and centrifuge the tubes at 2500 rpm for 5 minutes. Take out the serum by means of a Pasteur pipette, and transfer the serum into another test tube. Use the specimen within a few hours or store it in a refrigerator at 4°C.

Heat-inactivation of Serum Specimen

Serum-complement interferes with many tests and can be inactivated by heating (56°C) the serum sample in a waterbath for 30 minutes. Inactivated serum must be used within 2 hours.

Reagents

Reagents for serological testing are usually purchased as prepackaged kits. The antigen suspension is in a dropper bottle that delivers a requisite quantity of antigen suspension (0.03 ml). The kit also contains buffer and other accessories required for performing the test. Saline (0.85%, NaCl, w/v, in deionized water) is the only reagent prepared by the laboratory. Always store the reagent according to the manufacturer's directions. Most reagents are refrigerated except saline.

Control sera: Control sera with known positive and negative reactions must be run along with the test specimen from the patient. These are supplied by the manufacturer along with the serodiagnostic kits.

Equipment

Serological tests are carried out routinely by one of the two ways—the slide method and the tube method. The slide method requires plate to observe the agglutination or flocculation by the naked eye. The microscope is needed only in case of VDRL reaction. The tube method, on the other hand, requires small-sized test tubes which are placed in a test tube rack while performing the test. A list of most commonly required equipment is provided herewith. Large glass plates with rows of squares. Squares are and made with a wax pencil or diamond tip. Tiles with ceramic rings are commercially available. High intensity spot light for viewing.

Pipettes: Serological

1 ml with 0.05 ml graduation-plenty
2 ml with 0.05 ml graduation
10 ml with 0.1 ml graduation
Pasteur pipettes or droper with teats-uncalibrated and calibrated (to deliver 0.03 ml and 0.05 ml fluid).

Others: glass rods, applicator sticks (or toothpicks), timer, test tubes (12 m × 77 mm), test tube rack, centrifuge, waterbath, rotator (for VDRL), microscope.

AGGLUTINATION TESTS FOR SERODIAGNOSIS OF FEBRILE ILLNESSES

Widal Test for Enteric Fever

Use killed standardized suspension of *Salmonella typhi* 'O' (STO), *Salmonella typhi* 'H' (STH), *Salmonella* paratyphi A 'H' (SPAH), and *Salmonella* paratyphi B 'H' (SPBH) as antigens. Ready to use antigens are stored at + 4°C.

Make separate series of dilutions for each antigen to be tested, beginning with 1 to 10 and doubling through to 1 to 160, in four rows of five tubes.

Procedure

Arrange four rows of five test tubes each. Transfer 0.9 ml physiological saline into the first tubes of all four rows. Transfer 0.5 ml saline to all other test tubes. To the first tubes in each row, add 0.1 ml serum, to get a dilution of 1/10. Dilute serially by mixing and transferring. 0.5 ml diluted serum from first into second tube, and so on. Discard 0.5 ml from the fifth tube in all rows.

Include four tubes with 0.5 ml saline each, for control of antigen to be used, on each day. Bring antigens to room temperature and shake the suspensions to ensure even distribution. Add 0.5 ml of STO antigen to all tubes in the first row and one tube for control. Similarly add STH, SPAH and SPBH to second, third and fourth rows respectively, and to control tube. Mix well. (Final dilution therefore becomes 1/20 to 1/320). Incubate the tubes overnight at 37°C. Let tests stand at room temperature for 15 to 20 minutes before reading (Fig. 13.1).

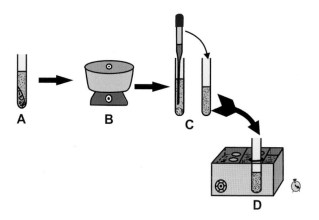

Fig. 13.1: Procedure of widal test

Interpretation

Look first at the control tubes. The antigen suspension should be evenly distributed in these tubes. Pick up individual tubes of the test and look for clearing of supernatant and clumping of bacterial suspension. When a complete, positive reaction occurs, practically all of the organisms are removed from suspension and the supernatant is entirely, or almost entirely, clear. When the reaction is negative, the suspension should look the same as the antigen control. When turbidity in the supernate is half that of the control the reaction is read as "2+".

Shake the tubes gently and note the characteristics of the deposits formed. In positive reactions with H antigens, deposits of agglutinated bacteria should appear fluffy or floccular and with the O antigen, finely granular. Agglutination should be at least "2+" to be considered positive. If end point is not reached with the 5th tube, higher dilutions are tested (Fig. 13.2).

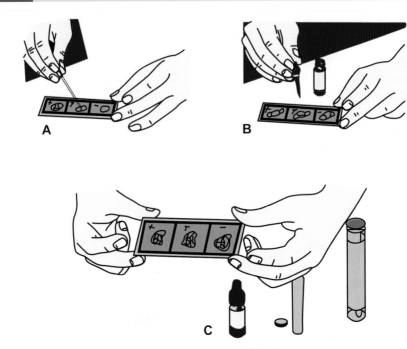

Fig. 13.2: Interpretation of widal text

Significance

Widal test becomes positive only about 7 to 10 day after onset of fever. STO antibodies rise first and titres of 80 or above can be significant. For H antigens, titres of 160 or above is considered significant.

Following is a general scheme of interpretation:

STO	STH	SPAH	SPBH	Interpretation
+	−	−	−	Enteric fever
+	+	−	−	Typhoid fever
−	+	−	−	Typhoid fever
+	−	+	−	Paratyphoid (A) fever
−	−	+	−	Paratyphoid (A) fever
−	+	+	−	TV vaccination
+	+	+	−	Recent TA vaccination
−	−	−	+	Sal Gr B infection
−	+	+	+	Anamnestic reaction

WEIL-FELIX TEST FOR RICKETTSIAL FEVERS

This is a non-specific agglutination test, where antibodies against rickettsiae are detected using proteus antigens. Certain antigens of non-motile strains of **Proteus** are found also in certain rickettsiae. Hence, antibodies to rickettsiae may be detected with antigens prepared from *Proteus.*

Procedure

Do not inactivate serum. Carry out serum dilutions as for Widal test, but make three rows of five tubes only. Killed standardised suspension of *Proteus* **OX 2**, *Proteus* **OX 19** and *Proteus* **OXK** are used as antigens. These are stored at + 4°C. Incubate test at 37°C overnight. Let stand at room temperature for 15-20 minutes before reading. Fine granular agglutination and clearing of supernantant indicate a positive test.

Results are interpreted as follows:

0 × 19	0 × 2	0 × k	Interpretation
+	–	–	Epidemic/endemic typhus
+	+	–	Indian tick typhus (spotted fever)
–	–	+	Scrub typhus

Titres of 160 or above are considered significant. However, lower titres can occur in early phases of infection.

TESTS FOR BRUCELLOSIS

Antibodies against *Brucella abortus* can be detected using a slide screening test or a tube dilution test.

SLIDE TEST

B. abortus undiluted coloured antigen is used. Store antigen at 4 to 10°C.

Procedure

Inactivate serum at 56°C for 30 minutes. Place 0.04 ml of the serum in a ring slide. Add 0.03 ml of the coloured *Brucella* antigen. Mix well with sterile stick. Allow to stand on the bench for 5 minutes. Tilt slide back and forth for about 30 seconds, and let it stand for further 3 minutes. Read reaction with a total magnification of 100 (10 × objective). Test is positive, if clumping in any degree is observed and is negative, if there is complete absence of agglutination. Include a known positive and a known negative control for each batch of tests.

STANDARD AGGLUTINATION TEST FOR BRUCELLOSIS

This is a specific test. *B. abortus* undiluted **Plain** antigen is used. The antigen is stored at 4-10°C.

Procedure

Inactive sera for 30 min at 56°C. Dilutions are made in 0.5 per cent phenol saline. Arrange a row of 7 tubes for each serum to be tested. Add 0.9 ml of 0.5

per cent phenolized physiological saline to the first tube and 0.5 ml there after. Add 0.1 ml serum to the first tube and dilute serially till the 7th tube in 0.5 ml amounts, to get dilutions from 1/10 to 1/640. Shake antigen to ensure even distribution and add 0.5 ml to each tube of diluted serum. (So final dilution becomes 1/20 to 1/1280). Set up one set of 5 tubes to serve as antigen controls. Label these as 0 per cent, 25 per cent, 50 per cent, 75 per cent, and 100 per cent.

Add reagents to these as follows:

Diluent	Antigen	Control
1 ml	1 ml	0%
1.25 ml	0.75 ml	25%
1.5 ml	0.5 ml	50%
1.75 ml	0.25 ml	75%
2.0 ml	0 ml	100%

Incubate all tubes at 37°C for 24 hours.

Reading

Do not shake the tubes. Make readings on the basis of the opacity, or clearance, of the supernatant fluids, and not by the sediment of agglutinated bacteria.
a. Compare the turbidity in each tube with that of the control tubes.
b. Take the highest dilution with 50 per cent clearing (matches with 50% control) as the end point.
c. Convert titre into International Units (IU) per ml of patient's serum, by multiplying by 2.

Example: If endpoint is 1/40, titre is 40 report is 80 IU.

COLD AGGLUTINATION TEST FOR *M. PNEUMONIAE* INFECTION

Sera of patients infected with *M. pneumoniae* frequently show the presence of cold agglutinins. These are non-specific antibodies which agglutinate human group O red cells in the cold but not at 37°C. Since cold agglutinins are not found usually in normal individuals their detection is a valuable aid in the diagnosis of infections of the respiratory tract. Blood to be tested for the presence of cold agglutinins should be kept at **room temperature until separation of the clot.**

Procedure

Inactivate serum at 56°C for 10 min. Prepare human Group 'O' red cells suspension as follows: Collect 3.5 ml blood from an 'O' group individual into a citrated tube. Wash 3 times in saline. Resuspend in saline and aliquot as

1 ml, suspension and store at 4 to 10°C for not more than 3 days. Prepare a 0.2 per cent RBC suspension on the day of testing. Arrange a row of 8 tubes in a rack. Add 0.7 ml of suspension to the first tube and 0.4 ml to the rest. Add 0.1 ml serum to the first tube. Mix and transfer 0.4 ml to the second tube and proceed with serial doubling dilutions through to the 7th tube to obtain dilutions of 1/8 to 1/512. The eighth tube serves as a cell control. Add 0.4 ml red cell suspension to all eight tubes. Additions of cell suspension results in further doubling of the serum dilution, i.e. 1/16 to 1/1024. Mix well and refrigerate at 4 to 10°C over right. Include RBC control and a known negative control serum with each batch.

Reading

Read the test **immediately on removal from the refrigerator.**

Look first at the cell control for a compact button which on shaking shows no clumping. Then examine the negative control serum, which should also show no agglutination. In the test, look for clumping of cells which does not break up on gentle shaking, as evidence of a positive reaction. If there is clumping, check for reversibility of agglutination by incubating at 37°C for 30 minutes. Record the highest dilution of the serum which produces reversible agglutination as the end point.

Report

Report the titre as the reciprocal of the end-point.

A titre of 64 is considered significant, but a 4-fold rise in titre between acute and convalescent serum sample, whatever the range, is also diagnostic. As in other tests, lower titres can occur if serum is collected very early in disease.

Specific tests (e.g. IgM, ELISA) are now available for the diagnosis of *M. pneumoniae* infection.

PAUL-BUNNELL TEST

This test is for heterophile antibodies occurring in infectious mononucleosis. These are produced against antigens common to Epstein-Barr virus (EBV), the agent of infectious mononucleosis, and sheep RBC.

Presumptive Test

Inactivate serum at 56°C for 30 minutes A non-specific antigen (Sheep RBC) is used on the basis of similar antigens being present on normal sheep erythrocytes and on Epstein-Barr virus.

Procedure

Set up a series of 10 tubes for each serum to be tested. Add 0.4 ml physiological saline to the first tube, and 0.25 ml to the other tubes. Add 0.1 ml serum to the first tube. Mix and then transfer 0.25 ml serially through to the 9th tube. Discard 0.25 ml from this tube. The 10th tube is an antigen control and does not contain serum. Use sheep blood not less than 24 hours and no more than 7 days after drawing. RBCs are stored at + 4°C. Wash these cells 3 times with physiological saline. Prepare a 2 per cent suspension in saline and mix well to ensure even distribution. Add 0.1 ml to each tube of diluted serum and to the control. Shake well. Final dilutions of serum begin with 1 to 7 and double thereafter. Incubate at room temperature for 2 hours.

Reading

Gently, shake antigen control (by tilting). No clumping should be visible to the naked eye or when seen against the concave mirror of a microscope. Gently shake the tubes of the test serum. Check again for clumping as with the control. Record the highest dilution with at least "2+" reaction (Clumping of 50% cells). Test higher dilutions, if necessary, to determine endpoint.

SERODIAGNOSIS OF SYPHILIS

Two groups of tests are available.

NON-TREPONEMAL TESTS

This group of tests known as "standard tests for syphilis (STS)" make use of cardiolipin antigen and hence are nonspecific. Commonly used test in this group is VDRL test (flocculation test). This is a good screening test especially for early syphilis (primary, secondary and early latent stages). This test is also used for diagnosing neurosyphilis, by testing CSF.

    Treponemal tests are specific tests using treponemal antigens. Elevated levels of antibodies against these antigens remain for long periods, even after effective treatment.

Non-treponemal Tests (Non-specific Tests)

These tests are simple to perform and economical. Hence, these are used for routine screening for syphilis. VDRL test is the most popular test in this group. Examples of other flocculation tests that can be used in field conditions are:

a. Rapid plasma reagin (RPR) test.

b. Unheated serum reagin (USR) tests, e.g.

Toluidine red unheated serum test (TRUST).

Currently, reagents containing cardiolipin are available commercially for detecting reagin antibodies. These are mostly supplied as ready to use reagents. Antigens are usually presented on carrier particles like carbon, which makes the reading macroscopic. Some of these tests are to be done on plasma, while others can be done on uninactivated serum. Control sera also come with the kit.

Venereal Disease Research Laboratory (VDRL) Test on Serum

This test is non-specific and depends on the presence of antibodies that will react with cardiolipin antigen. Antigen has to be prepared serum has to be inactivated.

Antigen Preparation

An alcoholic solution containing 0.03 per cent cardiolipin, 0.9 per cent cholesterol and purified lecithin sufficient to produce standard reactivity is commercially available. This is mixed with buffered saline for preparation of antigen suspension.

Procedure

Pipette 0.4 ml of buffered saline solution to the bottom of a 30 ml flat bottom ground glass or screw-cap-stoppered bottle. Add 0.5 ml antigen (from the lower half of a 1.0 ml pipette graduated to the tip or using a micropipette) directly onto the saline solution while continuously, but gently, rotating the bottle on a flat surface. Add antigen drop by drop but rapidly, so that approximately six seconds are allowed for each 0.5 ml antigen. Pipette tip should remain in upper third of bottle, and rotation should not be vigorous enough to splash saline solution onto pipette. Blow last drop of antigen from pipette without touching pipette to saline solution. Continue rotation of bottle for ten more seconds. Add 4.1 ml buffered saline solution using a 5.0 ml pipette. Place top on bottle and shake vigorously for approximately 30 times within 10 seconds. The antigen suspension is now ready for use.

Qualitative Slide Test for Sera

Inactivate sera at 56°C for 30 minutes immediately before testing. Test control sera to give expected reactions. Pipette 0.05 L (50 ml) each of inactivated serum onto a ring on a VDRL glass slide. Add one-drop (1/60 ml) antigen suspension to each of the sera. Rotate slide for 4 minutes using a VDRL rotator set at 180 rpm. Read tests immediately after rotation using a total magnification of 100.

Reading

Observation	*Interpretation*
No clumping	Non reactive (NR)
Small clumps	Weakly reactive (WR)
Medium and large clumps	Reactive (R)

All reactive, weakly reactive and those sera with doubtful reactions are tested after diluting.

Quantitative Serum Slide Test

Serum Dilutions

Pipette 0.5 ml of freshly prepared 0.9 per cent saline into each of 5 or more tubes. Add 0.5 ml of inactivated serum to first tube mix well, and transfer 0.5 ml to tube No. 2 Continue mixing and transferring until the last tube contains 1ml. Discard 0.5 ml from last tube. The dilutions will be 1-2, 1-4, 1-8, etc. Test each serum dilution as described under Qualitative Serum Test.

Reading

Read tests microscopically as described for the qualitative procedure.

Report

Reactive: 'Reactive' and add the titer in dilution, e.g. "VDRL reactive, 16 dilutions". Reactive only in undiluted serum: report as "VDRL reactive, 1 dilution". Weakly reactive samples are reported as 'VDRL weakly reactive, 0 dilution". Negative samples are reported as 'non-reactive'.

VDRL Test on Cerebrospinal Fluid (CSF)

Inspect the fluid specimens which are visibly contaminated or contain blood are unsatisfactory for testing. Centrifuge and decant if necessary to remove any extraneous particles and obtain clear fluid.

Preparation of antigen suspension for testing

CSF

Prepare suspension as described for serum test. Add one part of 10 per cent sodium chloride solution to one part of the antigen prepared for testing serum. Mix gently by rotating the bottle or inverting the tube. Let stand for at least 5 minutes, but not more than 2 hours, before use.

Serum Reaction

Expected Reaction with CSF antigen

1. NR   NR
2. 2+   3+
3. 1+   2+
4. WR   1+

Qualitative test for CSF: Pipette 0.05 ml spinal fluid into a concave VDRL slide. Add one drop (1/100 ml) of CSF VDRL antigen suspension. Rotate slide for 8 minutes (in a box containing a moistened blotter to prevent evaporation during rotation). Read immediately after rotation as for serum. Record definite clumping as "reactive"; no clumping, as "non-reactive".

Quantitative test for CSF: Pipette 0.2 ml (0.9% saline) into each of 5 or more tubes. Add 0.2 ml spinal fluid to first tube, mix well, and transfer 0.2 ml to the second on so on to get doubling dilutions. Test each dilution as described for Qualitative Test.

Reading

Read tests microscopically as for serum test. Record highest dilution that produces a positive reaction.

Report

As for serum.

TREPONEMAL TESTS (SPECIFIC TESTS)

Treponemal tests use antigens from cultivable *Treponema*. The treponemal tests are highly specific and sensitive and so used as confirmatory, tests when VDRL test is positive. In certain situations, e.g. tertiary syphilis specific tests may be positive when VDRL is negative. Of the tests available, *Treponema pallidum* Haemagglutination Test (TPHA) and *Treponema pallidum* particle agglutination (TPPA) are easy to perform and are reproducible. In this test, coloured gelatin particles coated with *T. pallidum* antigen when mixed with serum containing antibodies to *T.pallidum,* agglutination occurs. The test is done in microtitre plates. Manufacturer's instruction should be followed. A 'matt' is considered positive and 'button' negative. Routinely only qualitative tests are done.

SERODIAGNOSIS OF STREPTOCOCCAL INFECTION

Clinical Significance

A streptococcal infection by group A (*Streptococcus pyogenes*) causes a number of ailments like sore throat, rheumatic fever, and glomerulonephritis. The laboratory' diagnosis of a sore throat due to *S. pyogenes* infection is done by the laboratory culture of a throat swab specimen; the anti-streptolysin test or ASO is recommended for the diagnosis of rheumatic fever; the laboratory diagnosis of glomerulonephritis is based on urine culture and the recently introduced anti-deoxyribonuclease-B. (ADN-B) test. The ASO test is still the most popular test used in the diagnosis and monitoring of streptococcal infections. The latex agglutination test which utilises the principles of ASO testing is becoming more popular because of its rapidity and easy handling. Two newer tests for the diagnosis of streptococcal infection are the streptozyme test and the antihyaluronidase test.

ANTISTREPTOLYSIN O (ASO) TEST

The ASO test is based on the principle of neutralisation. Where the antibody neutralises the toxin and prevents it from its haemolytic activity on erythrocytes. The slide latex agglutination test is more simple and requires minimal equipment and reagent.

Principle

The basic reagent is the suspension of polystyrene latex particles which are coated with SLO reagent. These particles agglutinate in the presence of ASO present in patient's serum and the immunologic reaction is visible to the naked eye.

Reagent and Equipment

ASO latex agglutination kit: This contains SLO reagent, ASO coated latex suspension, and control sera ASO positive and ASO negative.

Qualitative

Reconstitute the reagents according to the manufacturer's directions and bring all reagents to room temperature. Note: Mix the SLO reagent gently while reconstituting or it may lose its activity. Add 0.1 ml of SLO reagent into the three small test tubes previously labelled as test (T), positive control (PC), and negative control (NC). Add 0.1 ml of the patient's serum to the first tube (T),

0.1 ml of ASO positive control serum to the second tube (PC) and 0.1 ml of ASO negative control serum to the third tube (NC). Mix by shaking and let them stand for 15 minutes at room temperature. Following incubation, add 1 drop (0.05 ml) of each mixture on three separated fields marked as T, PC and NC—on the black test slide. Shake the ASO latex reagent (coated with SLO) to obtain a uniform suspension and add one drop to each field containing the serum to be tested. Mix the serum with the latex suspension thoroughly with the help of a wooden applicator stick and spread mixture over most of the field. Tilt the slide back and forth and let the suspension go from end to end of the periphery. Examine for agglutination. Agglutination in test serum indicates that the patient has a significant ASO titre (higher than 200 Todd units). Agglutination should also be seen in the positive control (PC) serum but not in the negative control serum (NC).

Quantitative

Slide method is primarily for screening but can also be applied for quantitative testing. Tube method is preferred for titre determination.

ASO Tube Method

This is a quantitative procedure.

Reagent and Equipment

ASO kit: The kit contains buffer, streptolysin O (SLO) and control sera in dry form which are reconstituted according to the manufacturer's instructions. In addition, a suspension of defibrinated rabbit blood is needed. The cells are washed three times (or until the supernatant is clear) in normal saline with the aid of centrifugation (2000 rpm for 5 minutes) and finally a 5 per cent suspension is made in saline which is used on the same day or in buffer which can be stored at 4°C for a limited period. Citrated human blood (group 0) from the blood bank can also be used but rabbit blood is preferred.

Equipment: General equipment required in serodiagnostic tests has been presented earlier. This includes pipettes, test tubes (75 × 100 mm), test tube rack, glass marking pencil, and timer.

Procedure

Prepare initial serum dilutions of (1:100. and 1:500):
a. Take two test tubes and mark them as A and B.
b. Place them on one side of a test tube rack.
c. Add 9.9 ml of buffer in tube A and 8.0 ml buffer in tube B.

d. Using a 1 ml graduated serological pipette transfer 0.1 ml of the serum specimen into tube A (Procedure: Draw the test serum up to 0.9 ml mark, wipe the tip and blow the aliquot of serum specimen into 9.9 ml of buffer placed in tube A). Mix thoroughly by repeatedly blowing, into the pipette.

e. Then, with the help of a 2 ml serological pipette, transfer 2 ml of the diluted serum from tube A into tube B. Mix thoroughly by repeatedly blowing into the pipette.

Arrange 7 test tubes in a row in the test tube rack and number them from 1 to 7. Use tube nos. 1 to 5 for placing diluted serum and the subsequent tubes for red cell control (Tube no. 6) and Streptolysin or SLO control (Tube no. 7). Place the amounts of diluted serum from tubes A and B and the corresponding amounts of buffer in order to give a dilution gradient of the serum. Note that tube no. 6 (RBC control) gets 1.5 ml buffer and tube no. 8 (serum control) receives 1.5 ml of undiluted serum. The total volume of all tubes at the final stage, after adding red cells, will be 2.0 ml. Shake gently to mix. Incubate at 37°C for 15 minutes. Add 0.5 ml of 5 per cent red cell suspension. Incubate at 37°C for 15 minutes. Shake gently after 15 minutes and continue to incubate for another 45 minutes.

Setting up for ASO titration by tube method								
					Tube no.			
	1	2	3	4	5	6	7	8
						RBC Controls		
							SLO	Pt
Volume (ml) of diluted serum (from tube A/B)	1.0 A	0.5 A	0.33 A	0.25 A	1.0 B	0.0	0.0	1.5
Volume (ml) of buffer	0.0	0.0	0.67	0.75	0.00	1.5	1.0	0.0
Streptolysin (ml)	0.5	0.5	0.5	0.5	0.5	-	0.5	-
RBC (ml)	0.5	0.5	0.5	0.5	0.5	0.5	0.5	0.5
Dilution in Todd unit 1:	100	200	300	400	500*		Control	
Observation (example)	NH	NH	NH	NH*	H	NH	H	NH

* The titre is 400 (1:400); this may correspond to the same as the Todd units (IU/ml) or may be different. Compare the test serum against the Todd unit calculation of Todd unit:

$$\frac{Todd\ unit}{Serum\ dilution} = \frac{Total\ volume\ used}{Volume\ of\ serum\ used}$$

Example with tube No. 5: x/500 = 1/0 or x = 500 where, x= Todd Unit

Centrifuge at 2500 rpm for 1 minute and read haemolysis:

Supernatant pink-coloured = Haemolysis (H)

Supernatant clear = No haemolysis (NH)

Look first at the controls. Reagent (SLO) control should show complete haemolysis (Tube no. 7) and cell control (Tube No. 6) should show no hae-molysis. The patient's serum control (Tube no. 8) should also show no haemolysis (Fig. 13.3).

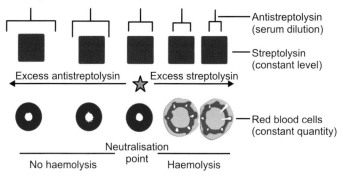

Fig. 13.3: Interpretation of ASO test

STREPTOZYME TEST

Clinical Significance

This immunologic test recognises the presence of antibodies in patients against streptococcal exoenzyme antigen.

Principle

The test is based on the principle that sheep red cells are sensitised with *Streptococcus.* A extracellular antigens which include streptolysin 0, DNAase, NAD ase, hyaluronidase and streptokinase. The antigen-coated sheep cells will react with antibodies to these antigens to give a positive agglutination reaction.

Special Reagent

Streptozyme reagent kit (cell suspension, positive control serum): Carefully read the instructions provided by the manufacturer before using the kit.

Accessories: Accessories required by the slide test as described earlier.

Procedure

The procedure described here is for qualitative testing. This can be modified for a quantitative determination by serially diluting the serum specimens as described earlier with other tests. Take 5 ml of saline in a test tube and add to this 0.05 ml of serum transfer 0.05 ml of the diluted serum onto the slide. Add a drop of the reagent cell suspension. Mix the fluids individually and thoroughly with the help of disposable stirrers (use separate stirrer for each mixture). Spread out the mixture on the entire space provided (2 cm × 2 cm).

Rock the slide back and forth, gently and evenly, for a period of 2 minutes at the rate of tilting 8 to 10 times per minute. Following the 2 minute period of immunologic reaction, gently place the slide on a flat surface and observe for agglutination reaction within 10 seconds. A direct light source above the slide facilitates reading of the agglutination reaction. A negative agglutination reaction is marked by a uniformly turbid or slightly granular appearance of the cell suspension. A positive agglutination reaction is recognised by the readily visible agglutination and clumping of the cells.

SERODIAGNOSTIC TESTS FOR MISCELLANEOUS DISORDERS

A few other disorders need to be mentioned here which are diagnosed in the laboratory by special immunologic techniques. Details of these procedures will not be presented here as they are not done routinely in the laboratories of developing countries.

Acquired Immunodeficiency Syndrome (AIDS)

It is a debilitating and terminal disease involving a defect in cell mediated (T-cell) immunity. The causative agent is retrovirus identified as HTLV-III (human T-cell lymphotropic virus type III) in North America and LAV (lymphadenopathy-associated virus) in Europe. This disease is transmitted through sexual contact (homosexual and/or bisexual), transfusion of contaminated blood products, and from infected mothers to foetuses. The disease progresses from vague symptoms, to generalised lymphadenopathy, weight loss, severe diarrhoea, recurring infections and susceptibility to malignant neoplasms. Laboratory diagnosis of AIDS is currently based on the detection of antibodies to the HTLV/LAV virus. Enzyme-linked immunosorbent assay (ELISA) is relatively easy to perform than the radionucleotide probe method (western blot technique).

Rubella

Rubella or German measles is caused by a virus which is spread to airborne microdroplets and enters its host via the respiratory tract. If a pregnant woman is infected during the firs trisemester, the foetus may be subject to cataract formation, deafness, stillbirth or spontaneous abortion. Haemagglutination inhibition test (HI) is applied in the laboratory diagnosis of *Rubella.*

Toxoplasmosis

Toxoplasmosis is caused by *Toxoplasma gondii*, an obligate intracellular protozoan; cats serve as its definitive host. A variety of clinical symptoms may be exhibited which can be diagnosed in the laboratory by direct and indirect.

IMMUNOLOGIC TEST FOR PREGNANCY

The pregnancy testing may be performed in clinical chemistry or in the serology department.

Clinical Significance

A pregnant woman discharges human chorionic gonadotrophin (hCG) in the urine which can be detected within 2 weeks after the first missed menstrual period by the latex agglutination test or haemagglutination inhibition test. Several commercial companies supply the rapid slide agglutination test kits for early diagnosis of pregnancy.

The pregnancy test can be done both for the qualitative estimation of hCG and the quantitative estimation of hCG. The former is used solely for the early detection and confirmation of pregnancy. The quantitative estimation of hCG in serum is of value in cases of pre-eclamptic Toxaemia, hydatidiform mole and choriocarcinoma.

Slide Test

Specimen

The first morning specimen of urine is desirable which is naturally concentrated. The urine specimen should be free of blood, pus and bacteria; collected in a clean container free from detergent.

Reagents and Accessories

All reagents and accessories to carry out the test are available in the commercially available kits. The kit should include latex antigen (polystyrene particles coated with hCG), anti-hCG serum (rabbit), control urine-positive and negative.

Procedure

Place one drop of the urine sample on the ring of the special slide provided by the manufacturer. Use the disposable plastic dropper or a Pasteur pipette for making the transfer. Add one drop of anti-hCG reagent to the urine specimen placed on the slide. Mix the two fluids well with the help of a disposable applicator stick or tooth pick and spread out to the entire area provided. Rock the slide gently for about 30 seconds, gently shake the vial with latex antigen and then add one drop of the latex antigen to the above mixture of urine and anti-hCG on the slide. Mix again with the applicator stick and spread out the pool of liquid uniformly within the entire area of the ring marked on the special slide. Now rock the slide gently back and forth and observe for the appearance of agglutination at 2 minutes. Examine under a bright light, preferably with a hand lens.

Observation and Reporting

Latex particles agglutinated within two minutes	Negative (normal)
Homogenous suspension of latex particles without any sign of agglutination	Positive (pregnant)

C-REACTIVE PROTEIN TEST (CRP)

The C-reactive protein or CRP is a normal alpha globulin found in low quantities in serum. As an acute phase reactant, CRP appears rapidly in the serum of patients. who have an inflammatory condition of either infectious or non-infectious origin and is absent in serum from normal individuals. It is almost always elevated following surgery and after myocardial infarction. The name of the protein is derived from the fact that CRP has the capacity for precipitating the somatic C-carbohydrate of *Pneumococcus*.

When CRP is injected into a laboratory animal (rabbit), a specific antibody reacting with the protein is produced which will react with the CRP present in the patient's serum. The immunologic reaction can be seen by a precipitation reaction or by a rapid latex agglutination slide test. This simple and inexpensive method can be used as a screening test to detect the presence of CRP.

Clinical Significance

The CRP test is a non-specific test, but the findings can be used as a simple index of disease activity and treatment status in such conditions as rheumatic fever, rheumatoid arthritis, viral meningitis, and bacterial infections. The CRP test can also help in determining post-surgical complications. Results of the CRP test frequently overlap with the erythrocyte sedimentation rate but the former is a valuable adjunct or substitute for the sedimentation test and is also more sensitive than the ESR test.

Principle

Latex particles are coated with anti-CRP that agglutinate in the presence of CRP. Thus in this test the presence of antigen in the patient's serum is detected rather than the antibody.

Procedure

Inactivate the test serum at 56°C for 30 minutes in a waterbath. Prepare a 1:5 dilution of serum sample by adding 0.1 ml of serum with 0.4 ml of glycine-saline buffer supplied in the kit. Place the diluted and undiluted serum on two sides of a divided microscope slide. Use a grease pencil for marking the dividing line on the slide. Two squares of about 2× 2 cm made on the slide are desirable. The same pipette or a capillary tube may be used for transferring both the samples of diluted and undiluted serum specimen, provided the pipette is used for the diluted one first and then emptied as completely as possible before it is used for the undiluted sample. Add 1 drop of latex anti-CRP reagent to each section holding the serum specimen. Mix the serum with the latex reagent (diluted one first) by means of a wooden applicator stick or glass rod or toothpick and spread it over the square. Tilt the slide slowly from side-to-side under sufficient illumination and observe for macroscopic clumping for 2 minutes. A dark background helps see the agglutination reaction clearly.

Interpretation: Visible agglutination in either or both sections indicates the presence of CRP in the patient's serum. Prozones may be encountered (false weak reaction). Hence, weak reacting specimens must be subjected to titre determination. A negative reaction appears as a smooth suspension with no visible agglutination in either section, and indicates that the patient's serum is devoid of abnormal quantities of CRP. The test results are recorded as follows:

  0 = Smooth suspension, no agglutination
1+ = Slight agglutination amidst smooth suspension of latex particles
2+ = Relatively heavier agglutination with some amount of smooth
       suspension of latex particles
4+ = Heavy agglutination

Example: Interpretation of slide agglutination reaction

Agglutination reaction with patient's serum

Undiluted	Diluted	Interpretation
0-2+	2+ - 4+	Strongly positive (prozone)
3+ - 4+	0 - 2+	Positive
1+ - 2+	Negative	Weakly positive
Negative	Negative	Negative

Add 1 drop of latex reagent with anti-CRP coating to each drop of serum dilution. Using a thin glass rod, or a wooden applicator, or a toothpick, mix the serum with the latex reagent and spread it over the area of the squares (approximately 2cm × 2 cm). Start mixing with the highest serum dilution (6) and proceed towards the lowest dilution (1) in sequence. Tilt the glass plate (or slide) slowly from side to side (use a dark background with good illumination from the top) for a period of 1 to 2 minutes. Observe for macroscopic agglutination. The reciprocal of the highest serum dilution showing visible agglutination is taken as the CRP titre of the specimen.

RHEUMATOID ARTHRITIS (RA) TEST

It is believed that during some infections of a joint (rheumatoid arthritis) an IgG antibody is produced. Subsequent to its production, the antibody becomes altered for some reason, and against this altered antibody a second antibody (anti-antibody is formed). Usually this second antibody is IgM. To demonstrate this antibody which is RF (rheumatoid factor), the serum is reacted with IgG-coated (purified human gammaglobulin) latex particles. Agglutination indicates the presence of RF.

Clinical Significance

The RA test is done in order to detect the presence of rheumatoid factor in the serum of patients with rheumatoid arthritis. The factor is non-specific and is found in other diseases as well, both related and non-related to rheumatoid arthritis.

Latex Agglutination Slide Test

Specimen

Serum which can be stored in a refrigerator or frozen (–20°C) for prolonged storage. Plasma should not be used. Heat inactivation of the serum is recommended (56°C for 30 minutes) but not essential.

Reagent and Equipment

Commercially available kits are available for this test. Other equipment required for the slide agglutination test is given earlier-graduated serological pipettes, calibrated pipettes (0.05 ml/drop) or capillary pipette (50 ul), and glass plates, plenty of test tubes (75 mm × 100 mm), waterbath, grease pencil and timer.

Procedure

1. Transfer the specimens in pre-labelled test tubes, and place them in a 56°C water bath for 30 minutes. This will inactivate the complement.
2. Allow all reagents to come to room temperature.

3. Arrange the heat-inactivated specimens on the front row of a test tube rack and number them. Enter the number in the record book.

4. Place an empty test tube behind each specimen tube and dispense 0.95 ml of buffer in each tube. To make the transfer of buffer, use, 1 1 ml serological pipette and draw the buffer to the 0.05 mark (the pipette now holds 1.95 ml of fluid).

5. Add 0.05 ml of test serum to the corresponding buffer solution in the rear test tube. Use separate 1 ml pipette (0.01 ml graduation) for each specimen. Mix each tube separately and thoroughly. This gives a 1:20 dilution of the serum.

   Alternatively: Steps, (4) and (5) can be modified in order to reduce the time for screening. In case of a positive reaction, proceed as follows: Using a calibrated dropper that, delivers 0.05 ml fluid per drop, mix one drop of undiluted serum specimen with 1 ml of buffer (dispense 1 ml buffer in all the empty tubes, using a 5 ml pipette). Mix well. The dilution is approximately the same as above, 1:20.

6. Place 1 drop, of this diluted serum in a space on the divided slide. Use a separate is posable plastic dropper for each serum specimen to be used in the test.

7. Include the positive and negative control along with the run. This is done only once a day.

8. Give each square a number corresponding to the specimen number on the tube, and mark the additional two squares as "+" and "–" corresponding to the positive and negative controls, when they are included, (once a day).

9. Shake the gammaglobulin reagent bottle gently then add one drop of latex suspension to each drop of diluted serum specimen placed in the square.

10. Mix the serum and latex reagent well with the applicator stick or tooth-pick provided in the kit. Spread the suspension over the entire square.

11. Tilt the slide gently to and fro for 1 minute.

12. Mix the serum and latex reagent well with the applicator stick or tooth pick provided in the kit. Spread the suspension over the entire square area.

13. Tilt the slide gently to and fro for 1 minute then note the reaction reactivity.

    Note: Do not examine after 2 minutes.

14. Reporting:

Negative (N):	Smooth homogeneous suspension
Weak positive (WP):	Finer agglutination and usually takes more than two minutes
Strong positive (SP):	Course agglutination and usually occurs within one minute

Quantitative

1. Prepare the serial dilutions of serum in the same way as done in other quantitative serological tests.
   a. Take 5 test, tubes in a test tube rack and add 1.95 ml of buffer in tube 1 and 1.0 ml in other tubes.
   b. Add 0.05 ml of undiluted serum in tube 1, mix and transfer to tube 2, and continue the process to tube 5. Keep 1 ml of the mixture in a separate empty tube for further dilution, if needed.
   c. The above process will lead to the following serial dilutions:
      (1) 1:40, (2)1:80, (3) 1:160, (4) 1:320, (5) 1:640
2. Perform the slide test as described earlier with different dilutions of the serum. Use the same pipette in transferring 1 drop of the diluted specimen (0.05 ml), starting from the highest dilution (tube 5, 1:640) until the lowest dilution (1:40) is reached in tube 1.

Additional Information

1. Store all reagents in the refrigerator. Do not freeze.
2. Avoid drying which may cause formation of flakes of' the latex gamma-globulin reagent; keep the bottle closed after use.
3. All glassware used must be thoroughly cleaned with detergent but they must be washed thoroughly; residual detergent will interfere in the test results.
4. During the rocking of the slide, cover the entire circle or else the periphery may start drying which may lead to erroneous results.
5. A quality control run is an essential part of all laboratory testing and should be done for RF testing also, using control sera (positive and negative). This should be done once a day, preferably during the first run, and for new batch of reagents.

SECTION

3

SYSTEMATIC BACTERIOLOGY

Chapter

14

Staphylococci

Staphylococci are gram-positive cocci arranged in grape-like clusters. The genus *Staphylococcus* contains various species but the medically important species are:
1. *Staphylococcus aureus*
2. *Staphylococcus epidermidis*
3. *Staphylococcus saprophyticus*

Staphylococcus Aureus

Alexander Ogstan in 1880 identified the pathogenic role of coccus in suppurative lesions. He gave the name *Staphylococcus* (*staphyle* meaning a bunch of grapes, *kokkos* meaning a berry) from the typical arrangement of the cocci in grape-like clusters.

Staphylococcus Epidermidis

This is a skin commensal.

Staphylococcus Saprophyticus

This acts as an opportunistic pathogen. Based on the production of enzyme coagulase, these are classified as coagulase positive and coagulase negative. The coagulase positive strains are pathogenic and the coagulase negative are generally non-pathogenic, non-toxigenic and forms white colonies, these are called *S. epidermidis* (also known as *S. albus*).

MORPHOLOGY

Staphylococcus aureus are gram-positive cocci arranged in grape-like clusters, approximately 1 μm in diameter. They may also be present singly or in pairs, tetrads or short chains.

Culture: They grow readily on ordinary culture medium.

Nutrient agar: The colonies are 2 to 4 mm in diameter, circular and smooth. Most of the strains produce golden yellow pigment.

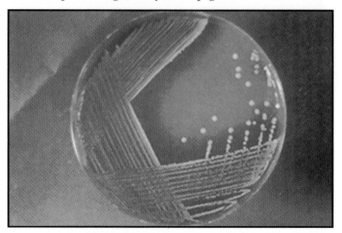

Fig. 14.1: Culture on blood sugar

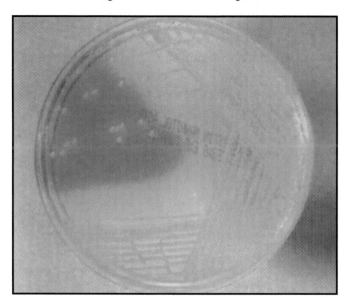

Fig. 14.2: Yellow colonies of *S. aureus*

Blood agar: Colonies of *Staphylococcus aureus* produce a beta type of haemolysis. Haemolysis is best observed with sheep or rabbit blood (Fig. 14.1).

MacConkey's agar: Colonies are very small and pink due to lactose fermentation.

Mannitol salt agar: Yellow coloured colonies (Fig. 14.2) are seen on the medium due to fermentation of mannitol. The mannitol fermentation leads to production of acid and lowers the pH of medium (Phenol red produce yellow colour in acidic pH).

Milk agar: Colonies of *S. aureus* are larger than those on nutrient agar and pigment production is enhanced.

Biochemical Reactions

S. aureus is catalase positive and oxidase negative. It ferments a number of sugars producing acid without gas.

Pathogenesis

S. aureus is a pyogenic organism and they may be classified as cutaneous and deep infections.

Cutaneous infections include pustules, boils, carbuncles, abscesses, styes, impetigo, pemphigus neonatorum, wound and burn infections.

Deep infections include osteomyelitis, tonsillitis, pharyngitis, sinusitis, pneumonitis, endocarditis, meningitis, bacteraemia, septicaemia. They may cause urinary tract infections also.

Various other infections like food poisoning, nosocomial infections, Exfoliative skin diseases and toxic shock syndrome are caused by *Staphylococcus aureus.*

Laboratory diagnosis of *Staphylococcus aureus* infections.

1. **The specimens include the following according to the nature of lesion**

Pus	Suppurative lesions
Sputum	Respiratory infection
Blood	Septicaemia or Puo
Urine	Urinary tract infection
CSF	Meningitis
Faeces	Food poisoning
Food of vomit	Food poisoning

   Nasal and perineal swabs—detection of carriers

2. **Collection and Transport**

Specimens should be collected in sterile containers under all aseptic conditions. In case of urine, midstream urine should be collected. Blood should be collected in blood culture bottles comprising of glucose broth and taurocholate broth.

Specimens should be transported immediately to the laboratory and processed.

3. Direct Microscopy

Direct microscopy with Gram's stain. Smears of pus or wound exudates may show gram-positive cocci in clusters.

4. Culture

The specimens are inoculated on

a. Blood agar
b. Peptone water
c. Salt agar salt milk agar and Robertson's cooked meat medium (RCM)— where specimens are expected to have scanty organisms.
d. Nutrient agar
e. MacConkey's agar

The inoculated media are incubated at 37°C for 18 to 24 hours.

The culture plates are examined for morphology of bacterial colonies and other characters. Uniform turbidity is produced in liquid medium (Peptone water).

On blood agar, colonies are 2 to 4 mm in diameter, circular, raised, opaque and produce golden-yellow pigment. Beta-haemolysis is seen around these colonies.

On nutrient agar, the colonies are large (2 to 4 mm diameter), circular, convex, smooth, shiny, opaque and easily emulsifiable. Most strains produce golden yellow pigment.

On Mac Conkey's medium, *Staphylococcus aureus* are pink due to lactose fermentation.

Biochemical Reaction

S. aureus is catalase positive and oxidase negative. It breaks down carbohydrates by fermentation and can be tested on high-Heifson medium.

The following features help to differentiate *Staphylococcus aureus* from other non-pathogenic strains.

1. Production of golden-yellow pigment
2. Beta type haemolysis on blood agar
3. Coagulase production
4. Mannitol fermentation
5. Gelatin liquefaction
6. Phosphatase production
7. Production of enzyme deoxyribonuclease
8. Tellurite reduction

Catalase Test (Fig. 14.3)

All staphylococci (pathogenic and non-pathogenic) are catalase positive. The enzyme catalase mediates the breakdown of hydrogen peroxide into oxygen and water. This results in the production of bubbles.

Procedure

1. Use a loop of sterile wooden stick to transfer a small amount of colony growth to the surface of a clean, dry glass slide.
2. Place a drop of 3 per cent hydrogen peroxide (H_2O_2) on to the inoculum.
3. Observe for the evolution of oxygen bubbles.

Inference

Catalase positive *Staphylococcus* produce copious bubbles, catalase-negative organisms, e.g. streptococci and enterococci yield no bubbles.

Fig. 14.3: Catalase test

Coagulase Test

It is positive in *S. aureus* but negative in other *staphylococci*.

Principle

This test is used to differentiate *Staphylococcus aureus* (positive) from coagulase negative staphylococci (negative). *S. aureus* produce two forms of coagulase-"bound and free bound coagulase, or clumping factor", is bound to the bacterial cell wall and reacts directly with fibrinogen.

Method

Slide test

1. Place a drop of coagulase plasma on a clean, dry glass slide.

2. Place a drop of distilled water or saline next to the drop of plasma as a control.
3. With a loop, straight wire or wooden stick, emulsify a portion of the colony.
4. Mix well with a wooden applicator stick.
5. Rock the slide gently for 5 to 10 seconds.

Inference

Positive: Macroscopic clumping in 10 seconds or less in coagulated plasma drop and no clumping in saline or water drop.

Negative: No clumping in either drop
   Note: Negative slide tests must be confirmed using the tube test.

Tube Test (Fig. 14.4)

1. Emulsify several colonies in 0.5 ml of rabbit plasma (with EDTA) to give a milky suspension.
2. Incubate tube at 35°C in ambient air for 4 hours.
3. Check for clot formation. Tests can be positive at 4 hours and then revert to negative after 24 hours.
4. If negative at 4 hours, incubate at room temperature overnight and check again for clot formation.

Inference

Positive: Clot of any size
Negative: No clot

Fig. 14.4: Tube coagulase test

Phosphatase Test

Most strains of *S. aureus* produce enzyme phosphatase for detection of this enzyme, *S. aureus* is grown on nutrient agar containing phenolphthalein diphosphate.

The enzyme phosphatase acts on phenolphthalein salt to release free phenolphthalein. The colonies turn pink when exposed to ammonia vapour due to the presence of free phenolphthalein.

Deoxyribonuclease Test (Fig. 14.5)

S. aureus produce enzyme deoxyribonuclease which hydrolyses DNA.

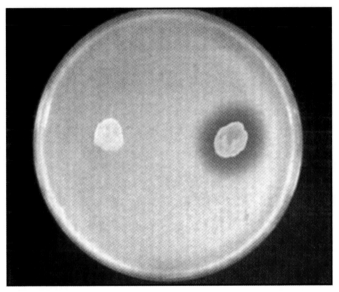

Fig. 14.5: DNAase test

Principle

The medium is pale green because of the DNA- methyl green complex. If the organism growing on the medium hydrolysis DNA, the green colour fades and the colony is surrounded by a colourless zone.

Method

1. Inoculate the deoxyribonuclease agar with the organism to be tested and streak for isolation.
2. Incubate aerobically at 35°C for 18 to 24 hours.

Inference

When DNA is hydrolysed, methyl green is released and combines with the highly polymerized DNA at a pH of 7.5 turning the medium colourless around the organism. Then the test is considered positive.

If there is no degradation of DNA, the medium remains green.

Tellurite Reduction

S. aureus reduces tellurite to tellurium producing black colonies, when grown in a medium containing potassium tellurite.

Other Biochemical Tests

Staphylococci hydrolyse urea, reduce nitrate to nitrites are indole negative and are MR and VP positive.

Bacteriophage Typing

This test is useful in food poisoning in large number of persons affected in the community. It is done for epidemiological purpose to find the source of *Staphylococcus aureus* infections. Other typing methods include antibiogram pattern, plasmid typing, ribotyping and DNA finger printing.

TREATMENT OF STAPHYLOCOCCAL INFECTION

Benzyl penicillin is the most effective antibiotic. Cloxacillins are used against beta-lactamase producing strains.

Vancomycin is used in treating infections with MRSA (Methicillin resistant *Staphylococcus aureus*).

COAGULASE-NEGATIVE STAPHYLOCOCCI

Staphylococcus Epidermidis

This is a skin commensal and is an opportunistic pathogen in prosthetic devices, e.g. prosthetic heart valves.

Intraperitoneal catheters, orthopaedic prostheses and vascular grafts. It may cause septicaemia and subacute bacterial endocarditis.

It is usually pathogenic in immunocompromised individuals.

Staphylococcus Saprophyticus

This acts as an opportunistic pathogen and may cause septicaemia and endocarditis.

Differentiating features of the three species of *Staphylococcus*			
Features	S. aureus	S. epidermidis	S. saprophyticus
1 Coagulase test	+	–	–
2 Mannitol fermentation	+	–	–
3 Production of DNAase	+	–	–
Phosphatase	+	–/+	–
a Lysin	+	–	–
4 Protein 4 in the cell wall	+	–	–
5 Novobiocin resistance	–	–	+

Micrococci

These are free living in the environment and are gram-positive cocci, catalase positive, coagulase-negative arranged in clusters. These appear in irregular clusters, groups of four or of eight. They are larger than staphylococci.

Colonies are white in colour and their staining is often not uniform. micrococci can be differentiated from staphylococci by high-Heifson's oxidation-fermentation (OF) test where micrococci show oxidative and staphylococci show fermentative breakdown of carbohydrate.

OXIDATION / FERMENTATION (OF) TEST

Principle

This test is used to determine whether an organism uses carbohydrate substrate to produce acid by products. Six carbohydrates (Glucose, xylose, mannitol, lactose, sucrose and maltose) are used of which glucose is used to determine whether an organism, ferments or oxidizes glucose.

Method

1. To determine whether acid is produced from carbohydrates, inoculate agar, with bacterial growth from on 18 to 24 hours. Culture by stabbing a needle 4 to 5 times into the medium to a depth of 1 cm.
2. Incubate the tubes at 35°C for upto 7 days.

Inference

Positive: Acid production (A) is indicated by the colour indicator changing to yellow.

Negative: Red or alkaline (K) colour with carbohydrates

Micrococci are saprophytes and commensals and may rarely cause opportunistic infection.

Chapter 15

Streptococcus

Streptococci are gram-positive cocci which are arranged in chains. They are part of the normal human flora and some of them are pathogens causing pyogenic infections.

The term streptococci means as follows, *streptos* meaning twisted or coiled. Streptococci are classified into:

1. Alpha (α) haemolytic streptococci
2. Beta (β) haemolytic streptococci
3. Gamma (γ) or non-haemolytic streptococci.

The above classification is based on the type of haemolysin, the organism produces on blood agar. Further the beta-haemolyic streptococci are classified by Lancefield as follows.

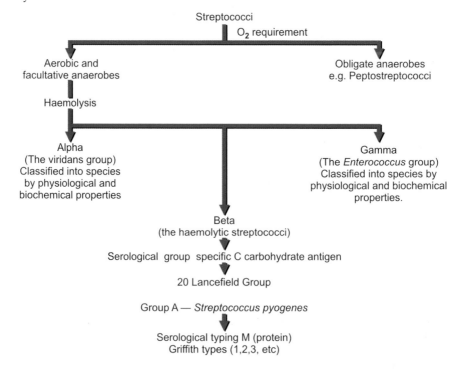

Streptococci

O₂ requirement

Aerobic and facultative anaerobes

Obligate anaerobes e.g. Peptostreptococci

Haemolysis

Alpha
(The viridans group)
Classified into species by physiological and biochemical properties

Gamma
(The *Enterococcus* group)
Classified into species by physiological and biochemical properties.

Beta
(the haemolytic streptococci)

Serological group specific C carbohydrate antigen

20 Lancefield Group

Group A — *Streptococcus pyogenes*

Serological typing M (protein)
Griffith types (1,2,3, etc)

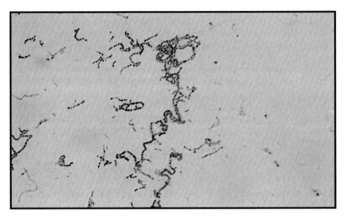

Fig. 15.1: *Streptococcus pyogenes*

STREPTOCOCCUS PYOGENES

Morphology

These are gram-positive cocci, non-motile and non-sporing. The individual cocci are spherical or oval. 0.5 to 1.0 mm in diameter and are arranged in chains. large chains are formed in liquid than in solid media (Fig. 15.1).

Culture

These are most exacting in nutritive requirements, growth occurs only in media containing blood, serum or sugars on blood agar, the colonies are small, circular, semitransparent, low convex with a wide zone of beta-haemolysis around them.

Selective media containing 1:500,00 crystal violet (crystal violet blood agar) permit growth of streptococci and inhibit other bacteria particularly staphylococci.

In liquid media, such as glucose broth, growth occurs as a granular turbidity with a powdery deposit.

BIOCHEMICAL REACTIONS

Streptococci are catalase-negative, unlike staphylococci which are catalase positive.

These are not soluble in 10 per cent bile (Bile solubility test).

Bile Solubility Test (Fig. 15.2)

Principle

The bile solubility test differentiates *Streptococcus pneumoniae* (Positive) from alpha-haemolytic streptococci (negative).

Fig. 15.2: Bile solubility test

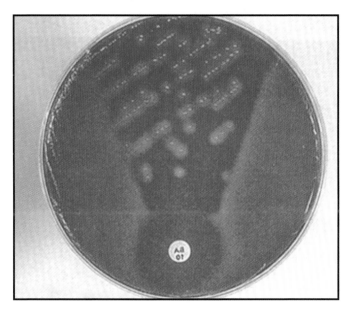

Fig. 15.3: Beta-haemolytic colonies

Bile or a solution of a bile salt, such as sodium desoxycholate rapidly lyses pneumococcal colonies (Fig. 15.3).

Method

1. Place 1 to 2 drops of 10 per cent sodium desoxycholate to the side of a well isolated colony grown on 5 per cent sheep blood agar.
2. Gently wash liquid over colony, dislodging colony from agar.
3. Incubate plates at 35°C in ambient air for 30 minutes.
4. Examine for lysis of colony.

Fig. 15.4: Lancefield grouping

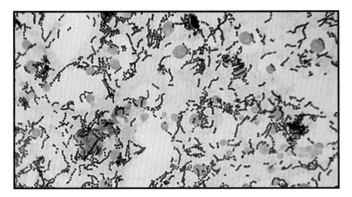

Fig. 15.5: *Streptococcus agalactiae*

Inference

Positive result shows a disintegration of colony (Fig. 15.4).

Negative result shows intact colonies (Fig. 15.5).

Several sugars are fermented by streptococci producing acid but no gas.

Pathogenesis

Streptococcus pyogenes produce pyogenic infections with a tendency to spread locally. Non-suppurative sequelae of local infection include acute glomerulonephritis and rheumatic fever.

Pyogenic infections include sore throat tonsillitis and pharyngitis. The organisms may spread to surrounding tissues in the throat and result in complication like cervical adenitis, otitis media, quinsy, Ludwig's angina and mastoiditis.

Scarlet Fever

It consists of sore throat and a generalised erythematous rash. The skin infections are erysipelas and impetigo. Erysipelas is an acute spreading lesion with massive oedema and erythema.

Impetigo is found mainly in young children and leads to acute glomerulonephrits.

Other pyogenic infections of *streptococcus* are
1. Puerperal sepsis
2. Wound and burns infections
3. Pyaemia along with septicaemia, abscess in brain, lung, liver and kidney.

Non-suppurative Complications

Streptococcus pyogenes infections are followed by non-suppurative sequelae:
1. Acute rheumatic fever
2. Acute glomerulonephritis

Rheumatic fever is often preceded by sorethroat while acute glomerulo-nephritis is by the skin infection. These sequelae or complications are believed to occur due to hypensensitivity to some streptococcal components. The mechanism by which streptococci produce rheumatic fever is still not clear. four common cross-reacting antigen may exist in streptococci and heart, therefore, antibodies to streptococci could cross-react with myocardial and heart valve tissue, causing cellular destruction.

Acute glomerulonephritis is probably due to antibodies formed against streptococci which cross react with glomerular basement membrane and result in damage.

LABORATORY DIAGNOSIS

Acute Suppurative Infections

Specimens

Specimen includes according to the site of lesion such as swab, pus, blood or CSF.

Collection and Transport

Specimens should be collected in sterile container under aseptic conditions and plated immediately or sent to the laboratory in Pike's transport medium (Blood agar containing 1 in 1,000,000 crystal violet and 1 in 10,000 sodium azole).

Gram Stain of Smears

Gram-positive cocci in chains are seen and is of importance in smears collected from pus and CSF.

Culture

The specimen is inoculated on blood agar medium and incubated at 37°C for 18 to 24 hours.

The colonies of streptococci are small 0.5 to 1.0 mm size, circular, low convex with a zone of Beta-haemolysis around them.

Biochemical Reaction

Streptococci are catalase-negative and are not soluble in 10 per cent bile. Streptococci ferment sugars producing acid but no gas.

Non-suppurative Complications

In rheumatic fever and glomerulonephritis, serological tests are done. The routine test done is anti-streptolysin O (ASO) titration. A titre of 200 units or more is significant, in rheumatic fever and is indicative of prior streptococcal infection.

ASO test is a neutralisation reaction where antibodies to streptolysin 'O' (ASO) are neutralised with streptolysin 'O' antigen. ASO titre is usually found in high levels in rheumatic fever but in glomerulonephritis, titres tend to be low, therefore anti-deoxyribonuclease. B (anti-DNAase B) estimation is more reliable. Titres higher than 300 or 350 are significant. Anti-hyaluronidase is another useful test for pyoderma infection of streptococci.

Treatment

Penicillin G is the drug of choice in patients allergic to penicillin, erythromycin or cephalexin is used.

OTHER HAEMOLYTIC STREPTOCOCCI

Group B Streptococci

Streptococcus agalactiae (Fig. 15.5) is an important pathogen causing neonatal septicaemia and meningitis. It may also cause septic abortion and puerperal sepsis. It is a commensal of female genital tract from where bacterial infection in neonates occur. Other group B infection in neonates include osteomyelities, arthritis, conjunctivitis, respiratory infection, endocarditis and peritonitis.

The identification of this depends on their ability to hydrolyse hippurate. Hippurate positive bacteria produce a deep purple colour whereas hippurate negative organisms produce a slightly yellow pink colour or fail to produce any colour.

The other test to identify group B is the CAMP reaction (Christie, Atkins, and Munch-Peterson) which can be demonstrated by the production of a zone of haemolysis resembling butterfly appearance when inoculated perpendicular to a streak of *S. aureus* grown on blood agar.

CAMP TEST

Certain organisms produce a diffusible extracellular protein (CAMP factor) that acts synergistically with the beta-lysine of *S. aureus* to cause enhanced lysis of red blood cells.

Method

1. Streak a beta-lysine producing strain of *S. aureus* down the centre of a sheep blood agar plate.
2. Streak test organisms across the plate perpendicular to the *S. aureus* streak.
3. Inoculate overnight at 35°C in ambient air.

Inference

Positive: Enhanced haemolysis is indicated by an arrowhead shaped zone of beta haemolysis at the juncture of the two organisms.

Negative: No enhancement of haemolysis.

Group C Streptococci

These are predominantly animal pathogens. The species *Streptococcus equisimilis* is isolated from human infection like pharyngitis, rarely endocarditis, septicaemia and meningitis.

Streptococcus equisimilis produces streptolysin O, streptokinase and other extracellular substances. The streptokinase is used for thrombolytic therapy in patients.

Group D Streptococci

Streptococci belonging to group 'D' are classified as enterococci (Faecal streptococci) and non-enterococci (non-faecal streptococci). They are usually non-haemolytic. The enterococci group has been reclassified as a separate genus called enterococci.

Enterococci are normal inhabitants of human intestinal tract and possess some distinctive properties as follows.

1. They are relatively heat resistant and can withstand heat at 60°C for 30 minutes (heat test or heat resistance test).

2. Their ability to grow in the presence of (i) 40 per cent bile, (ii) 6.5 per cent sodium chloride.
3. Their ability to grow at 45°C and at pH 9.6.
4. They are PYR test positive.

On MacConkey's medium they grow as tiny deep pink colonies. On Grams staining enterococci appear as pairs of oval cocci and short chains. The identification of the species is based on biochemical reactions. *Enterococcus faecalis* is the most commonly isolated *Enterococcus* from human sources.

E. faecalis is frequently isolated from cases of urinary tract infection and wound infection like subacute bacterial endocarditis, septicaemia, peritonitis and infection of biliary tract. Vancomycin is the primary alternative drug to penicillin resistant strains.

VIRIDANS STREPTOCOCCI

This group of streptococci produce alpha haemolysis on blood agar. They are known as viridans streptococci due to greenish discoloration (α-haemolysis) on blood agar (Fig. 15.6). They are commensals of mouth and upper respiratory tract. This group contains

Streptococcus salivarius
Streptococcus mutans
Streptococcus sanguis and
Streptococcus miller

These are non-pathogenic but are occasionally associated with dental caries and subacute bacterial endocarditis in persons with predisposing factor such as valvular disease of the congenital heart disease and cardiac surgery. Prophylactic antibiotics is of use in such patients.

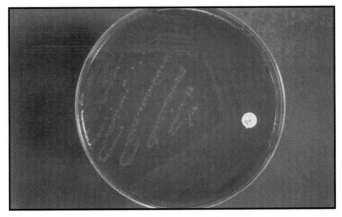

Fig. 15.6: *S. viridans* on blood agar—alpha-haemolysis and optochin

Chapter

16

Pneumococcus

Pneumococci are normal commensals of the upper respiratory tract. These are reclassified as *Streptococcus pneumoniae.*

Morphology

Pneumococci are gram-positive, small and slightly elongated cocci arranged in pairs. Each coccus has one end broad or rounded and other end pointed (flame-shaped diplococci). These are capsulated and the capsules are demonstrated by India-Ink preparation (Fig. 16.1).

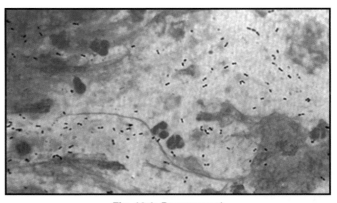

Fig. 16.1: Pneumococci

CULTURE

Pneumococci have complex nutritional requirement.

On blood agar: The colonies are usually dome shaped, with an area of greenish discolouration- alpha-haemolysis around them.

Pneumococci are typically alpha-haemolytic but under anaerobic conditions colonies show beta haemolysis due to oxygen labile pneumolysin O.

In liquid medium such as glucose broth, pneumococci produce uniform turbidity.

Biochemical Reaction

Pneumococci ferment several sugars with production of acid only. Pneumococci are also soluble in bile and bile solubility test is an important diagnostic test to differentiate *Pneumococcus* from other streptococci.

Pneumococci are catalase and oxidase negative.

BILE SOLUBILITY TEST

Principle

This test differentiates *Streptococcus pneumoniae* from alpha-haemolytic streptococci.

Bile or a solution of a bile salt, such as sodium desoxycholate rapidly lyses pneumococcal colonies.

Procedure

Place 1 to 2 drops of 10 per cent sodium desoxycholate to the side of a isolated colony growing on 5 per cent sheep blood agar. Incubate plate at 35°C in ambient air for 30 minutes examine for lysis of colony.

Observation and Result

Positive: Colony disintegrates; e.g. *Streptococcus pneumoniae.*

Negative: Intact colonies; e.g. *Enterococcus faecalis.*

Pathogenesis

Streptococcus pneumoniae is one of the most common bacteria causing pneumonia.
1. Lobar pneumonia
2. Bronchopneumonia
3. Meningitis: *Streptococcus pneumoniae* is the second most important cause of pyogenic meningitis after *Neisseriae meningitidis*
4. Other infections: *Pneumococcus* may also produce empyema, pericarditis, otitis media, sinusitis, conjunctivitis peritonitis and suppurative arthritis usually as complications of pneumonia.

Laboratory Diagnosis

Specimens

Clinical samples are:
- Sputum
- Cerebrospinal fluid (CSF)

- Pleural exudates
- Blood.

Collection and Transport

Specimens should be collected in sterile containers and processed immediately. In case of delay, CSF specimen should never be refrigerated but kept at 37°C.

Microscopy

Gram staining of smear reveals a large number of polymorphs along with gram-positive diplococci which may be intracellular as well as extracellular in CSF smear.

Culture

On blood agar, typical a haemolytic colonies are isolated (Fig. 16.2).

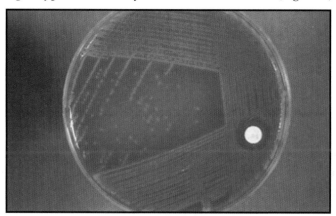

Fig. 16.2: Pneumococci on blood agar

Biochemical Reactions

Inulin fermentation and bile solubility tests are significant in diagnosing pneumococcal infection.

Treatment

The antibiotic of choice is parenteral penicillin. In case of penicillin resistant strains, cephalosporin is used.

Chapter 17

Neisseria and Branhamella

The genus *Neisseria* consists of two pathogenic species namely *N. meningitidis* and *N. gonorrhoeae*. The non-pathogenic species include *N. flarescens, N. sicca, N. subflava* and other species.

NEISSERIA MENINGITIDIS

Morphology

These are gram-negative, spherical or oval cocci arranged in pairs with the adjacent sides flattened. These are non-motile and are generally intra-cellular.

Culture

Blood agar, chocolate agar and Müeller-Hinton agar are commonly used. Thayer-Martin medium with antibiotics like vancomycin, colistin, nystatin and trimethoprim are selective media used for isolation of these cocci.

The colonies are small, round, convex, grey and translucent.

Biochemical Reactions

N. meningitidis is catalase and oxidase positive. The prompt oxidase reaction helps in the identification of *N. meningitidis* and gonococci.

Glucose and maltose are fermented with acid production but no gas. They do not ferment lactose or sucrose.

Indole and H_2S are not produced and nitrates are not reduced.

Pathogenesis

N. meningitidis causes pyogenic meningitis and the infection is acquired by droplet spread via nasopharynx.

Laboratory Diagnosis

Specimens

- CSF
- Blood
- Nasopharyngeal swab.

Collection and Transport of Specimens

CSF is collected by lumbar puncture and blood by venipuncture under strict sterile conditions. Nasopharyngeal specimen is collected by using a sterile swab. This swab is transported using the transport medium namely Stuart's medium.

Direct Microscopy and Culture

CSF is divided into three portions. One portion is centrifuged and is used for Gram staining where the typical structure of menigococci is observed.

The second portion of the CSF is used for direct culture and the third portion is incubated overnight along with equal volume of glucose broth and then subcultured on blood agar or chocolate agar (Figs 17.1 to 17.3).

Cultural Characteristics

On solid media, colonies are small, round, convex, grey and translucent.

Serological Diagnosis

Direct slide agglutination of the organism may be done with specific antisera. Specific antibodies to capsular polysaccharide may also be demonstrated by haemagglutination test.

Fig. 17.1: *Neisseria meningitides*——chocolate agar

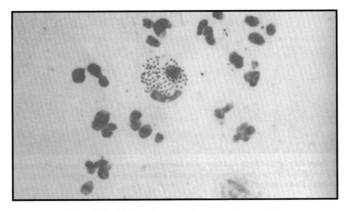

Fig. 17.2: *Neisseria gonorrhoeae*—Gram stain

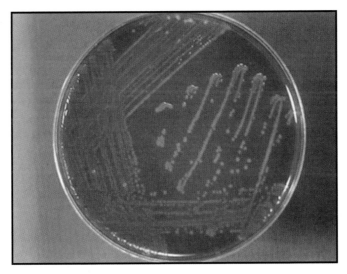

Fig. 17.3: *Neisseria gonorrhoeae*—culture

Treatment

Penicillin is the drug of choice and chloramphenicol is used in persons who are allergic to penicillin. Cefotoxime or ceftriaxone are as effective as chloramphenicol.

Neisseria Gonorrhoeae

This causes gonorrhoea, the sexually transmitted disease.

Morphology

These are gram-negative oval cocci arranged in pairs with adjacent sides concave (diplococci-pear or bean shaped). Gonococci possess pili on their surface which helps in the adhesion of the cocci to the host surface.

Culture

These grow on selective medium like chocolate agar. Thayer-Martin medium with antibiotics (vancomycin, colistin, nystatin and trimethoprim). These also grow well on Müeller-Hinton agar.

Colony Morphology

Colonies are small, round, grey, translucent, convex with tiny granular surface.

Biochemical Reactions

N. gonorrhoeae is oxidase positive and ferments glucose with acid production only (Fig. 17.4).

Pathogenesis

Gonorrhoeae: This is a sexually transmitted disease infecting both sexes. In males, it starts as acute urethritis and extends to the prostate, testes, seminal vesicles and epididymis. In females, the primary infection starts from the urethra and extends to the cervix, Bartholin's glands, uterus, fallopian tubes, ovaries and results in pelvic inflammatory disease.

Ophthalmia Neonatorum

This is a non-venereal gonococcal conjunctivitis in the newborn through infected birth canal.

Fig. 17.4: *Neisseria gonorrhoeae* — oxidase test

Laboratory Diagnosis

Specimens include urethral discharge and cervical discharge in females.

Transport

All specimens should be transported using Stuart's transport medium and should be processed at the earliest.

Microscopy

Gram's staining shows gram-negative intracellular diplococci.

Cultural Characteristics

Chocolate agar and Thayer-Martin medium are used for culture. The colonies are small, round grey translucent and convex. The colonies are confirmed by Gram staining.

Serology

Serological tests like complement fixation test, immunofluorescence and ELISA tests are used to detect antibodies.

Treatment

Penicillin and doxycycline are used to treat gonococcal infection.

Non-gonococcal Urethritis

Causative agents:
a. Bacterial:
- *Chlamydia trachomatis*
- *Ureaplasma urealyticum*
- *Mycoplasma hominis*
- *Gardnerella vaginalis*
b. Viral
- Herpes virus
- Cytomegalovirus
c. Fungal: *Candida albicans*
d. Protozoal: *Trichomonas vaginalis.*

Treatment

Tetracycline is effective for both *C. trachomatis* and *Ureaplasma urealyticum* infections.

Branhamella

Neisseria catarrhalis is now reclassified as *Branhamella catarrhalis*.

Morphology

These are gram-negative diplococci. Oval with adjacent sides flattened.

Culture

This grows on ordinary medium like nutrient agar.

Biochemical Reactions

It does not ferment any carbohydrates but hydrolyses tributyrium. It is catalase and oxidase positive.

Pathogenesis

Branhamella causes lower respiratory tract infections. Otitis media and rarely meningitis, endocarditis and sinusitis.

Treatment

Penicillin is the drug of choice. There are some strains of *B. catarrhalis produce* beta-lactamase and are resistant to penicillin. In such cases other higher antibiotics are used.

Chapter

18

Corynebacterium Diphtheriae

Morphology

These are thin slender, gram-positive bacilli and are also pleomorphic. These are club-shaped with metachromatic granules at one or both ends. These granules are also called as volutin or Babe's. Ernst granules. Special stains like Alberts and Neisser are used for staining the bacilli (Fig. 18.1).

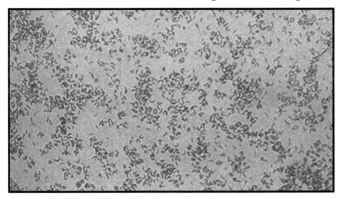

Fig. 18.1: *C. diphtheriae*—Bacilli

Cultural Characteristics

C. diphtheriae grow best on media enriched with blood, serum or egg.

Loeffler's Serum Slope

The colonies are small, circular white and glistening.

Tellurite Blood Agar Medium

The potassium telluride in this medium inhibits other bacteria. *C. diphtheriae* grow slowly on this medium and form grey or black colonies due to reduction of potassium tellurite to tellurium (Fig. 18.2).

Fig. 18.2: Culture on tellurite blood agar

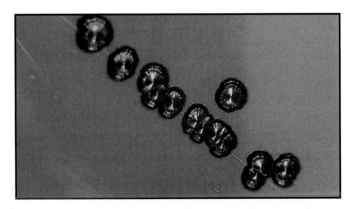

Fig. 18.3: *C. diphtheriae,* gravis type

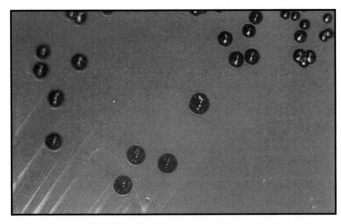

Fig. 18.4: *C. diphtheriae,* mitis type

Based on colony morphology on tellurite medium, three main biotypes of *C. diphtheriae* are distinguished namely.

C. *diphtheriae*-gravis, intermedius and mitis (Figs 18.3 and 18.4).

Biochemical Reactions

These ferment glucose and maltose with the production of acid but no gas. They do not ferment lactose, mannitol or sucrose.

Pathogenesis

Diphtheria affect the following sites:
1. Faucial
2. Laryngeal
3. Nasal
4. Conjunctival
5. Otitic
6. Vulvovaginal
7. Cutaneous around the mouth and nose.

The bacilli multiplies in the above sites and causes toxic effect. The toxin also causes necrosis and results in pseudomembrance. This toxin also diffuses into the bloodstream and results in heart failure, polyneuropathy and degenerative change in adrenals, kidney and liver.

Laboratory Diagnosis

Collection of Specimen

Swabs are taken from the lesions, namely throat, nose, larynx, ear and conjunctiva.

Vagina or Skin

Two swabs are taken one swab is used for smear examination and the other for culture.

Microscopy

Smears are stained with Gram's stain and Albert's stain. Diphtheria bacilli show beaded slender green rods in typical Chinese letter pattern on Albert's staining.

Culture

Swabs are inoculated on the following culture media.

1. Loeffler's serum slope growth appears in 6 to 8 hrs and subculture from Loeffler's serum slope is made on tellurite blood agar.
2. Tellurite blood agar: These plates have to be incubated at 37°C for 48 hours.
3. *Blood agar:* This helps to distinguish mitis biotype which shows haemolysis.

VIRULENCE TEST

These tests demonstrate the production of toxin by bacteria isolated on culture. Virulence testing done by *in vivo* or *in vitro* methods.

IN VIVO TESTS

Subcutaneous Test

The growth from an overnight culture on Loeffler's serum slope is emulsified in 2.5 ml broth and 0.8 ml of this emulsion is injected subcutaneously into two guinea pigs, one of which has received an intramuscular injection of 500 units of diphtheria antitoxin 18 to 24 hours previously. If the strain is virulent, the unprotected animal will die within 2 to 3 days with evidence of haemorrhage in the adrenal glands.

Intracutaneous Test

Two guinea pigs (or Rabbits) are injected intracutaneously with 0.1 ml emulsion from growth on Loeffler's serum slope, one of these animals is protected with 500 units antitoxin the previous day. Control and the other is given 50 units of antitoxin intraperitoneally four hours after the skin tests in order to prevent death. If the strain is toxigenic, the inflammatory reaction at the site of injection, progresses to necrosis in 48 to 72 hours in the test animal but there is no change in the control animal.

IN VITRO TESTS

Elek's Gel Precipitation Test (Fig. 18.5)

This is an immunodiffusion test. A rectangular strip of filter paper soaked in diphtheria antitoxin is placed on the surface of a 20 per cent horse serum agar plate while the medium is still fluid when the agar solidifies, the test strain is streaked at right angle to the filter paper strip. The positive and negative controls are also put up. The plate is incubated at 37°C for 24 to 48 hours. The toxin produced by the bacterial growth diffuses in the agar and produces a line of precipitation where it meets the antitoxin at optimum concentration. Nontoxigenic strains will not produce any precipitation line.

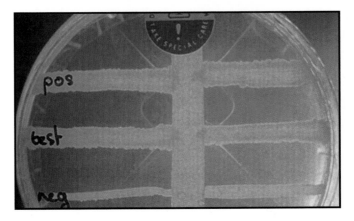

Fig. 18.5: Elek plate test

Treatment

C. diphtheriae is sensitive to penicillin, erythromycin and other antibiotics.

Prophylaxis

Immunisation is with toxoid in a combined form namely DPT.

Chapter

19

Bacillus Anthracis

This is the causative agent of anthrax. *Bacillus anthracis* is a gram-positive, non-acid-fast, non-motile, rectangular, large, spore forming bacillus. When Blood films containing anthrax bacilli are stained with polychrome methylene blue for 10 to 20 seconds, an amorphous purple material is noticed around the bacilli. This is the capsular material of the anthrax Bacillus and the test is known as McFadyean's reaction (Fig. 19.1).

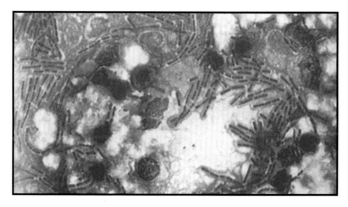

Fig. 19.1: McFadyean's reaction

Culture

On nutrient agar, colonies are round 2 to 3 mm in diameter, raised, opaque and grayish-white. Under microscopic examination the colony is composed of long interlacing chains of bacilli, resembling locks of matted hair—known as the medusa head appearance (Fig. 19.2).

Blood Agar

The colonies are non-haemolytic but occasional strains produce a narrow zone of haemolysis (Fig. 19.3).

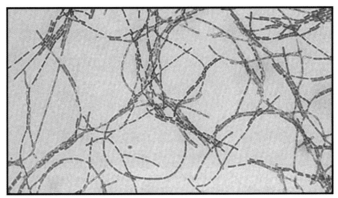

Fig. 19.2: Long interlacing chain of bacilli

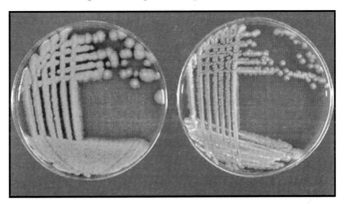

Fig. 19.3: Bacillus on blood agar

Gelatin Stab Culture

Gelatin is liquefied at the top portion of the culture medium. This is known as inverted fir tree appearance.

A selective medium (PLET medium) consisting of heart infusion agar with polymyxin, lysozyme, ethylene diamine tetra aceticacid (EDTA) and thallos acetate is used for the isolation of *B. anthracis*.

Biochemical Reactions

Glucose, maltose and sucrose are fermented with production of acid only catalase is formed and nitrates are reduced to nitrites.

Pathogenesis

Human anthrax: There are three clinical type of disease namely. Cutaneous anthrax, pulmonary anthrax and intestinal anthrax. All types lead to septicaemia.

Laboratory Diagnosis

Specimens

Swabs, fluid or pus from pustules; sputum and blood from pulmonary and septicaemic anthrax are collected.

Microscopy

Gram stained smear from the specimen shows chains of large gram-positive bacilli. India- Ink staining capsule appears as a clear hole around the bacterium.

McFadyean's reaction is demonstrated with the polychrome methylene blue stain.

Culture

Medusa head colonies appear on culture and inverted fir tree appearance on gelatin stab culture.

Treatment

Penicillin is the drug of choice.

Clostridium Tetani

Clostridium tetani is the causative agent of tetanus.

Morphology

It is a gram-positive, slender bacillus (measuring 4-8 mm × 0.5 mm) with spherical terminal spores giving the bacillus, the characteristic drumstick appearance.

Culture

It is an obligate anaerobe which grows on ordinary media. It can grow well in Robertson cooked meat broth (RCM), thioglycollate broth, nutrient agar and blood agar. In RCM broth growth occur as turbidity and there is also some gas formation. The bacilli produce a swarming (thin spreading film) growth on blood agar. However, on horse blood agar they produce alpha-haemolytic colonies which subsequently develop into beta-haemolytic, due to the production of a haemolysin (Figs 20.1 and 20.2).

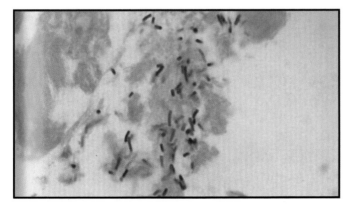

Fig. 20.1: *Clostridium perfringens (C. welchii)*

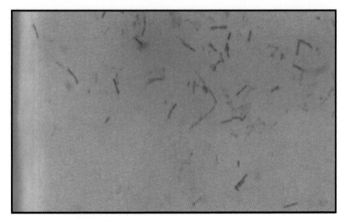

Fig. 20.2: *Clostridium tetani*, Gram's stain

Biochemical Reactions

It does not ferment any sugar. It forms indole, H_2S not formed. Nitrates are not reduced.

Pathogenesis

Tetanus develops following the contamination of wound with *C. tetani*. Infection stricty remains localised in the wound. Pathogenic effects are mainly due to tetanospasmin (neurotoxin) of *C. tetani*. This results in muscle rigidity and spasms.

Laboratory Diagnosis

The diagnosis of tetanus should always be made clinically and laboratory test are done only to confirm it (Figs 20.3 to 20.6).

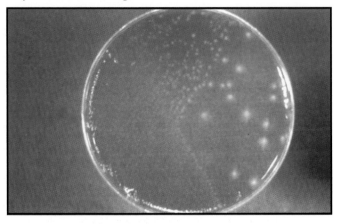

Fig. 20.3: *Clostridium perfringens*

Fig. 20.4: *Clostridium perfringens,* blood agar

Fig. 20.5: *Clostridium perfringens,* Robertson's cooked meat medium

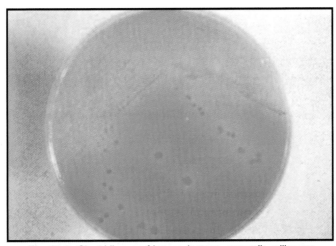

Fig. 20.6: *Clostridium perfringens,* lactose, egg yolk, milk agar

Microscopy

Gram staining may show gram-positive bacilli with drumstick appearance (Fig. 20.7).

Culture

Specimen is inoculated on freshly prepared blood agar and incubated at 37°C for 24 to 28 hours under anaerobic conditions. The specimen is also inoculated in three tubes of Robertson's cooked meat broth (see Fig. 20.5). One of these tubes is heated at 80°C for 15 minutes. The second tube for five minutes and the third left unheated heating for different periods to kill vegetative bacteria, while leaving tetanus spores undamaged. These cooked meat broths are incubated at 37°C and subcultured on blood agar places daily for up to four days.

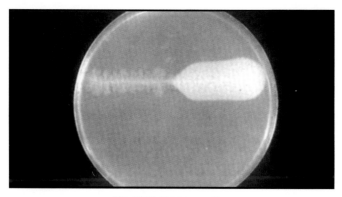

Fig. 20.7: Nagler reaction

Immunisation

Tetanus toxoid. Three doses at 0.5 ml tetanus toxoid (TT) each are given intramuscularly with an interval at 4 to 6 weeks between first two doses and 6 to 19 months between the second and third doses. A booster dose of toxoid is recommended after 10 years.

Tetanus toxoid is given along with diphtheria toxoid and pertusis vaccine (DPT) in children.

Treatment

Tetanus patients are treated in special isolated units. Treatment consists of controlling spasms. Maintaining airway by tracheostomy and attention with penicillin or metronidazole should be started and continued for a week or more.

Chapter

21

Escherichia Coli

Morphology

E. coli is a gram-negative *Bacillus* measuring 1 to 3 mm. Most strains are motile by peritrichate flagella. It is non-sporing and non-capsulated.

Culture

It is an aerobe and facultative anaerobe and grows on ordinary culture medium at optimum temperature of 37°C. Colonies of some strains show beta-haemolysis on blood agar. On MacConkey's medium, colonies are pink due to lactose fermentation. In general colonies are circular, moist smooth with entire margin and non-mucoid.

Biochemical Reactions (Fig. 21.1)

They ferment most of the sugars (glucose, lactose, monnitol, maltose) with production of acid and gas. Indole and methyl red (MR) reaction are positive but voges proskauer (VP) and citrate utilisation tests are negative (imvic + + −) urea is not split, gelatin is not liquefied, H_2S is not turned and growth does not occur in medium. The important biochemical reactions are summarised as follows.

Glucose	Lactose	Mannitol	Sucrose
AG	+	+	−

Indole	Urease	Citrate	MR	VP
+	−	−	+	−

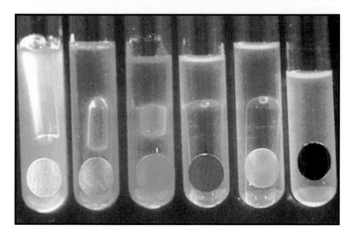

Fig. 21.1: Biochemical tests of *E. coli*

Pathogenesis

E. coli forms a part of normal intestinal flora of man and animal. There are four major types of clinical syndromes which are caused by *E. coli.*
1. Urinary tract infection
2. Diarrhoea
3. Pyogenic infections
4. Septicaemia.

Urinary Tract Infection

E. coli is the commonest organism responsible for urinary tract infection (UTI). Route of infection to reach urinary tract is either the ascending route or the haematogeneous route. The ascending route is through faecal flora spreading to the perineum and from there they ascend into the bladder.

Diarrhoea

E. coli causing, diarrhoeal diseases are of five groups.

Enteropathogenic E. coli (EPEC)

They cause enteritis in infants especially in tropical countries. Common serogroups of EPEC are 026, 086, 055, 086, 0111, 01114, 01119, 0125, 0126, 0129, 0128 and 0148.

Enterotoxigenic E. coli (ETEC)

They are a major cause of diarrhoea in children in developing countries and are the most important cause of travellers diarrhoea. The name travelers diarrhoea refers to diarrhoea in persons from the developed countries within a few days of their visit to one of the developing countries. ETEC sometimes causes a diarrhoea similar to that produced by *Vibrio cholerae*.

Enteroinvasive E. coli (EIEC)

Some strains of *E. coli* involves the intestinal epithelial cells as do dysentery bacilli and produce disease identical to *Shigella dysentery*. These belong to serogroups: 028, 0112, 0124, 0186, 0143, 0144, 0152 and 0164.

Enterohaemorrhagic E. coli (EHEC) or Verocytotoxin-producing E. coli (VTEC)

These strains cause haemorrhagic colitis (HC) and haemolytic uraemic syndrome (HUS). It is most common in infant and young children but can occur in all ages.

Enteroaggregative E. coli (EAEC)

They form a heat stable enterotoxin called enteroaggregative heat stable enterotoxin-1 (EAST-1).

Pyogenic Infections

E. coli may cause wound infection peritonitis, cholecystitis and meningitis. It is an important cause of neonatal meningitis.

Septicaemia

E. coli is a very common cause of septicaemia in many hospitals and leads to fever, hypotension and disseminated intravascular coagulation (endotoxic shock).

Laboratory Diagnosis

Urinary tract infection.

Specimen Collection

Midstream Urine Specimen (MSU)

It is collected preferably prior to administration of antibiotics. Specimen is collected in a sterile container. Before collecting a sample, genitalia should be cleaned with soap and water and men are instructed to retract the foreskin of glans penis whereas women should keep the labia apart. The first portion of urine is allowed to pass, without interrupting the urine. Now, mid portion of the stream is collected and the first portion of urine adequately flushed.

Catheter Specimen

Urine should be collected directly from the catheter and not from the collection bag. The catheter should not touch the container.

Urine Specimens from Infants

A clean catch specimen after cleaning of genitalia is preferred. Another procedure of collecting specimen in infants is by suprapubic aspiration.

Transport

As urine is a good culture medium, specimens after collection should reach the laboratory with minimum delay, if it is not possible the specimen is to be refrigerated at 4°C.

Laboratory Methods

Part of the specimen is used for bacteriological culture and the rest is examined immediately under the microscope.

Microscopy

Urine is certrifuged and deposit is examined under the microscope for detecting pus cells. Erythrocytes, epithelial, cells and bacteria.

Culture

Most laboratories use a semiquantitative method (Standard loop technique) for culture of urine specimens.

Standard loop technique: A standard calibrated loop is used to culture a fixed volume of uncentrifuged urine. Blood agar and MacConkey's agar are used and incubated at 37°C for 24 hours. Next day, the number of colonies grown is counted and total count per ml is calculated (Figs 21.2 and 21.3).

Interpretation of results: Significant bacteriuria when bacterial count is more than 10^5/ml of a single species.

Doubtful significance-between 10^4 to 10^5 bacteria per ml. Specimen should be repeated for culture. No significant growth – $<10^3$ bacteria per ml and are regarded as contaminant.

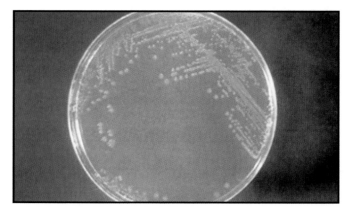

Fig. 21.2: *E. coli,* sorbitol MacConkey's medium

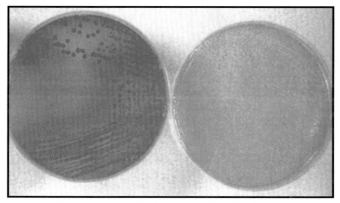

21.3: *E. coli,* MacConkey and sorbitol MacConkey

Identification

The organisms are identified by colony characters, Gram's staining, motility, biochemical reactions and slide agglutination test.

Diarrhoea

Faeces or rectal swab is placed directly on blood agar and MacConkey's medium. After overnight incubation. Growth on culture media is identified by colony morphology, Gram staining, motility and biochemical reactions. *E. coli* colonies are emulsified in saline on a slide and tested by agglutination with polyvalent and monovalent αβ antisera against enteropathogenic (EPEC) serotypes.

Special indicator media may be employed for specific serotypes with distinctive properties. A modified MacConkey's medium in which sorbitol is incorporated instead of lactose, has been employed for the detection of serotypes O55: B5 and 0111: B4 as these do not ferment sorbitol unlike other *E. coli* strains.

Pyogenic Infections

The specimens are usually pus and wound swab. Cultures are made on MacConkey's agar and the isolate is identified by colony morphology, staining motility and biochemical reactions.

Septicaemia

Diagnosis depends on the isolation of the organism by blood culture and its identification by colony morphology, staining, motility and biochemical reactions.

Chapter

22

Klebsiella

The genus *Klebsiella* consists of gram-negative rods.

Property	K. pneumoniae	K. ozaenae	K. rhinoscleromatis
Gas from glucose	+	V	–
Acid from lactose	+	V	–
Urease	+	V	–
Citrate	+	V	–
Malonate	+	–	+
MR	–	+	+
VP	+	–	–
Lysine decarboxylase	+	V	–

Gram-negative capsulated, non-sporing, non-motile bacilli that grow well on ordinary media produce pink mucoid colonies on MacConkey's agar.

Morphology

These are short plump, gram-negative capsulated non-motile bacilli. They are about 1 to 2 μm × 0.5 to 0.8 μm in size.

Culture

Klebsiellae grow well on ordinary media at optimum temperature at 37°C in 18-24 hours. On MacConkey's agar, the colonies appear large mucoid and pink to red in colour (Fig. 22.1). Mucoid nature of colonies is due to capsular material produced by the organism.

Biochemical Reactions

They ferment sugars (glucose, lactoses, sucrose, mannitol) with production of acid and gas. They are urease positive, indole negative, MR negative VP positive and citrate positive (Imvic –++). These reactions are typical of *K. pneumoniae*.

Glucose	Lactose	Mannitol	Sucrose
AG	+	+	+

Indole	Urease	Citrate	MR	VP
–	+	+	–	+

Pathogenesis

It is responsible for severe bronchopneumonia, urinary tract infections, nosocomial infections, wound infections, septicaemia, meningitis and rarely diarrhoea, *Klebsiella pneumonia* is a serious disease with high case fatality. It is very important pathogen causing nosocomial infections of the lower respiratory tract.

Laboratory Diagnosis

Laboratory diagnosis is done by culture of appropriate specimens on blood agar and MacConkey's agar (Fig. 22.1). The isolate is identified by colony morphology, Gram staining, test for motility and biochemical reactions.

Treatment

Klebsiellae are usually sensitive to cephalosporins, trimethoprim, nitrofurantoin, co-trimoxazole and gentamicin.

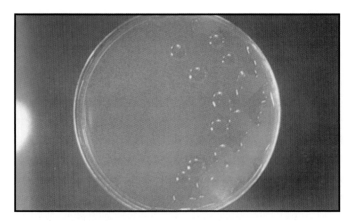

Fig. 22.1: *Klebsiella pneumoniae*, MacConkey's medium

Proteus

These are motile, gram-negative bacilli characterised by swarming growth on agar.

Morphology

These are gram-negative bacilli measuring 1 to 3 × 0.5 μm in size. They are non-capsulated, non-sporing and actively mobile. They possess peritrichate flagella.

Culture

They are aerobic and facultatively anaerobic. They grow on ordinary media and culture emits a characteristic putrefactive (fishy or seminal) odour. When grown on nutrient agar or blood agar. *P. vulgaris* and *P. mirabilis* exhibit swarming. Swarming of *Proteus* appears to be due to vigorous motility of the bacteria (Fig. 23.1).

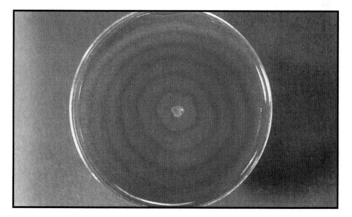

Fig. 23.1: *Proteus mirabilis* blood agar

Fig. 23.2: *Proteus mirabilis*, MacConkey's medium

They form smooth, pale or colourless colonies on MacConkey's agar (Fig. 23.2) and do not swarm this medium. In liquid medium. (Peptone water). *Proteus* produce uniform turbidity with a slight powdery deposit and an ammoniacal odour.

Biochemical Reactions (Fig. 23.3)

The distinctive characters of this genus are:

i. Deamination of phenylalanine to phenylpyruvic acid (PPA) test. It is always positive.
ii. Hydrolysis of use by enzyme urease. It is another characteristic feature of *Proteus.*

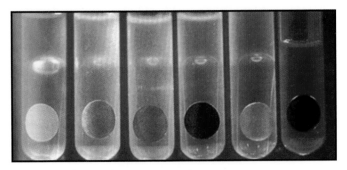

Fig. 23.3: Biochemical test of *Proteus mirabilis*

They ferment glucose by producing acid and gas. Lactose is not fermented. Indole is formed by *P. vulgaris* but is negative in *P. mirabilis*. H_2S is produced by *P. vulgaris* and *P. mirabilis*. They are MR positive and VP negative.

Pathogenesis

Proteus species are saprophytic and widely distributed in nature. They also occur as commensals in the intestine. They are opportunistic pathogens and may cause many types of infections such as urinary tract infection (UTI), Pyogenic lesions, infection of ear, respiratory tract infection and nosocomial infections.

Laboratory Diagnosis

1. Specimens
    a. Mid stream urine sample in UTI
    b. Pus in pyogenic lesions
2. Collections of specimen should be collected in sterile container, under all aseptic conditions and transported immediately.
3. *Culture:* It is cultured on MacConkey's agar or blood agar with 6 per cent agar to inhibit swarming. Culture media are incubated at 37°C for 18-24 hours. WLF colony are seen on MacConkey's agar. Peptone water is also inoculated.
4. *Gram's staining:* Gram-negative bacilli which are non-capsulated and non-sporing.
5. *Hanging drop preparation:* Actively motile bacilli are observed.
6. *Biochemical reactions:* Most important are PPA and urease tests. PPA test is positive in all *Proteus* species. Urease test is also positive in all *Proteus* species.
7. *Agglutination test:* Strain may be agglutinated with polyvalent group specific sera to confirm it.

Morphology

Shigellae are short, gram-negative bacilli measuring about 1 to 3 mm × 0.5 mm. They are non-motile, non-capsulated and non-sporing.

Culture

After overnight incubation, colonies are about 2 mm in diameter, circular, convex and smooth. Colonies in MacConkey's agar and deoxycholate citrate agar (DCA) are colourless (non-lactose fermenting-NLF). DCA is a useful selective medium to isolate these organisms from faeces. However, Xylose Lysine Deoxycholate (XLD) agar is a better selective medium than DCA.

Salmonella-Shigella (SS) agar is a highly selective medium for the isolation of *Salmonella* and *Shigella*. Colonies of *Shigella* on this medium are colourless (due to non-lactose fermentation) with no blackening, while those of *Salmonella* are colourless with black centres. This medium contains salts like sodium citrate, ferric citrate, lactate and neutral red (indicator). The high bile salt concentration and sodium citrate in this medium, inhibit all gram-positive bacteria and coliforms.

Enrichment Medium

Selenite F broth: Sodium solvent in this medium inhibits coliform bacilli while permitting salmonellae and shigellae to grow. It is recommended for the isolation of these organisms from faeces.

Pathogenesis

Shigellae cause bacillary dysentery. Mode of infection is by ingestion of contaminated food.

The food or drink is contaminated by faces of cases or carriers. The organisms infect the epithelial cells of the terminal ileum and colon and multiply inside them.

Inflammatory reaction develops and there is necrosis of surface epithelial cells.

S. dysentriae type 1 causes toxaemia due to production of exotoxin. The toxin causes accumulation of fluid leading to severe diarrhoea (enterotoxic action) and may cause complications like polyneuritis, arthritis, conjunctivitis and parotitis, haemolytic uraemic syndrome may occur as a rare complication in severe cases.

The term shigellosis has been employed to include the whole spectrum of the disease caused by shigellae.

Laboratory Diagnosis

Diagnosis depends upon isolating the bacillus from faeces.

Specimens

Fresh stool is collected. Rectal swabs are not satisfactory. The ideal specimen is a direct swab of an ulcer taken under sigmoidoscopic examination.

Transport

Specimens should be transported immediately and inoculated without delay. specimens should be transported in a suitable medium such as buffered glycerol saline, pH 7.0 to 7.4.

Direct Microscopy

Saline and iodine preparation of faeces show large number of pus cells, erythrocytes and macrophages.

Culture

Specimen is inoculated on selective media like MacConkey's agar, DCA or XLD agar. Selective F broth (0.4%) is used as enrichment medium (Figs 24.1 to 24.3).

Colony Morphology and Staining

Colourless (NLF) colonies appear on MacConkey's agar which are further confirmed by Gram's staining, hanging drop preparation and biochemical reactions. Shigellae are gram-negative bacilli and are non-motile.

Biochemical Reactions

Any non-motile *Bacillus* that is urease, citrate, H_2S and KCN negative should be further confirmed by various biochemical test.

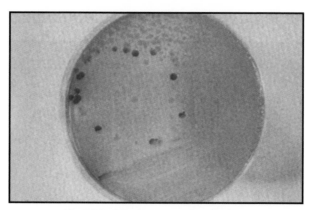

Fig. 24.1: *Shigella sonnei,* MacConkey

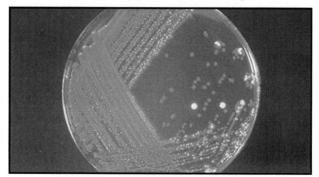

Fig. 24.2: *Shigella sonnei, XLD medium*

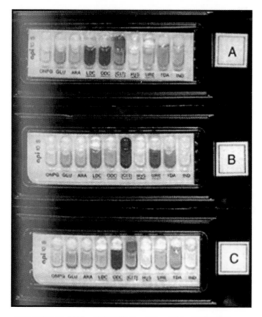

Fig. 24.3: API – 105 kit for *Enterobacteriaceae*

Slide Agglutination Test

Identification of *Shigella* is confirmed by slide agglutination with polyvalent antisera and monovalent sera. Then type specific antisera belonging to subgroups A, B, or C is used for agglutination test.

Treatment

Dehydration has to be corrected promptly particularly in infants and young children.

In very serious infections, nalidixic acid and has been life-saving.

Chapter

25

Salmonella

Salmonella produce three main types of disease in man.

Enteric Fever

The causative agents for enteric fever are *Salmonella typhi* (causing typhoid fever) or *S. paratyphi* A, B and C (causing paratyphoid fever). The term enteric fever includes both typhoid and paratyphoid fever.

Gastroenteritis

Salmonellae under this group are *S. typhimurium, S. enteritidis, S. newport, S. dubin* and *S. thompsun.*

Septicaemia

The commonly associated *Salmonella* is *S. choleraesuis* but other species may also cause septicaemia.

Morphology

Salmonellae are gram-negative bacilli measuring 1 to 3 mm × 0.5 mm. They are motile, non-sporing and non-capsulated. Motility is due to the presence of peritrichous flagella except *S. gallinarium* and *S. pullorum* which are non motile.

Table 25.1: Biochemical reactions of S. typhi and S. paratyphi							
	Glucose	Mannitol	Lactose	Sucrose	Indole	Citrate	MR
S. typhi	A	A	–	–	–	–	+
S. paratyphi A	AG	AG	–	–	–	–	+
S. paratyphi B	AG	AG	–	–	–	+	+
S. paratyphi C	AG	AG	–	–	–	+	+

VP	H₂S	Xylose	D-tarture	Mucate
–	+	D	A	D
–	–	–	–	–
–	+	AG	–	AG
–	+	AG	AG	–

Culture

Salmonellae grow on ordinary culture media at optimum temperature of 37°C range 15 to 41°C, pH 6-8 aerobic and facultatively anaerobic. They produce colonies of 2-3 mm in diameter, circular, translucent, low convex and smooth. On MacConkey's agar and deoxycholate citrate agar (DCA), colonies are colourless due to non-lactose fermentation (NLF). On Wilson and Blair Bismith sulphate medium (Selective medium for salmonellae) jet black colonies with metallic sheen are formed due to formation of hydrogen sulphide. *S. paratyphi* A and other species which do not form H₂S produce green colonies. Xylose lysine deoxycholate (XLD) agar is another medium used for isolation of this organism. Most strains of salmonellae produce red colonies with black centres, when grown on this medium. H₂S negative serotypes of *Salmonella* produce red colonies without black centres.

Selenite F broth and tetrathionate broth (TTB) are commonly used as enrichment media for inoculation of specimens especially faeces.

Biochemical Reactions

Salmonellae ferment glucose, mannitol and maltose fermenting acid and gas except *S. typhi* which produce only acid and no gas. They do not ferment lactose or sucrose. Indole is not produced. Most salmonellae produce H₂S in

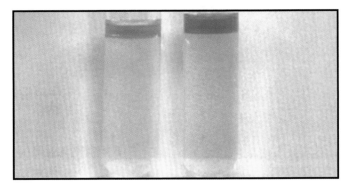

Fig. 25.1: Tube indole test

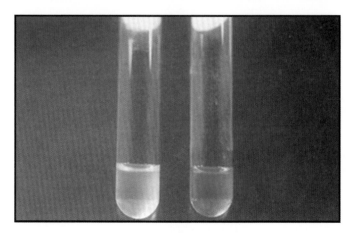

Fig. 25.2: Methyl indole test

Fig. 25.3: Voges -Proskauer (VP) test

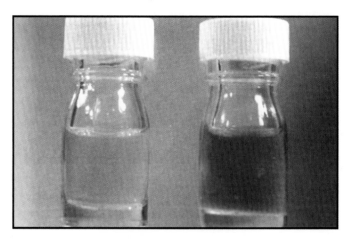

Fig. 25.4: Koser's citrate medium

triple sugar iron (TSI) agar except *S. paratyphi* A and *S. choleraesuis*. They utilise citrate (except *S. typhi* and *S. paratyphi* A) and are MR positive and VP negative. Urea is not hydrolysed. Common biochemical reactions of salmonellae are shown in Table 25.1 and Figures 25.1 to 25.4.

Pathogenesis

S. typhi, S. paratyphi A and usually *S. paratyphi* B are confined to human beings.

Salmonellae cause three types of clinical syndrome in human beings, enteric fever, septicaemia and gastroenteritis.

ENTERIC FEVER

Typhoid Fever

The injection is acquired by ingestion through contaminated food and water.

Salmonellae multiply abundantly sets in the gallbladder as bile is a good culture medium for the bacilli. These bacteria are discharged continuously into the intestine involving the Peyer's patches and typhoid follicles of the ileum.

The clinical course may vary from a mild pyrexia to a fatal fulminating disease. The illness is usually gradual, with headache, anorexia and congestion of mucous membranes. The characteristic features are hepatosplenomegaly, step-ladder pyrexia with relative bradycardia and leucopenia. Skin rashes known as rose-spots may appear during the second or third week. The infecting organisms appear in stool during third week and in urine during third to fourth week. 'Rose-spots' appear on the skin during the second or third week.

Paratyphoid Fever

Paratyphoid fever resembles typhoid fever but is milder. *S. paratyphi* A, B and C cause paratyphoid fever. *S. paratyphi* C more often leads to a frank septicaemia with suppurative complications.

Septicaemia

Salmonella septicaemia is commonly caused by *S. choleraesuis* or *S. paratyphi* C and occasionally by other salmonellae. Infections occurs through oral route. It may cause osteomyelitis, pneumonia, pulmonary abscess, meningitis or endocarditis.

Gastroenteritis

Salmonella gastroenteritis or food like meat, milk, egg contaminated by certain salmonellae which are primarily animal pathogens, eggs and egg products are of great concern.

The illness is characterised by fever, vomiting, abdominal pain and diarrhoea, *Salmonella* food poisoning is of infective type in which the organisms not only grow in the food before ingestion but also in the intestine.

Laboratory Diagnosis

Bacteriological diagnosis of enteric fever consists of:
1. Isolation of bacilli
2. Demonstration of antibodies
3. Demonstration of circulating antigen
4. Other laboratory tests.

Isolation of Bacilli

This may be done by culture of specimens like blood, faeces, urine, aspirated duodenal fluid, etc.

Blood Culture

10 ml of blood is collected by venipuncture under aseptic conditions and transferred into blood culture bottles (glucose broth and taurocholate broth). Blood should be transferred through a hole in a cap by inserting the needle of the syringes rather than opening the bottle (Fig. 25.5).

Blood culture bottles are incubated at 37°C for overnight. The glucose broth is subcultured on blood agar and the taurocholate (bile) broth on MacConkey's agar. Pale non-lactose fermenting (NLF) colonies appear on MacConkey's agar and are picked out for biochemical reactions and motility. Salmonellae will be gram-negative, motile bacilli and fermenting glucose, mannitol but not lactose or sucrose. *S. typhi* will be anaerogenic but paratyphoid bacilli will

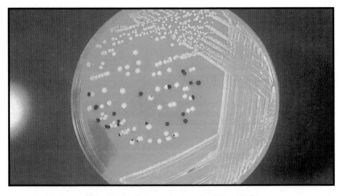

Fig. 25.5: *Salmonella enteritidis*, XLD

form acid and gas from carbohydrate (sugars). Final identification of the isolate is by slide agglutination with O and H antisera.

Clot Culture

It is an alternative to blood culture. 5 ml of blood is withdrawn aseptically into a sterile container and allowed to clot. The serum is separated and used for the widal test. The clot is broken up with a sterile glass rod and added to bile broth containing streptokinase (100 units/ml) which digest the clot causing its lysis and thereby the bacteria are released from the clot.

Faeces Culture

Salmonellae are shed in the faeces throughout the disease. Hence, faecal cultures may be helpful in patients as well as for the detection of carriers. Faecal samples are inoculated into one tube each of selenite and tetrathionate broth (both enrichment media) and are also plated directly on MacConkey's agar, DCA, XLD and Wilson-Blair media, salmonellae appear as pale yellow (NLF) colonies on MacConkey's agar and DCA media. On Wilson-Blair medium, *S. typhi* forms large black colonies with a metallic sheen whereas *S. paratyphi* A produces green colonies due to the absence of H_2S production.

Enrichment media (selenite F and tetrathionate broths) are incubated for 6-8 hours before subculture on to selective media such as MacConkey's agar and DCA. These selective media are then incubated at 37°C for overnight.

Urine Culture

Urine culture is less frequently positive than the culture of blood or faeces. Cultures are generally positive only in the second and third weeks. Urine samples are centrifuged and the deposit is inoculated into enrichment and selective media.

Other Specimens for Culture

Bone marrow culture is valuable as it is positive even when blood cultures are negative. Culture of bile is usually positive and may be useful in detection of chronic carriers. Other materials which may be used for culture are rose spots discharge, pus from suppurative lesions, CSF and sputum.

Colony Morphology and Staining

On MacConkey's agar or DCA, salmonellae grow as pale yellow, non-lactose fermenting (NLP) colonies. Gram, staining from these colonies show gram-negative bacilli and on hanging drop preparation these are motile bacilli.

Biochemical Reactions

Salmonellae are catalase positive, oxidase, negative, nitrate reduction positive and ferment glucose, mannitol but not lactose or sucrose. *S. typhi* ferments glucose, mannitol but not lactose or sucrose. *S. typhi* ferments glucose and mannitol with production of acid only but paratyphoid bacilli (*S. paratyphi* A, B and C) form acid and gas.

Slide Agglutination Test

A loopful of the growth from a nutrient agar slope is emulsified in two drops of saline on microscopic slide. One emulsion acts as a control and other as a test. Control is to show that the strain is not autoagglutinable. Agglutination is first carried out with the polyvalent O and the polyvalent *H. antisera*. Positive agglutination indicates that the isolate belongs to genus *Salmonella* (Fig. 25.6).

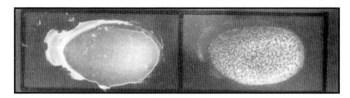

Fig. 25.6: *Salmonella* identification by O serotyping

DEMONSTRATION OF ANTIBODIES

Widal Test

It is an agglutination test for detection of agglutinins (H and O) in patients with enteric fever. *Salmonella* antibodies start appearing in the serum at the end of first week and rise sharply during the third week of enteric fever. Two specimens of sera at an interval of 7 to 10 days are preferred to demonstrate a rising antibody titre (Fig. 25.7).

Procedure

Two types of tubes were originally used for the test-Dreyer's tube (narrow tube with a conical bottom) for the H agglutination and felix tube (round bottomed tube), for the O agglutination. Equal volumes (0.4 ml) of serial dilutions of the serum (1:10 to 1:640) and the H and O antigens are mixed and incubated in a water bath at 37°C for 4 hours and read after overnight refrigeration at 4°C. Control tubes containing the antigen and normal saline are included to check for autoagglutination. H agglutination leads to the formation of loose, cotton-wool clumps, while O agglutination appears as a granular deposit at the bottom of the tube. The highest dilution of the serum showing agglutination indicates the antibody titre against that particular antigen. Control tubes show a compact deposit (Button formation).

The antigens used in the test are the H and O antigens of *S. typhi* (TH and TO antigens) and H antigens of *S. paratyphi* A (AH antigen) and *S. paratyphi* B (BH antigen).

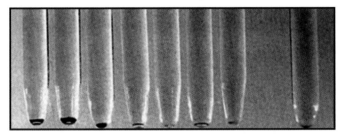

Fig. 25.7: Widal test for serological diagnosis of typhoid fever

Preparation of Widal Antigen

The H agglutinable suspension of bacteria is prepared by adding 0.1 per cent formalin to 24 hours broth culture or saline suspension of an agar culture. For preparation of O suspension of bacteria, the bacillus is cultured on phenol agar (1,800). The growth is scraped off in a small volume of saline and mixed with 20 times its volume of absolute alcohol. It is then heated in a water bath at 40 to 50°C for 30 minutes, centrifuged and the deposit resuspended in saline to the appropriate density. Chloroform is then added as a preservative. *S. typhi* 901, O and H strains are used for preparation of antigens. Each batch of prepared antigen should be compared with a standard. Widal kits of stained antigens are available commercially.

Result

The highest dilution (titre) of patient's serum in which agglutination occurs is noted, i.e. if the dilution is 1 in 160, the titre is 160.

Interpretation of Widal Test

a. The agglutinin (antibody) titre depends on the stage of the disease. Agglutinins usually start appearing in the serum by the end of the first week, so that blood specimen taken earlier than first week may give a negative result. The titre rises steadily till the third or fourth week after which it declines.

b. *Rising titre:* Demonstration of a four fold or greater rise in titre of both H and O antibodies at an interval of 7 to 10 days is more meaningful and diagnostic than a single test.

c. *Single test:* O titre of 1:100 or more and H titre of 1:200 or more signifies presence of active infection.

Other Serological Tests

ELISA is a sensitive method of measuring antibody against the lipopolysaccharide of salmonellae. Haemagglutination and CIEP are other serological methods of diagnosis.

Demonstration of Circulating Antigen

Typhoid bacilli antigens are present in the blood in the early phase of the disease, and also in the urine of patients. The antigen can be detected by coagglutination test. *S. aureus* (Cowan 1 strain) containing protein A, is first stabilised with formaldehyde and then coated with *S. typhi* antibody. These sensitised staphylococcal cells (1% suspension) are mixed on slide with patient serum. The typhoid antigen present in the serum combines with the antibody coated on staphylococci producing visible agglutination within two minutes. The test is rapid, sensitive and specific, but is positive only during the first week of the disease.

Counterimmunoelectrophoresis (CIEP) and ELISA have also been employed to detect typhoid antigens in blood and urine.

Other Laboratory Tests

1. Total leucocyte count (TLC)
   Leucopenia with a relative lymphocytosis is found.
2. Diazole test in urine
   This test becomes positive generally between 5th and 14th days of fever and remains positive till the fever subsides.

Procedure

Equal volumes of patient's urine and the diazo reagent are mixed and a few drops of 30 per cent ammonium hydroxide are added. On shaking the mixture, a red or pink froth develops, if the test is positive.

Diazo reagent

Solution A –	Sulphanilic acid
	Conc. H_2SO_4
	Distilled water
Solution B –	Sodium nitrate
	Distilled water

For use, 40 parts of solution A are added to one part of solution B.

Treatment

Chloramphenicol has been the antibiotic of choice for enteric fever. Amoxycillin, ampicillin, furazolidone and cotrimoxazole are also effective.

Vaccination

Vaccine is indicated for travelers or who live in endemic areas.

TAB Vaccine

It is heat-killed whole cell vaccine which contains *S. typhi* 1,000 millions, *S. paratyphi* A and B, 750 millions each per ml and preserved in 0.5 ml at an interval of 4-6 weeks followed by booster dose every three years.

Live Oral (Ty 21a) Typhoid Vaccine

Avirulent mutant strain of *S. typhi* (Ty21a) lacking the enzyme UDP-galactose-4-epimerase (Gal E mutant) has been used as a live oral vaccine.

Dose Schedule

Three doses are given on alternate day. The oral vaccine (typhoral) is available in capsule form containing 10^9 viable lyophilised mutant bacilli.

Purified Vi Polysaccharide Vaccine (typhin-Vi)

It contains purified Vi antigen. Dose schedule: It is injected intramuscularly in a single dose of 25 μg. The efficacy is about 75 per cent.

Vibrios are gram-negative, oxidase positive, rigid, curved rods that are actively motile by a polar flagellum. The most important member of the genus is *Vibrio cholerae*. It is the causative agent of cholera.

Morphology

V. cholerae is gram-negative, curved or comma-shaped rod, non-sporing, non-capsulated about 1.5 × 0.2 to 0.4 µm in size (Fig. 26.1).

The organism is very actively motile with a single polar flagellum and movement is named as darting motility.

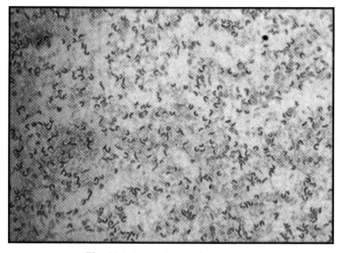

Fig. 26.1: *V. cholerae* – Gram's stain

CULTURE

Nutrient Agar

After overnight incubation, the colonies are moist, translucent, round disks, 1 to 2 mm in diameter, with a bluish tinge in transmitted light.

MacConkey's Agar

The colonies are colourless or pale at first but becomes reddish or pink on prolonged incubation due to late fermentation of lactose.

Blood Agar

V. cholerae, does not produce haemolysis although some strains produce greenish discolouration around colonies which later becomes clear due to haemodigestion.

Peptone Water

It grows as a surface pellicle because of its aerobic nature.

Gelatin Stab Culture

At first a white line of growth appears along the track of inoculation. Gelatin liquefaction begins at the top which spreads downward in a funnel-shaped form (infundibuliform or napiform) in 4 days at 22°C.

Special Media

The special media are classified as follows:
1. Transport or holding media.
2. Enrichment media.
3. Plating media.

Transport or Holding Media

Venkatraman-Ramakrishnan (VR) Medium

It contains 20 gm common salt and 5 gm peptone in one litre of distilled water and pH is adjusted to 8.6 to 8.8. It is dispensed in screw-capped bottles in 10 to 15 ml amount. About 1 3 ml faeces is added to each bottle. Vibrios do not multiply, but remains viable for several weeks.

Cary-Blair Medium

This medium is prepared by adding disodium phosphate, sodium thioglycollate, sodium chloride and calcium chloride and pH is adjusted at 8.4. It is a suitable transport medium for *Salmonella, Shigella* and *Vibrio.*

Enrichment Media

Alkaline paptone water (APW).

It is a peptone water at pH 8.6. Besides enrichment medium. It is also an excellent transport medium.

Monsur's taurocholate tellurite peptone water (Fig. 26.2): It contains peptone, sodium chloride, sodium taurocholate in one litre of distilled water and pH is adjusted at 9.2. To this medium, sterile potassium tellurite solution is added to give a final concentration of 1:200,000 like APW, it is not only a good enrichment medium but is transport medium as well.

Fig. 26.2: Culture of *Vibrio cholerae*, alkaline peptone water

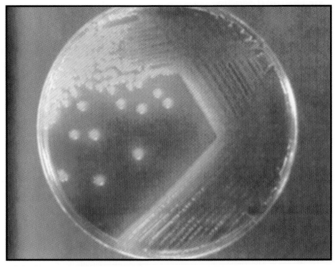

Fig. 26.3: *Vibrio* culture—TCBS agar

PLATING MEDIA

Alkaline Bile Salt Agar (BSA); pH 8.2

It is modified nutrient agar medium containing 0.5 per cent sodium taurocholate (bile salt). The colonies are similar to those on nutrient agar.

Monsur's Gelatin Taurocholate Tryptcase Tellurite Agar (GTTTA) Medium; pH 8.5

The colonies are small, translucent with a greyish-black centre and a turbid halo around the colonies due to hydrolysis of gelatin. The colonies become 3 to 4 mm in size after 48 hours incubation.

Thiosulphate Citrate Bile Sucrose (TCBS) Agar; pH 8.6 (Fig. 26.3)

It is the most widely used selective medium for isolation of vibrios. It contains sodium thiosulphate, sodium citrate, bile salts, sucrose, bromothymol blue (indicator), yeast extract, peptone, sodium chloride, ferric citrate and water. *V. cholerae* forms yellow colonies due to sucrose fermentation, while non-sucrose fermenters such as *V. parahaemolyticus* produce green colonies. The colonies of *V. cholerae* are large, yellow, convex and turns green on continued incubation.

Biochemical Reactions

It is catalase and oxidase-positive. It ferements glucose, mannitol, sucrose, maltose and mannose, but not lactose, though lactose may be split very slowly. It is indole positive and reduces nitrates to nitrates. These two properties contribute to the *cholera red reaction* which is tested by adding a few drops of concentrated sulphuric acid to peptone water culture. In case of *V. cholerae*, a reddish-pink colour is developed due to the formation of nitrosoindole. It is methyl red (MR) and urease negative. Gelatin is liquefied. It decarboxylates lysine and ornithine but do not utilise arginine. Voges-Proskauer (VP) reaction and haemolysis of sheep erythrocytes are positive in ELTor biotype and both these tests are negative in classical biotype.

Pathogenesis

V. cholerae causes an acute diarrhoeal disease known as cholera and it occurs only in man. The human infection occurs by ingestion of contaminated foods and drink.

LABORATORY DIAGNOSIS

Specimens

1. Watery stool
2. Rectal swab.

Collection and Transport

Specimens should be collected preferably prior to start of antibiotics. Specimens should be immediately transported to the laboratory for processing. In case of delay, stool samples may be preserved in holding media such as VR fluid or Cary-Blair medium for long periods. If the specimen can reach the laboratory in a few hours, enrichment media such as alkaline peptone water or Monsur's medium may be used as transport media. When transport media are not available, strips of blotting paper may be soaked in watery stool and sent to the laboratory after proper packing in plastic envelopes. If possible, specimens should be plated at bedside and the inoculated plates sent to the laboratory.

Direct Microscopy

For rapid diagnosis, the characteristic darting motility of the *Vibrio* and its inhibition by adding antiserum can be demonstrated under the dark field or phase contrast microscope, using cholera stool.

Culture

Stool sample is directly cultured on following media.
a. Selective media (BSA, TCBS or Monsur's GTTA) and non-selective media (blood agar and MacConkey's agar) are inoculated. These plates are incubated at 37°C for overnight.
b. Enrichment media such as alkaline peptone water or Monsur's liquid media are inoculated. These media are incubated at 37°C for 6 to 8 hours before subculturing on to selective media.

Colony Morphology and Staining

After overnight incubation, culture media are examined for typical colonies of *V. cholerae*. On MacConkey's agar colonies are pale and on Monsur's medium the colony has a black centre with a turbid halo around the colony. TCBS shows yellow colonies and on BSA, translucent colonies are present.

Gram staining from colony shows typical gram-negative comma shaped bacilli. These show darting motility on hanging drop preparation. Further confirmation is done by biochemical reactions and agglutination test.

Biochemical Reactions

V. *cholerae* ferments glucose, mannitol, sucrose, maltose, mannose with acid production. Lactose is usually not fermented. Catalase, oxidase and cholera red reactions are positive. The ELT or biotype is usually haemolytic, VP positive, agglutinate chick erythrocytes and is resistant to polymyxin B and group IV cholera phage.

Agglutination Test (Fig. 26.4)

Colonies are picked up with a straight wire and tested with *V. cholerae* 01 antiserum. If positive, the test of repeated with monospecific Ogawa and Inaba sera for serotyping.

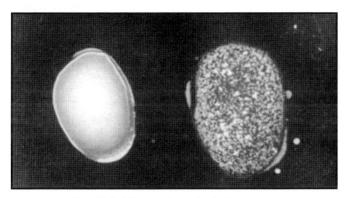

Fig. 26.4: *Vibrio cholerae:* Agglutination test

BACTERIOLOGICAL EXAMINATION OF WATER AND SEWAGE

1. *Water sample:* 900 ml of water is added to 100 ml of ten-fold concentrated peptone water (pH 9.2) and incubated at 37°C for 6 to 8 hours. After incubation, it is inoculated on selective media.
2. *Sewage:* It is diluted in saline and filtered through gauze. The filtrate is treated as water.

Treatment

1. Oral rehydration therapy
   The most important is prompt water and electrolyte replacement to correct the severe dehydration and sale depletion.
2. Antibiotics
   It is of secondary importance tetracycline is useful.

Chapter

27

Helicobacter Pylori

Morphology

Helicobacter pylori is a short curved, spiral or S-shaped gram-negative bacterium measuring 3×0.5 to 0.9 µm. It is motile by a tuft of unipolar sheathed flagella.

Culture

It is strict microaerophilic and grows in an atmosphere of 5 per cent O_2, 10 per cent CO_2 and 85 per cent N_2. It does not grow anaerobically. High humidity and 5 to 10 per cent CO_2 are required for growth. The optimum temperature for growth is 37°C and not 42°C. It can grow in moist freshly prepared chocolate agar and campylobacter selective media. After incubation at 37°C under microaerophilic atmosphere for 3 to 5 days, *H. pylori* produces circular, convex and translucent colonies.

Biochemical Reactions

H. pylori is biochemically inactive in most conventional tests but produces urease, catalase and oxidase.

Pathogenesis

The organisms are present in large number in mucous overlying mucosa and also deep in gastric glands. *H. pylori* is implicated to cause chronic active gastritis, duodenitis and peptic ulcer.

Laboratory Diagnosis

Specimens

Biopsy of gastric mucosa.

Direct Microscopy

a. *Gram staining:* Gram staining of the specimen shows short curved, spiral gram-negative bacteria
b. *Haematoxylin-Eosin staining:* It can be used in histological sections.
c. *Warthin-Starry silver staining:* Organisms can be seen most clearly with this method.

Culture

Specimen is inoculated on chocolate agar and *Campylobacter* selective medium followed by incubation as mentioned above. The organism is identified on the basis of colony morphology, Gram's staining and biochemical reactions (Figs 27.1 and 27.2).

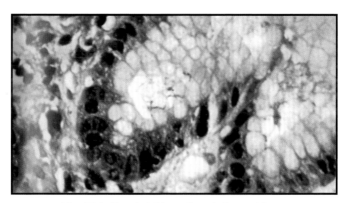

Fig. 27.1: *H. pylori,* Giemsa's stain from a biopsy

Fig. 27.2: *H. Pylori,* blood agar

Biopsy Urease Test

The biopsy material is placed in urea solution with an indicator and incubated at 37°C. If the tissue contains *H. pylori*, pH changes within a few minutes to 2 hours due to production of ammonia. Urea is broken down to ammonia by action of enzyme urease produced by the organism.

Urea Breath Test

Urea labelled with an isotope of carbon is given orally to the patient. If the patient's stomach contains *H. pylori*, urea is converted into ammonia and CO_2. This labelled CO_2 appears in the breath where it can be measured.

Serology

Antibodies to *H. pylori* can be detected in patient serum by ELISA.

28

Pseudomonas Aeruginosa

MORPHOLOGY

It is slender, gram-negative *Bacillus*, 1.5 to 3 × 0.5 µm, non-capsulated, non-sporing and is actively motile by a polar flagellum. Most strains possess pili. It is non-capsulated. Occasionally strains have two or three polar flagella.

CULTURE

It is a strict aerobe and grows well on ordinary media like nutrient broth and nutrient agar.

Nutrient Agar

Colonies are smooth, large, translucent, low convex, 2 to 4 mm in diameter. The organism produces a sweetish aromatic odour. This is due to the production of 2-aminoacetophenone. There is greenish-blue pigment which diffuses into the medium (Fig. 28.1).

Blood Agar

Colony characters are similar to those on nutrient agar. Many strains are haemolytic on blood agar.

MacConkey's Agar

Colonies are pale or colourless (non-lactose fermenters, NLF) (Fig. 28.2).

Cetrimide Agar

It is a selective medium for *P. aeruginosa*.

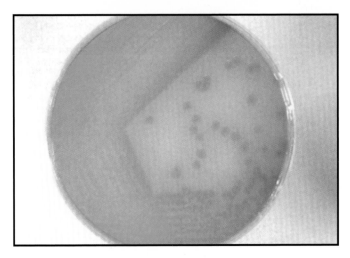

Fig. 28.1: *Pseudomonas* culture

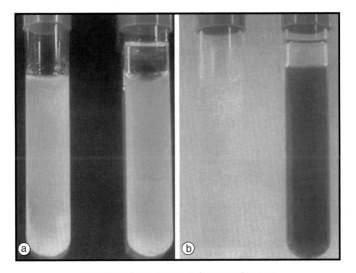

Fig. 28.2: Oxidation and fermentation test

Peptone Water

It forms a turbidity with a surface pellicle. *Pseudomonas* being a strict aerobe tends to collect at the surface for more oxygen hence forming surface pellicle.

PIGMENT PRODUCTION

P. aeruginosa produces a number of pigments which diffuse into surrounding medium).

BIOCHEMICAL REACTIONS

P. aeruginosa derives energy from carbohydrates by oxidative breakdown rather than a fermentative metabolism. Special OF medium of Hugh and Leifson must be used to find out oxidative metabolism. It utilises only glucose oxidatively with acid production. Lactose and maltose are not utilised. All strains of *P. aeruginosa* are oxidase positive and utilise citrate as the sole source of carbon. They are catalase positive and indole, MR, VP and H_2S tests negative. They reduce nitrates to nitrites and further to gaseous nitrogen. Important biochemical reactions are summarised below.

OF test	Catalase	Oxidase	Nitrate reduction	Glucose	Lactose
Oxidative	+	+ +	+	A –	–
Mannitol	Sucrose	Citrate	Indole	Urease	H_2S
–	–	+	–	–/+	–

Laboratory Diagnosis

Specimens

Pus, wound swab, urine, sputum, blood or CSF.

Culture

Specimens may be inoculated on nutrient agar, blood agar or MacConkey's agar and incubated at 37°C for 18 to 24 hours. On nutrient agar, there is bluish-green pigment diffused in the medium. On MacConkey's agar they grow as pale colonies (NLF).

Gram Staining and Motility

They are gram-negative bacilli and are actively motile.

Biochemical Reactions

The oxidase test is positive within 30 seconds. They are non-fermenter. They breakdown glucose oxidatively with acid production only.

Treatment

Ciprofloxacin, piperacillin, ticarcillin, aziocillin, cefotaxime, ceftazidime, gentamicin and tobramycin are used in treatment of *P. aeruginosa* infections.

Chapter

29

Haemophilus Influenzae

MORPHOLOGY

It is a small (1.5 × 0.3 mm), gram-negative, non-motile *Bacillus* showing considerable pleomorphism.

CULTURE

H. influenzae has fastidious growth requirements. It grows better in aerobic than in anaerobic conditions. It requires enriched media such as blood agar or chocolate agar because the accessory growth factors known as X and V present in blood are essential for growth Figure 29.1.

Satellitism

After inoculating suspected *H. influenzae* on a blood agar plate, *S. auerus* is streaked across the same blood agar plate and incubated at 37°C for 18 to 24 hours. The colonies of *H. influenzae* will be large and well developed alongside the streak of staphylococci while those further away from staphylococcal streak

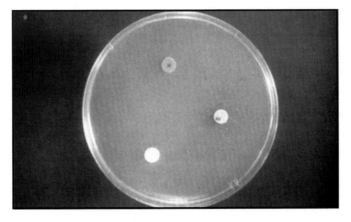

Fig. 29.1: X and V of *H. influenzae*

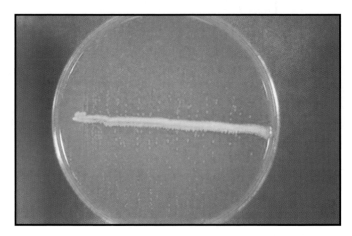

Fig. 29.2: Satellitisms

are smaller. This phenomenon is called satellitism and demonstrates that V factor is available in high concentration near the staphylococcal growth and only in smaller quantities away from it (Fig. 29.2).

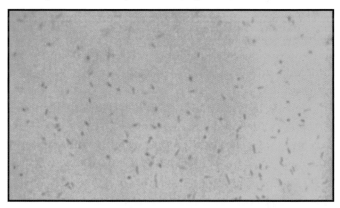

Fig. 29.3: *H. influenzae*, Gram stain

BIOCHEMICAL REACTIONS

H. influenzae is catalase and oxidase positive, ferments glucose and galactose, reduces nitrate to nitrite. It does not ferment sucrose, lactose and mannitol. On the basis of production of indole, urease, ornithine decarboxylase, it is divided into eight (I-VIII) biotypes (Fig. 29.3).

Pathogenesis

The following infections are caused by *H. influenzae*.

Meningitis, acute epiglottis, pneumonia, bronchitis, suppurative lesions.

LABORATORY DIAGNOSIS

Specimens

Depending upon the type of lesion, the following specimens may be collected.
1. Cerebrospinal fluid (CSF)
2. Blood
3. Throat swab
4. Sputum
5. Pus
6. Aspirates from joints, middle ears or sinuses, etc.

Collection and Transport

Specimens should be collected in sterile containers and under all aseptic conditions. As *H. influenzae* is very sensitive to low temperature, therefore, clinical specimens should never be refrigerated.

Direct Microscopy

i. Gram staining: In meningitis, Gram stained smear of CSF shows pleomorphic gram-negative coccobacilli (Fig. 29.4).
ii. Immunofluorescence and quellung reaction: These can be employed for direct demonstration of *H. influenzae* after mixing with specific type b antiserum.
iii. Type b capsular antigen can be also detected in patient's serum, CSF, urine or pus by the following methods:
   a. *Latex agglutination:* Latex particles coated with antibody to type b antigen are mixed with the specimen. In positive test, agglutination occurs.

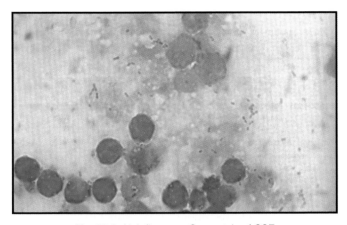

Fig. 29.4: *H. Influenzae,* Gram stain of CSF

Fig. 29.5: *H. Influenzae*, chocolate agar

b. *Coagglutination:* Instead of latex particles, *S. aureus* is coated with antibody to type b antigen and mixed with specimen. If positive, agglutination occurs.

c. *Counterimmunoelectrophoresis (CIE):* Specific antiserum is put in one well of agarose gel and specimen is put in other well. Current is passed. In positive test, precipitation line occurs in between the two wells.

Culture

i. *CSF culture:* CSF should be plated promptly on blood agar or chocolate agar (Fig. 29.5). A strain of *S. aureus* should be streaked across the blood agar plate on which the specimen has already been inoculated. Plates are then incubated at 35-37°C, aerobically with 5 to 10 per cent CO_2, overnight. The isolate is then identified by its colony morphology, satellitism, Gram staining and serotyping.

ii. *Blood culture:* It is usually positive in cases of epiglottitis and pneumonia.

iii. *Sputum culture:* Sputum should be homogenised by treatment with pancreatin or by shaking with sterile water and glass beads for 15 to 30 minutes.

Colony Morphology and Staining

After overnight incubation, small opaque colonies appear that show satellitism. A smear is made from colony and stained with Gram's stain. It shows small gram-negative bacilli or coccobacilli.

Serotyping

Typing may be done with type specific antisera.

Treatment

H. influenzae is suspectable to sulphonamides, chloramphenicol, trimetho-prism, ampicillin, tetracycline, ciprofloxacin, cefuroxime, cefotaxime and ceftazidime.

Mycobacterium Tuberculosis

INTRODUCTION

Mycobacteria are slender bacilli that sometimes show branching filamentous forms resembling fungal mycelium (*Myces* meaning fungus). They are difficult to stain, but once stained, resist decolourisation with dilute mineral acids and are therefore called *acid-fast bacilli* or *AFB*. These organisms are aerobic, non-motile, non-capsulated and non-sporing. Growth is generally slow.

MYCOBACTERIUM TUBERCULOSIS

Morphology

M. tuberculosis is a slender, straight or slightly curved bacilli with rounded ends, occurring singly, in pairs or in small clumps. It measures, 1 to 4 × 0.2 to 0.8 μm (average 3 × 0.3 μm) in size. These bacilli are acid-fast, non-sporing, non-capsulated and non-motile. Ziehl-Neelsen staining is useful to study the morphology of these organisms. With this stain, *tubercle bacilli* are seen bright red (acid-fast), while the tissue cells and other organisms are stained blue. *Tubercle bacilli* may also be stained with the fluorescent dyes (auramine O, rhodamine) and appear yellow luminous bacilli under the fluorescent microscope.

Culture

Tubercle bacilli can grow on a wide range of enriched culture media but Lowenstein-Jensen (LJ) medium is most commonly used. This medium consists of beaten eggs, asparagines, mineral salts, malachite green and glycerol or sodium pyruvate. It is solidified by heating. In this medium egg acts as a solidifying agent. Malachite green inhibits the growth of organisms other than mycobacteria and provides a colour of the medium. The addition of glycerol improves the growth of human type of *M. tuberculosis.* Sodium pyruvate improves the growth of both *M. tuberculosis* and *M. bovis.* Colonies of *M. tuberculosis* are dry, rough, buff coloured, raised with a wrinkled surface

(Figs 30.1 and 30.2). They are tenacious and not easily emulsified. In contrast, the colonies of *M. bovis* are flat, smooth, moist and white, breaking up easily when touched. *M. tuberculosis* has a luxuriant growth in culture as compared to sparsely grown *M. bovis*.

In liquid media, the bacilli grow as surface pellicle due to hydrophobic properties of their cell wall. Diffuse uniform growth can be obtained by addition of a detergent Tween 80 (polyoxyethylene sorbitan mono-oleate) in Dubo's medium. Tween 80 wets the surface and permits them to grow diffusely.

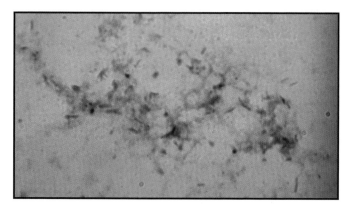

Fig. 30.1: Ziehl-Neelsen staining

Fig. 30.2: Culture on LJ medium

BIOCHEMICAL REACTIONS

Niacin Test

Niacin is produced as a metabolic by product by all mycobacteria. When 10 per cent cyanogen bromide and 4 per cent aniline in 96 per cent ethanol are added to a suspension of bacterial culture, a canary yellow colour shows a positive reaction. The human tubercle bacilli give a positive test, while the bovine type is negative.

Arylsulphatase Test

Arylsulphatase is an enzyme formed by certain atypical mycobacteria. The organisms are grown in a medium containing 0.001 *M. tripotassium* phenol-phthalein disulphate. If aryl sulphatase enzyme is produced it liberates free phenolphthalein from tripotassium, phenolphthalein disulphate. This can be detected by adding 2N NaOH dropwise to the culture. A pink colour develops in a positive reaction.

Neutral Red Test

Virulent strains of tubercle bacilli can bind neutral red in alkaline buffer solution, whereas avirulent strains are unable to do so.

Catalase-peroxidase Tests

A Mixture of equal volumes of 30 per cent H_2O_2 and 0.2 per cent catechol in distilled water is added to a 5 ml test culture and left for few minutes. Catalase production is indicated by effervescence while browning indicates peroxidase activity.

Amidase Tests

A 0.00165 M solution of amide is incubated with the bacillary suspension at 37°C and to this is added 0.1 ml of $MnSO_4$. $4H_2O$, 1.0 ml of phenol solution and 0.5 ml of hypochlorite solution. The tubes are placed in boiling waterbath for 20 minutes. A blue colour indicates a positive test.

Nitrate Reduction Test

This test depends on reduction of nitrate to nitrite by an enzyme nitroreductase. This test is positive with *M. tuberculosis* and negative with *M. bovis*.

The test organism is suspended in a buffer solution containing nitrate and incubated at 37°C for 2 hours. Then sulphanilamide and n-naphthylethylene diamine dihydrochloride solutions are added. Development of pink or red colour indicates positive reaction.

Tween 80 Hydrolysis

Certain mycobacteria possess an enzyme lipase that splits Tween 80 into oleic acid and polyoxyethylated sorbitol which modifies the optical characteristics of the test solution from a yellow to pink. A pink colour indicates hydrolysis of Tween 80.

PATHOGENESIS

Primary Tuberculosis

Inhaled tubercle bacilli are engulfed by alveolar macrophages in which they replicate to form a lesion called *Ghon focus*. It is frequently found in the lower lobe or lower part of the upper lobe.

The Ghon focus together with the enlarged hilar lymph nodes is called the primary complex.

Secondary (Post-primary) Tuberculosis

It is caused by reactivation of the primary lesion (endogenous) or by exogenous reinfection.

Presence of caseous necrosis and cavities are two special features of secondary tuberculosis. Cavities may rupture into blood vessels, spreading mycobacteria throughout the body, and break into airways, releasing the organisms in aerosols and sputum.

TUBERCULIN TEST

Principle

It is delayed or type IV hypersensitivity reaction.

Reagents

Old Tuberculin (OT)

It was originally described by Robert Koch. It is a crude preparation of 6 to 8 weeks culture filtrate of tubercle bacilli, concentrated by evaporation on a heated water bath.

Purified Protein Derivative (PPD)

A purified preparation of the active tuberculoprotein was prepared by Seibert (1941). It was prepared by growing *M. tuberculosis* in a semisynthetic medium. It is called purified protein derivative (PPD). The dosage of PPD is expressed in tuberculin unit (TU). One TU is equal to 0.01 ml of OT or 0.00002 mg of PPD-S.

Method

Mantoux Test

0.1 ml of PPD containing 5 IU is injected intradermally into flexor aspect of forearm. A PPD dose of 1 TU is used when extreme hypersensitivity is suspected.

Heaf Test

This is done with a multiple puncture apparatus that pricks the skin. A drop of undiluted PPD is spread on the area of skin. The multiple puncture apparatus is pressed against this area of skin.

Result

Positive Test

In a positive reaction, there is induration (local oedema) of 10 mm diameter or more surrounded by erythema at the site of inoculation.

False-Negative

The test may become negative in following conditions:
a. Military tuberculosis
b. When anergy develops following overwhelming infection of measles, Hodgkin's disease, sarcoidosis, lepromatous leprosy, malnutrition, administration of immunosuppressive agents and corticosteroids.

False-Positive

This is observed in presence of related mycobacteria such as a typical mycobacteria.

Uses

i. To measure prevalence of infection in a community.
ii. To diagnose active infection in young children.
iii. It is used an indicator of successful BCG vaccination.

Laboratory Diagnosis

Bacteriological diagnosis can be established by microscopy, culture examination or by animal inoculation test.

Specimen

Pulmonary Tuberculosis

Sputum is the most common specimen. It is collected in a clean wide-mouthed container. A morning specimen may be collected on three consecutive days. If sputum is scanty, a 24 hours specimen may be collected. When sputum is not available, laryngeal swab or bronchial washings are collected. In children, gastric washings may be examined as they tend to swallow sputum.

Meningitis

Cerebrospinal fluid (CSF) from tuberculous meningitis (TBM) often forms a spider web clot on standing, examination of which may be more useful than of fluid.

Renal Tuberculosis

Three consecutive days morning samples of urine are examined.
Bone and joints tuberculosis: Aspirated fluid.

Tissue

Biopsy of tissue
    Smear is made from the specimen on a new glass slide and stained by the Ziehl-Neelsen technique. It is examined under oil immersion lens. The acid-fast bacilli (AFB) appear as bright red bacilli against a blue background. Grading of smears is done according to number of bacilli seen:
    1+ – 3-9 bacilli in the entire smear
    2+ – ≥ 10 bacilli in the smear
    3+ – ≥ 10 bacilli seen in most oil immersion fields.
    If a large number of smears are to be examined, fluorescent microscopy is more convenient. Smears are stained with fluorescent dyes such as auramine 'O' or auramine rhodamine and examined under ultraviolet light. The bacilli appear as bright bacilli against dark background.

Concentration of Specimens

Concentration of a specimen is done to achieve:
a. Homogenisation of the specimen.
b. Decontamination, i.e. to kill other bacteria present in the specimen.
c. Concentration, i.e. to concentrate the bacilli in the small volume without inactivation.

Several Concentration Methods are in Use:

Petroff's Method

It is a simple and widely used technique.

Sputum is mixed with equal volume of 4 per cent sodium hydroxide and is incubated at 37°C with frequent shaking for about 30 minutes. It is then centrifuged at 3,000 rpm for 30 minutes. The supernatant fluid is poured off and the deposit is neutralized by adding 8 per cent hydrochloric acid in presence of a drop of phenol red indicator. The deposit is used for smear, culture and animal inoculation.

Other Methods

Dilute acids (5% oxalic acid, 3% hydrochloric acid or 6% sulphuric acid), mucolytic agents such as N-acetyl-L-cysteine with sodium hydroxide and pancreatin are used for concentration of specimens. In urine and CSF specimens centrifugation is done to concentrate the specimen. Centrifuged deposit is used for smear and culture examination.

Culture

Culture is a very sensitive method for detection of tubercle bacilli. It may detect as few as 10 to 100 bacilli per ml. The concentrated material is inoculated on two bottles of Lowenstein-Jensen medium. In case of gastric washings, alkali is added to neutralise the acid present in the specimen and then inoculated on culture medium. Urine is centrifuged and then from deposit, culture medium is inoculated. In case of CSF, it is centrifuged and the deposit is used for culture and smear examination.

The tubercle bacilli usually grow in 2 to 8 weeks.

In a positive culture, characteristic colonies appear on culture medium. Smear in prepared from isolated colony and stained with Ziehl-Neelsen technique.

In radiometric method such as BACTEC, the growth may be detected in about a week by using ^{14}C labelled substrates. A growth index of >10 is considered as positive.

Animal Inoculation

0.5 ml of the concentrated specimen is inoculated intramuscularly into the thigh of two tuberculin negative healthy guinea pigs. Inoculation by subcutaneous route is avoided as it causes local ulcer which may be infectious. The animals are weighed prior to inoculation and thereafter at weekly interval. They are tuberculin tested after 3 to 4 weeks. There is progressive loss of weight and tuberculin test becomes positive in animals that develop tuberculosis. Animal is killed after six weeks.

Autopsy shows:

1. Caseous lesion at the site of inoculation.
2. Enlarged caseous inguinal lymph nodes. The infection may spread to other lymph nodes such as lumbar, portal, mediastinal and cervical lymph nodes.
3. Tubercles may be seen in spleen, lungs, liver or peritoneum.
4. Kidneys are unaffected.

Serology

Serology includes detection of antimycobacterial antibodies in patient serum. Various methods such as enzyme linked immunosorbent assay (ELISA), radio-immunoassay (RIA), latex agglutination assay have been employed.

Molecular Methods

Polymerase chain reaction (PCR) is a rapid method in diagnosis of tuberculosis. It is based on DNA amplification and has been used to detect *M. tuberculosis* directly in clinical specimens. The restriction fragment length polymorphism (RFLP) is used to type different strains for epidemiological purposes. The principal of this technique is that restriction endonuclease treatment yields nucleic acid fragments of different lengths, the patterns of which are strain specific. DNA probes have been used to identify mycobacterial species isolated on solid culture media or from broth culture.

Treatment

The antitubercular drugs include bactericidal agents such as rifampicin (R), isoniazid (H), pyrazinamide (Z), streptomycin and bacteriostatic agents include ethambutol (E), thiacetazone, ethionamide, para-aminosalicylic acid (PAS) and cycloserine. Short course regimens of 6 to 7 months are used.

Chapter

31

Mycobacterium Leprae

MORPHOLOGY

M. leprae is slender, slightly curved or straight curved or straight bacillus, 1 to 8 m × 0.2 to 0.5 μ and shows considerable morphological variations. It is acid-fast, but less so than the tubercle bacillus. The bacilli are seen singly and in groups, intracellularly or lying free outside the cells. They are present as bundles bound by a lipid like substance, the Glia. The masses are known as globi. The parallel rows of bacilli in the globi give appearance of a cigar-bundle.

CULTIVATION

Lepra bacilli have not yet been grown on artificial culture media or tissue culture.

Following animals have been used for experimental infection with *M. leprae.*
1. Mouse (foot-pad)
2. Nine-banded armadillos
3. Chimpanzees
4. Monkeys
5. Slender loris
6. Indian pangolin
7. Chipmunks
8. Golden hamsters
9. European hedgehog.

Among these animals, the most important are mouse and nine-banded armadillos.

Leprosy is a chronic granulomatous disease of humans and the only source of infection is patient.

LABORATORY DIAGNOSIS

Specimens

Specimens are collected from the nasal mucosa, skin lesions and ear lobules. The specimens from skin are obtained by *slit and scrape* method. Samples from skin should be obtained from the edges of the lesion. The skin is cleaned with spirit in order to remove any saprophytic acid-fast bacilli that may be present on the skin surface. The skin is pinched up tight to minimise bleeding. A cut about 5 mm long is made with a scalpel, deep enough to get into the infiltrated layers. After wiping off blood or lymph that may have exuded, the blade of the scalpel is then turned at right angle to the cut (slit). The bottom and the sides of the slit are scraped with the point of the blade so as to obtain a little tissue pulp which is smeared uniformly on a slide. Abut six different areas should be sampled, including the skin over the buttocks, chin, cheek, forehead and ears.

Smears from the nose are made by scraping a little material from the nasal septum with a small blade knife.

Skin biopsy is collected from active edges of the patches and nerve biopsy from thickened nerves. Histological examination of skin biopsy is useful in the diagnosis and accurate classification of leprosy lesion.

Acid-fast Staining

Slit-skin smears from skin patches and ear lobes and nasal mucosal scrapings are stained by Ziehl-Neelsen method using 5 per cent sulphuric acid as decolourising agent (Fig. 31.1). Acid-fast bacilli (AFB) arranged in parallel bundles within macrophages (Lepra-cell) confirm the diagnosis of lepromatous leprosy. The viable bacilli stain uniformly and the dead bacilli are fragmented, irregular or granular.

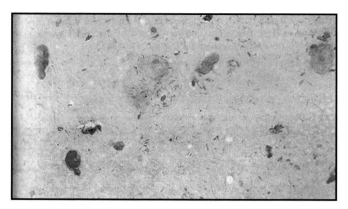

Fig. 31.1: *M. leprae*, Ziehl-Neelsen stain

The smear are graded, based on the number of bacilli as follows:

1 to 10 bacilli in 100 fields	= 1+
1 to 10 bacilli in 10 fields	= 2+
1 to 10 bacilli per field	= 3+
10 to 100 bacilli per field	= 4+
100 to 1000 bacilli per field	= 5+
More than 1,000 bacilli, Clumps and globi in every field	= 6+

Bacteriological Index (BI)

It is defined as number of total bacilli in a tissue. The bacteriological index is calculated by totalling the grades (number of pluses, +s scored in all the smears and divided by number of smears. Thus if seven smears examined have a total fourteen pluses (14+), BI will be 2. For calculating BI, a minimum of four skin lesions, a nasal swab and both the ear lobes are to be examined.

Morphological Index (MI)

It is defined as the percentage of uniformly stained bacilli out of the total number of bacilli counted. It provides a method for assessing the progress of patients of chemotherapy.

Skin and Nerve Biopsy

These are required for histological confirmation of tuberculoid leprosy when acid-fast bacilli cannot be demonstrated in direct smear.

Animal Inoculation

Injection of ground tissue from lepromatous nodules or nasal scrapings from leprosy patient into the foot pad of mouse produces typical granuloma at the site of inoculation within 6 months.

Lepromin Test

It is not a diagnostic test but is used to assess the resistance of patient to *M. leprae* infection.

Serological Test

Serodiagnosis of leprosy may be carried out by detection of anti-phenolic glycolipid-1 (anti-PGL-1) antibodies. Various serological tests like latex agglutination, *Mycobacterium leprae* particle agglutination (MLPA) and ELISA have been described.

Treatment

Chemotherapy

Dapsone monotherapy was the standard treatment for all types of leprosy for many years. Due to emergence of dapsone resistance, WHO recommended multiple drug therapy (MDT) for all leprosy cases.

Chapter

32

Spirochaetes

MORPHOLOGY

The spirochaetes have gram-negative type cell wall composed of an outer membrane, a peptidoglycan layer and a cytoplasmic membrane.

A characteristic feature is the presence of varying number of endoflagella.

DISEASES

Diseases caused by spirochaetes are shown in table below:

Genera	Species	Diseases
Treponema	pallidum	Syphilis
	pertenue	Yaws
	carateum	Pinta
	endemicum	Endemic syphilis
Borrelia	recurrentis	Relapsing fever
	vincentii	Vincent's angina
	burgdorferi	Lyme disease
Leptospira	interrogans	Leptospirosis
	biflexa	Saprophytes

TREPONEMA PALLIDUM

Morphology

It is a thin, delicate spirochaete with tapering ends, having about ten regular spirals. It is about 1 µm long (range 4 to 14 µm) and 0.1 to 0.2 µm wide.

Culture

Pathogenic treponemes cannot be grown in artificial culture media but are maintained by subculture in suspectible animals.

Pathogenesis

Natural infection with *T. pallidum* occurs only in human beings. Venereal syphilis is acquired by sexual contact.

Primary Syphilis

A papule appears on the genital area that ulcerates forming a classical chancre of primary syphilis, called *hard chancre.*

Secondary Syphilis

The secondary lesions are due to widespread multiplication of the treponemes and their dissemination through the blood. These are characterised by appearance of widespread macular rash on mucous membranes and skin which contain numerous treponemes. There may be retinitis, meningitis, periostitis and arthritis. Secondary lesions usually undergo spontaneous healing, in some instances taking as long as four or five years.

After the secondary lesions disappear, 30 per cent of cases remain dormant (latent) for many years without any clinical symptoms but with positive serology.

Tertiary Syphilis or Late Stage

Tertiary lesions contain few spirochaetes. There may be cardiovascular lesions including aneurysms, chronic granulomata (gummata) and meningovascular manifestations.

Congenital Syphilis

The treponemes can cross the placental barrier. Infection in foetus usually occurs from primary and secondary infection of the mother.

LABORATORY DIAGNOSIS

Demonstration of Treponemes

Demonstration of treponemes by microscopy is applicable in primary and secondary stages and in cases of congenital syphilis with superficial lesions.

Darkground Microscopy

The surface of chancre is cleaned with saline, a drop of exudates is collected on a slide by applying gentle pressure at the base of the lesion. The wet film is

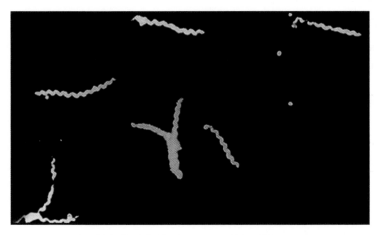

Fig. 32.1: Dark-field microscopy of *T. pallidum*

covered with a coverslip and examined under darkground microscope. *Treponema pallidum*, appears as a slender, spiral organism showing rotational as well as flexion and extension movements (Fig. 32.1). Dark-field examination should be repeated on three consecutive days before declaring it negative. The negative results do not exclude the diagnosis of syphilis because of its low sensitivity. A treponemal concentration of 10^4 per ml is required for the test to become positive.

Diagnostic Methods in Syphilis

Demonstration of treponemes	1.	Darground microscope
	2.	Direct fluorescent antibody staining for *T. pallidum* (DFA-TP)
	3.	Treponeme in tissue
		(i) Silver impregnation method Levaditi stain)
		(ii) Immunofluorescence staining

Serological tests

Non-treponemal tests	1.	VDRL
	2.	RPR
Treponemal tests	1.	FTA, FTA-ABS (using killed *T. pallidum*)
	2.	TPHA MHA-TP(using *T. pallidum* extract)
	3.	TPI test (using live *T. pallidam*)

Direct Fluorescent-antibody Staining for T. pallidum (DFA-TP)

Smear of the material to be tested is made on a glass slide. It is stained with fluorescent-labelled monoclonal antibody against *T. pallidum*. The treponemes appear distinct, sharply outlined and exhibit an apple green fluorescence. It is a better and safer method for microscopic diagnosis.

Serological Tests

Two types of antibodies are produced in syphilis, non-specific antibody (reagin antibody) and specific anti-treponemal antibody. Depending upon the antigen used, serological tests for syphilis are divided into *non-treponemal tests* (cardiolipin or lipoidal antigen is used) and *treponemal tests* (treponemes are used as the antigen).

Non-treponemal Tests

In the standard tests for syphilis (STS), reagin antibodies are detected by cardiolipin antigen. Cardiolipin antigen is an alcoholic extract of beef heart tissue to which lecithin and cholesterol are added. The STS includes venereal diseases research laboratory (VDRL) test, rapid plasma regain (RPR), Kahn test and Wassermann reaction. All these tests are *flocculation tests* except Wassermann reaction which is a complement fixation test (CFT). The Wassermann reaction is no longer in use. Similarly Kahn test is rarely done.

Flocculation Tests

Cardiolipin antigen with regain antibody in syphilitic serum resulting in formation of visible clumps or floccules. Results can be read in a few minutes. VDRL and RPR tests are equally sensitive.

*a. VDRL (Venereal Disease Research Laboratory) test:*It is the most widely used simple and rapid serological test. Small quantity of serum is needed. It can be also be used to detect antibodies in cerebrospinal fluid (CSF). It is a slide flocculation test. The VDRL antigen (ardiolipin antigen) must be prepared fresh daily.

Method
1. The test is done in a specially prepared slide with depressions of 14 mm diameter each.
2. 0.05 ml of inactivated serum (at 5°C for 30 minutes) is taken on a slide, to which one drop of freshly prepared cardiolipin antigen is added by a syringe delivering 60 drops in one ml.

3. The slide is then rotated at 180 revolutions per minute for four minutes. It can done by VDRL rotator or manually.
4. The slide is examined under low power objective of microscope. Presence of clumps signifies positive reaction while uniformly distributed crystals indicate a negative result.
5. If the test is positive, it is quantitiated by performing the test with serial dilutions (1:4, 1:8 and so on) of serum.

VDRL test sometimes may give false-negative reaction to high titres of antibody in patient's serum (prozone phenomenon). In such cases the test is performed with diluted serum and it becomes positive.

b. RPR (Rapid Plasma Reagin) test (Fig. 32.2): It is almost similar to VDRL test. Finely divided carbon particles are added to cardiolipin antigen. RPR test has got the following advantages over VDRL test:

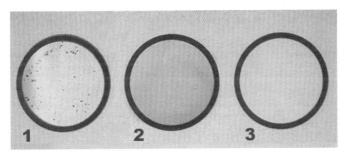

Fig. 32.2: RPR test

Unheated serum or plasma can be used.

A fingerprick sample of blood is sufficient.

It does not require microscope and can be done in the field.

It is available commercially as a kit.

The only *disadvantage* of the test is that it cannot be used with cerebrospinal fluid (CSF).

VDRL and RPR tests are useful in surveys because of their low cost. VDRL test or RPR test is positive in about 70 per cent of primary and 100 per cent of secondary syphilis. Automated RPR test (ART) is available for large scale tests. Similarly, an automated VDRL-ELISA test has also been developed.

Treponemal Tests

These tests may be divided as follows:
1. Tests using reiter treponeme
   Reiter protein complement fixation (RPCF) test.

2. Tests using *T. pallidum* (Nichol's strain)
   i. Using live *T. pallidum*
      Treponema pallidum immobilisation (TPI) test.
   ii. Using killed *T. pallidum.*
      Treponema pallidum agglutination (TPA) test.
      Treponema pallidum immune adherence (TPIA) test.
      Fluorescent treponemal antibody (FTA) test.
   iii. Using and extract of *T. pallidum*
      Treponema pallidum haemagglutination assay (TPH) test enzyme immunoassay (EIA).

CHOICE OF SEROLOGICAL TESTS

1. VDRL or RPR tests are used for screening or for diagnostic purposes of large number of sera. These tests are also used for quantitative measurement of reagin titre for assessment of clinical activity of syphilis.
2. Treponemal tests (TPHA or FTA-ABS) are used to confirm the diagnosis with a positive reagin test.

Treatment

1. *Early syphilis:* Primary, secondary and latent infection of two years duration or less are include in early syphilis.
   i. Benzathine benzyl penicillin 24 lacs units intramuscularly in a single dose after sensitivity test.
   ii. Alternatively, doxycycline 100 mg twice a day, orally for 15 days.
2. *Late syphilis:* Infection more than two years duration is included in late syphilis.
   Benzathine benzyl penicillin 24 lacs units, intramuscularly, once weekly for three weeks.

Chapter 33

Leptospira

Members of the genus *Leptospira* are actively motile, delicate spirochaetes possessing numerous closely wound spirals and characteristic hooked ends. They cannot be seen under light microscope due to its thinness (*leptos* meaning fine or thin). They do not stain readily. They may be observed ground illumination.

LEPTOSPIRA INTERROGANS

Morphology

These are spiral bacteria, 5 to 20 × 0.1 µm with numerous closely set coils and hooked ends. They are actively motile. They stain poorly with aniline dyes but can be observed by fluorescent antibody and silver impregnation techniques. Because of narrow diameter, they are best observed by dark ground, phase-contrast or electron microscopy. Leptospires rotate rapidly about their long axis and bending or flexing sharply.

Culture

They are aerobic and microaerophilic. Optimum temperature for their growth is 28 to 32°C and optimum pH 7.2 to 7.5. They can be grown in media enriched with rabbit serum. Several media, such as Korotkoff's, Stuart's and Fletcher's media have been described. Semisynthetic medium, such as, EMJH (Ellinghausen, McCulough Johnsom, Harris) is now commonly used. In semisolid media, growth occurs a few millimeters below the surface. Growth is detected usually after 6 to 14 days of incubation. The generation time of leptospires in laboratory media is 12 to 16 hours and 4 to 8 hours in inoculated animals.

Leptospires may be grown on chorioallantoic membrane (CAM) of chick embryos. They can be also be grown in guinea pigs.

Pathogenesis

L. interrogans causes a zoonotic disease named leptospirosis in rodents and sometimes in domestic animals. It is transmitted to humans by direct or indirect contact with water contaminated by urine of carrier animals. Leptospirosis is an established cause of aseptic meningitis. Severe leptospirosis (*Weil's disease*) associated with fever, conjunctivitis, albuminurea, jaundice and haemorrhage is usually caused by *L. icterohaemorrhagic* serogroup. It is a fatal illness with hepatorenal damage.

Laboratory Diagnosis

Demonstration of Leptospires in the Blood or Urine

Leptospires can be observed in the blood by darkground microscopy (Fig. 33.1). Blood examination is useful in first week as leptospires disappear from blood after 8 days.

Fig. 33.1: Dark field microscopy of *Leptospira*

Culture

Specimen of blood during first week and urine in the second week (up to six weeks) can be cultured in modified Korthof's medium or Fletcher's semisolid medium. Centrifuged deposit of urine is cultured. Media are incubated at 28 to 32°C aerobically and examined by darkground microscopy every third day up to six weeks before discarding it as negative.

Animal Inoculation

The blood or urine from the patient is inoculated intraperitoneally into young guinea pigs. From the third day after inoculation, the peritoneal fluid is examined daily for leptospires by darkground illumination. Heart blood of animal is inoculated into the culture media for isolation of organism.

Serological Tests

It is very useful method of diagnosis. Antibodies begin to appear at the end of first week and continue to rise till the fourth week and then begin to decline.

Treatment

Leptospires are sensitive to penicillin, tetracycline and erythromycin.

Chapter

34

Mycoplasma

MORPHOLOGY

Mycoplasmas are smallest free living microorganisms. They can pass through bacterial filters. They are most pleomorphic and may present as small spherical shapes (125 to 250 nm in diameter) to longer branching filaments (500 to 1000 nm in size). They lack cell wall but have a triple layered cell membrane which is rich in cholesterol and other lipids. They are gram-negatives, but are better stained by Giemsa's stain.

CULTURE (FIG. 34.1)

A medium widely used for the isolation of mycoplasmas is PPLO broth which contains bovine heart infusion broth enriched with 20 per cent horse serum

Fig. 34.1: Culture of Mycoplasma

and 10 per cent fresh yeast extract along with glucose, and phenol red as indicator. This medium can be made solid by the addition of agar. Penicillin, ampicillin and polymyxin B may be added in the medium to inhibit contaminating bacteria and amphotericin B to inhibit fungi.

Colonies usually appear after incubation for 48 to 72 hours. Typical tiny fried egg colonies appear in the medium. Colony consists of a central opaque granular area of growth, surrounded by a flat, translucent peripheral zone. Colony size varies from 200 to 500 μm for 'large colony' Mycoplasmas and 15 to 60 μm for the ureaplasmas. Colonies can be examined by hand lens but are best studied after staining by *Dienes method.* For this, a block of agar containing the colony is cut and placed on a microscopic slide. It is covered with a coverslip on which an alcoholic solution of methylene blue and azure has been dried.

As the colonies are too small, they can not be picked with platinum loops. Subcultures are done by cutting out an agar block with colonies and rubbing it on fresh culture plates. Most of the *Mycoplasma* colonies are haemolytic.

BIOCHEMICAL REACTIONS

Mycoplasmas are mainly fermentative. Most species utilise glucose or aginine as the main source of energy. Urea is not hydrolysed, except by ureaplasmas. They are generally not proteolytic.

PATHOGENS

Mycoplasma pneumoniae.

Laboratory Diagnosis

Isolation

The mycoplasmas can be recovered from throat swab, nasophryngeal swab, respiratory secretions, sputum (*M. pneumoniae)* or urethral secretions, prostatic secretions, cervical swabs, urine (*M. hominis* and *Ureaplasma urealyticum).*

Culture media should be inoculated immediately after collection of specimen. If inoculation is not possible immediately then specimen may be kept at 4°C up to 24 hours. In case of delay more than 24 hours, the specimen should be frozen at –70°C. *Mycoplasma* broth medium containing penicillin, polymyxin B, amphotericin B glucose and phenol red (indicator) is inoculated with the specimen and incubated at 37 °C. If specimen contains *M. pneumoniae,* growth is detected by turbidity and a colour changed to yellow of phenol red indicator, due to fermentation of glucose.

Serological Tests

These are of two types: (i) detection of antigen or nucleic acids, (ii) detection of antibody.

Detection of antigen or nucleic acids
a. Antigen can be detected by direct immunofluorescence test, counter-immunoelectrophoresis (CEIP) and enzyme immunoassay (EIA).
b. Specific DNA can be detected by hybridisation technique and PCR in respiratory secretions.

Detection of antibody: It can be done by using specific mycoplasmal antigens or by non-specific methods. Among the former are immunofluorescence, haemagglutination inhibition, complement fixation, enzyme immunoassay (EIA) and indirect haemagglutination assays (IHA).

MORPHOLOGY

Antinomycetes are gram-positive, non-motile, non-spring, non-acid-fast organisms. They often grow in mycelial forms and break up into coccal and bacillary forms. Most of them show true branching.

The organisms appear in the pus as *granules.* When these granules are crushed between two slides and examined by Gram staining, they consist of a central filamentous gram to positive mycelium surrounded by a peripheral zone of swollen radiating club shaped structures, presenting a *sun-ray appearance.* These clubs are gram-negative, acid-fast and are host origin.

The sulphur granules are white yellowish in colour and found only in tissues. Their size varied from minute specks to about 5 mm.

CULTURE

They grow best under anaerobic or microaerophilic conditions at the optimum temperature of 37°C under 5 to 10 per cent CO_2. They can be grown on brain heart infusion agar, blood agar or thioglycollate broth. Most species show good growth after 2 to 4 days.

PATHOGENESIS

The *Actinomyces* causes the disease known as *actinomycosis.* It is a granulomatous disease characterised by multiple abscesses, tissue destruction, fibrosis and formation of multiple sinuses.

Actinomycosis occurs in four clinical forms:

1. *Cervicofacial*: This is the commonest type and it occurs mainly in cheek and submaxillary regions.
2. *Thoracic*: It involves lungs.
3. *Abdominal:* It occurs usually in the ileocaecal region.
4. *Pelvic:* Pelvic actinomycosis has been reported in association with the use of intrauterine devices.

Macroscopically, it is a painless indurated swelling with multiple discharging sinuses. The pus contains usually yellow coloured sulphur granules. Actinomycosis may also present as mycetoma.

LABORATORY DIAGNOSIS

Specimens

- Pus from lesion or sinuses
- Discharges from fistula
- Sputum in pulmonary disease
- Tissue or biopsy.

Microscopy

Pus is shaken along with some saline in a test tube and the mixture is allowed to settle. The sulphur granules sediment is withdrawn with a capillary pipette. Granules are crushed between two slides and smears are prepared. One smear is stained by Gram's stain and other by acid-fast stain (decolorisation with 1per cent sulphuric acid). Gram's staining shows a dense network of thin gram-positive filaments, surrounded by a peripheral radiating gram-negative 'clubs' presenting a sun-ray appearance. Acid-fast staining shows central part as non-acid fast surrounded by acid-fast 'clubs'. In absence of sulphur granules, Gram's staining of pus shows gram-positive branching filaments.

In tissue sections, sulphur granules and mycelia are detected by using fluorescein conjugated specific antisera.

Culture

The sulphur granules or pus containing *'Actinomycetes'* are washed and inoculated into thioglycollate broth and streaked on brain heart infusion agar (BHI agar) and blood agar. Cultures are incubated anaerobically and aerobically with 5 per cent CO_2 at 37°C for at least 2 weeks.

Biopsy

Haematoxylin-eosin stained section shows mycelial mass surrounded by pus cells and chronic inflammatory cells.

Treatment

Surgical removal of affected tissue along with penicillin therapy are effective.

Chapter 36

Nocardia

Nocardiae resemble *Actinomyces* morphologically but are strictly aerobic. They are non-motile, gram-positive bacteria. They are acid-fast when decolourised with 1 per cent sulphuric acid.

MORPHOLOGY

Nocardiae are gram-positive bacteria and form a mycelium, that fragments into rod shaped and coccoid elements. *Nocardia* resembles *Actinomyces*, but some species are acid-fast, and a few are non-acid-fast. Differentiating feaures of *Actinomyces* and *Nocardia* as follow: .

Differentiating features of *Actinomyces* and *Nocardia*		
Property	*Actinomyces spp*	*Nocardia spp*
O$_2$ requirement	Anaerobic or microaerophilic	Strict aerobe
Temperature range for growth	35-37 °C	Wide range
Habitat	Oral commensals	Saprophytes of environment
Acid-fastness	Non-acid-fast	Weakly acid-fast
Mode of infection	Endogenous	Exogenous

CULTURE

Nocardiae readily grow in ordinary media. They are strict aerobes. They are slow growing (require 5 to 14 days). They can be grown on nutrient agar, Sabouraud dextrose agar (SDA) and brain heart infusion agar (BHI agar). The culture plates weeks. They can grow at wide range of temperature. Nocardiae form dry, granular, wrinkled colonies with pigmentation (white, yellow, pink or red).

PATHOGENESIS

Nocardiae produce opportunistic pulmonary disease known as nocardiosis in immunocompromised individuals including those with AIDS.

It causes systemic nocardiosis which manifests primarily as pulmonary disease, pneumonia, lung abscess or other lesions resembling tuberculosis. Metastasis may involve the brain, kidneys and other organs.

It may also cause mycetoma. It is a chronic granulomatous lesion involving the subcutaneous and deeper tissues mainly localised in the region of foot.

LABORATORY DIAGNOSIS

Specimens

Pus or purulent sputum.

Microscopy

The smears are stained with Gram's staining and Ziehl-Neelsen (ZN) technique using decolourisation with 1 per cent sulphuric acid. Gram-positive filamentous bacteria can be seen on Gram's staining. Acid-fast bacilli are detected on ZN technique though some species are non-acid-fast.

Culture

The specimens are inoculated on nutrient agar, SDA and BHI agar and incubated at 36°C for 3 weeks. Colony morphology is seen and bacteria are identified by staining.

TREATMENT

Nocardiae are sensitive to sulphonamides. They are also susceptible to nalidixic acid, amikacin, tobramycin and vancomycin.

Chapter
37

Chlamydiae

Chlamydiae are obligate intracellular parasites which are small, non-motile and gram-negative. They cause psittacosis, lymphogranuloma venerum (LGV) and trachoma in man and diverse diseases in birds and mammals.

MORPHOLOGY

Chlamydiae exist in two forms: the elementary body and the reticulate (initial body).

Elementary Body (EB)

It is a spherical particle measuring 200 to 300 nm in diameter with an electron dense necleioid. It is the extracellular infective form.

Reticulate (Initial) Body (RB)

Reticulate body is non-infectious in nature. It is the intracellular growing and replicative form.

Chlamydiae are gram-negative but they stain better with other methods like Castaneda, Machiavello or Gimenez stains. The inclusion bodies are basophilic and are present in cytoplasm. The inclusion bodies of *C. trachomatis* can be stained with Lugol's iodine because of the presence of glycogen matrix. Chlamydiae can also be demonstrated by immunofluorescent technique using antibody tagged with fluorescein isothiocyanate (FITC).

CULTURE

Chlamydiae can be isolated by:
1. Animal inoculation
2. Yolk sac inoculation
3. Tissue culture.

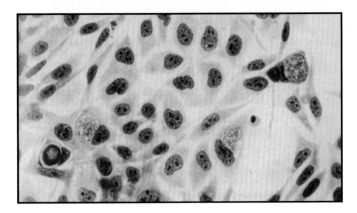

Fig. 37.1: *Chlamydia* on McCoy cells

Animal Inoculation

Mice are inoculated by intranasal, intraperitoneal or intracerebral inoculation.

Yolk Sac Inoculation

Yolk sac of chick embryo is inoculated and the organisms can be detected in impression smears stained by the Giemsa or Gimenez methods.

Tissue Culture

McCoy cells treated with cycloheximide are the most commonly used cell line (Fig. 37.1). Mouse fibroblast cell lines, Hela or monkey kidney cells can also be used for isolation of chlamydiae. The organisms in the tissue culture can be detected by staining for inclusions or elementary bodies.

PATHOGENESIS

Chlamydial infections in man occur in three forms.
1. Ocular infections
2. Genital infections
3. Respiratory infections.

LABORATORY DIAGNOSIS

Laboratory diagnosis depends on:
1. Direct detection of antigens
2. Isolation
3. Serology for antibody detection
4. Skin test.

Ocular, urethral, vaginal and cervical specimens are best collected by scraping the mucosa. In addition, blood, sputum, respiratory secretions, and other tissues can also be collected. In case of LGV, pus or discharge from bubo should be collected.

Direct Detection of Antigens

The following methods may be used for detection of chlamydial antigens in specimens.

Light Microscopy

C. trachomatis infections of conjuctiva, urethra and cervix may be diagnosed by demonstrating inclusion bodies in the smears stained with Giemsa, Castanaeda or Machiavello methods. The inclusion bodies of trachoma and inclusion conjunctivitis is named *Halberstaedter Prowazek* or HP bodies.

Immunofluorescence (IF)

It is FITC- labelled monoclonal antibodies and fluorescence can be detected in positive smears.

ELISA

It detects soluable genus-specific anitigen. Sensitivity and specificity of ELISA is similar to that of immunofluorescence test.

DNA Probes

DNA hybridisation can be used for detection of DNA of *C. trachomatis* in conjunctival and cervical smears. Specific DNA probes are used.

Polymerase Chain Reaction (PCR)

DNA is amplified and detected by PCR. This method is more sensitive than culture.

Isolation

Chlamydiae may be isolated by inoculation into mice, yolk sac of chick embryo or in tissue cultures.

Serology for Antibody Detection

1. Complement fixation test (CFT)
2.. Microimmunofluorescence

3. Immunoperoxidase test
4. ELISA test.

Skin Test

Frei's test is available for diagnosis of LGV.

TREATMENT

Tetracycline is given topically as well as systematically for several weeks. In young children, erythromycin is given.

SECTION

4

VIROLOGY

Chapter

38

Viruses: General Properties and Classification

Viruses are the smallest known infective agents. These are approximately 100 to 1000–fold smaller than the cells they infect. The smallest Viruses are 20 nm in diameter (1 nm = 10^{-9}m) whereas the largest viruses have a diameter approximately 300 nm. Most forms of life like animals, plants and bacteria are susceptible to infection with appropriate viruses.

The main properties of viruses are as follows:

i. *Small size:* Viruses are smaller than other organisms and vary in sizes from 10 nm to 300 nm.

ii. *Genome:* The genome of viruses may be either DNA or RNA; viruses contain only one kind of nucleic acid.

iii. *Metabolically inert:* Viruses have no metabolic activity outside the host cells; they do not possess active ribosomes or protein synthesizing apparatus and therefore cannot multiply in inanimate media but only inside living cells.

STRUCTURE OF VIRUSES

Viruses consist basically of a core of nucleic acid surrounded by a protein coat as in Figure 38.1.

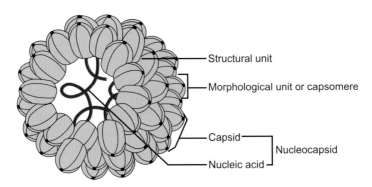

Fig. 38.1: Basic structure of virus

The structures which make up a virus particle are known as:

Virion	:	The intact virus particle
Capsid	:	**The protein coat**
Capsomers	:	The protein structural units of which the capsid is composed of nucleic acid
Envelope	:	The particles of many viruses are surrounded by a lipoprotein envelope containing viral antigens.

Virus particles show three types of symmetry.

Cubic	:	In which the particles are icosahedral protein shells with the nucleic acid contained inside
Helical	:	In which the particle contains an elongated nucleocapsid
Complex	:	In which the particle does not confirm to either cubic or helical symmetry.

Table 38.1: Classification of viruses		
Family	**Viruses**	**Diseases**
DNA Viruses		
Poxviruses	Variola, Molluscum	Smallpox, molluscum contagiosum
Herpesviruses	Herpes simplex, Varicella-zoster, Cytomegalovirus, EB Virus, HHV-6 and HHV-7	Herpes, chickenpox, Shingles, infectious mononucleosis Exanthum subitum
Adenoviruses	Adenoviruses	Sore throat, conjunctivitis
Hepadnavirus	Hepatitis B	Hepatitis
Papovaviruses	Papilloma JC virus	Warts, progressive multifocal leucoencephalopathy
Parvoviruses	B19	Erythema infectiosum, aplastic crises
RNA Viruses		
Orthomyxoviruses	Influenza	Influenza
Paramyxoviruses	Parainfluenza Respiratory syncytial viruses, Measles, mumps	Respiratory infection Measles, mumps
Coronaviruses	Coronavirus	Respiratory infection
Rhabdoviruses	Rabies	Rabies
Picornaviruses	Enteroviruses, Rhinoviruses Hepatitis A	Meningitis, paralysis, Colds, hepatitis
Calciviruses	Norwalk-like viruses	Gastroenteritis

Contd...

Contd...

Family	Viruses	Diseases
Togaviruses	Alphaviruses Rubiviruses (Group A arboviruses)	Encephalitis, haemorrhagic fevers, rubella
Flaviruses	Flaviruses (Group B arboviruses)	Encephalitis, haemorrhagic fevers
Bunyaviruses	Hepatitis-C Bunya, arboviruses, Hantan virus	Hepatitis encephalitis, H-fever, renal involvement
Reoviruses	Rotavirus	Gastroenteritis
Arenaviruses	Lymphocytic choriomeningitis, Lassa virus,	Meningitis
Retroviruses	HTLV 1	T-cell leukaemia-lymphoma, Paresis
Filoviruses	HIV-1, 2 Ebola virus Marburg virus	AIDS Marburg and ebola, Haemorrhagic fever

Viruses are assigned to groups mainly on the basis of the morphology of the virus particle, but also of their nucleic acid and method of RNA transcription.

A simplified scheme of classification of the main groups of medically important viruses and the diseases they cause is shown in Table 38.1.

Chapter
39

Cultivation of Viruses

As viruses are obligate intracellular parasites, they cannot be grown on any inanimate culture medium. Hence, there are three methods for the cultivation of viruses.
1. Animal inoculation
2. Embryonated eggs
3. Tissue cultures.

ANIMAL INOCULATION

The earliest method for the cultivation of viruses was inoculation in to human volunteers. Due to the serious risk involved, human volunteers are used only when no other method is available and when the virus is relatively harmless. Monkeys were used for the isolation of the polio virus by Landsteiner. However, due to their cost and risk to handlers, monkeys find only limited application in virology. The use of white mice, extended the scope of animal inoculation. Infant (suckling) mice are very susceptible to coxsackie and arboviruses, Mice may be inoculated by several routes intracerebral, subcutaneous, intraperitoneal or intranasal. Other animals such as guinea pigs, rabbits and ferrets are used in some situation.

EMBRYONATED EGGS

The embryonated hen's egg was first used for the cultivation of viruses by Goodpasture (1931) and the method was further developed by Burnet. The embryonated egg's offers several sites for the cultivation of viruses (Fig. 39.1). Inoculation on the chorioallantoic membrane (CAM) produces visible lesions (Pocks). Different viruses have different pock morphology (Fig. 39.2). Inoculation into the allantoic cavity provides a rich yield of influenza and some paramyxoviruses. Inoculation into the amniotic sac is employed for the primary isolation of the influenza virus. Yolk sac inoculation is used for the cultivation of some viruses, chlamydiae and rickettsiae.

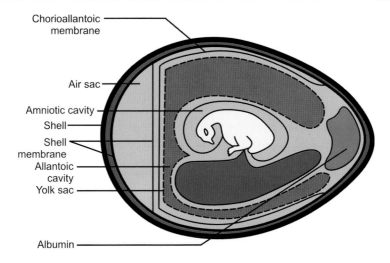

Chorioallantoic membrane

Air sac

Amniotic cavity

Shell

Shell membrane

Allantoic cavity

Yolk sac

Albumin

Fig. 39.1: Egg inoculation

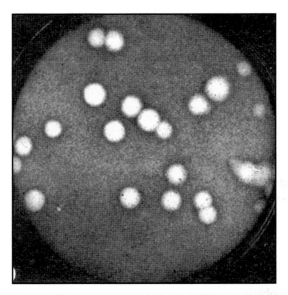

Fig. 39.2: Viral plaques on monolayer of chick embryo cells

Allantoic inoculation is employed for growing the influenza virus for vaccine production. Other chick embryo vaccines in routine use are the yellow fever (17D strain) and rabies (Flury strain) vaccines. Duck eggs are bigger and have a longer incubation period than hen's eggs. They, therefore, provide a better yield of rabies virus and were used for the preparation of the inactivated non-neural rabies vaccine.

TISSUE CULTURE

There are three types of tissue culture. They are:

Organ Culture

Small bits of organs can be used for the isolation of some viruses which appear to be highly specialised parasites of certain organs. For example, the tracheal ring organ culture is employed for the isolation of coronavirus, a respiratory pathogen.

Explant Culture

Fragments of minced tissue can be grown as explants embedded in plasma clots. They may also be cultivated in suspension. This was originally known as 'tissue culture'. Adenoid tissue explant cultures were used for the isolation of adenoviruses.

Cell Culture

Viruses are strict intracellular parasites, requiring a living cell for multiplication and spread. To detect virus using living cells, suitable host cells, cell culture media, and techniques in cell culture maintenance are necessary, Host cells referred to as cell cultures, originate as a few cells, and grow into a monolayer on the sides of glass or plastic test tubes. Cells are kept moist and supplied with nutrients by keeping them continuously immersed in cell culture medium. Once inoculated with specimen, cell cultures are incubated for 1 to 4 weeks, depending on the viruses suspected. Periodically the cells are inspected microscopically for the presence of virus, indicated by areas of dead or dying cells called **Cytopathic effect (CPE)**. TWO KINDS OF MEDIA—Growth medium and maintenance medium, are used for cell culture. Both are prepared with Eagle's minimum essential medium (EMEM) in Earle's balanced salt solution (EBSS) and include antimicrobials to prevent bacterial contamination. Usual antimicrobials added are vancomycin (10 microgram/ml), gentamicin (20 microgram/ml) and amphotericin (2.5 microgram/ml). Growth medium in a serum-rich (10% fetal, newborn, or agammaglobulinemic calf serum) nutrient medium designed to support rapid cell growth. This medium is used for initiating growth of cells. Maintenance medium is similar to growth medium but contains less serum (0 to 4%) and is used to keep cells in a steady state of metabolism. Several kinds of cell cultures are routinely used for isolation of viruses. A cell culture becomes a cell line once it has been passed or subcultured *in vitro*. Cell lines are classified as primary, low passage, or continuous. **Blind passage:** refers to passing cells and fluid to a second cell culture tube. Blind passage is used to detect viruses that may not produce CPE in the initial culture tube but will when the "beefed-up" inoculum is passed to a second tube. Cell cultures that show non-specific or ambiguous CPE are also passed to additional cell culture tubes (Figs 39.3 and 39.4). Passage in both instances, is performed by scraping the monolayer off the sides of the tube with a pipette or disrupting the monolayer by vortexing with sterile glass beads added to the culture tube, followed by inoculating 0.25 ml of the resulting suspension into new cell cultures.

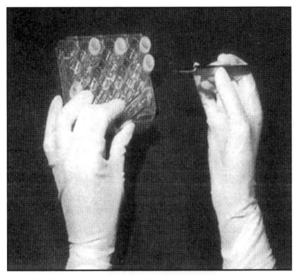

Fig. 39.3: Preparation of shell vial culture tubes

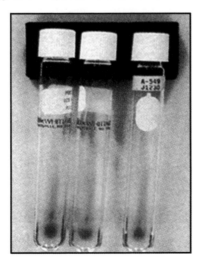

Fig. 39.4: Cell culture tubes

Some Cell Cultures in Common Use

a. Primary cell cultures
   1. Rhesus monkey kidney cell culture
   2. Human amnion cell culture
   3. Chick embryo fibroblast cell culture
b. Diploid cell strains
   1. WI-38   Human embryonic lung cell strain
   2. HL-8   Rhesus embryo cell strain
c. Continuous cell lines
   1. HeLa   Human carcinoma of cervix cell line
   2. HEP-2   Human epithelioma of larynx cell line
   3. KB   Human carcinoma of nasopharynx cell line
   4. McCoy   Human synovial carcinoma cell line
   5. Detroit-6   Sternal marrow cell line
   6. Chang C/I/L/K Human conjunctiva ©
   7. Vero   Intestine (I), Liver (L) and kidney (K) cell lines
                     Fervet monkey kidney cell line
                     Baby hamster kidney cell line.

LABORATORY METHODS IN VIROLOGY

The clinical virology laboratory must be familiar with cell culture, enzyme immunoassay and immunofluorescence methods, in addition to other common laboratory techniques.

The equipments required in setting up of a virology laboratory include *laminar flow, biosafety cabinet, fluorescence microscope, inverted bright-field microscope (Fig. 40.1), refrigerated centrifuge, incubator, refrigerator/freezer,* and *roller drum* for holding cell culture tubes during incubation.

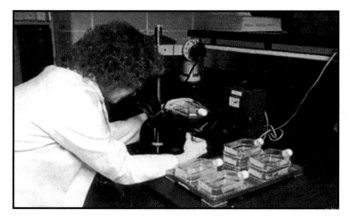

Fig. 40.1: Inverted microscope

SPECIMEN SELECTION AND COLLECTION

Specimen selection depends on the specific disease syndrome. Viral aetiologies suspected and time of year. Specimens for the detection of virus should be collected as early as possible following the onset of disease. Serum for serologic testing may be necessary, and some viral disease needs only to be considered. Recommendations for collection of common specimens are summarised in Table 40.1.

Table 40.1: Specimens for the diagnosis of viral diseases	
Respiratory infections	
1. Adenoviruses	Throat/Nasopharynx–Swab
2. Influenza viruses	Throat/Nasopharynx–Swab
3. Parainfluenza virus	Throat/Nasopharynx–Swab
Respiratory syncytial virus (RSV)	Throat/Nasopharynx–Swab
Rhino viruses	nasal–Swab
Dermatologic and mucus membrane	
Vesicular	Vesicle fluid or scraping
	Throat/Nasopharynx–Swab
Enterovirus	Stool
Herpes simplex	Throat/Nasopharynx–Swab
Varicella-zoster	Vesicle fluid or scraping
Adenoviruses	Throat/Nasopharynx–Swab
Influenza virus	Throat/Nasopharynx–Swab
Parainfluenza virus	Throat/Nasopharynx–Swab
Respiratory syncytial virus	Throat/Nasopharynx–Swab
Vesicular	Throat/Nasopharynx/Stool
Enterovirus	Throat/Nasopharynx/ Stool
Herpes simplex	Throat/Nasopharynx–Swab
Varicella-zoster	Throat
Exanthematous	Throat/Nasopharynx/Stool
Enterovirus	Throat/Nasopharynx/Stool
Measles	Throat/Nasopharynx/Urine
Rubella	Urine
Parvovirus	Amniotic fluid, liver tissue
Enteroviruses	Throat/Nasopharynx/Stool/Urine
Herpes simplex	Stool
Adenoviruses	Stool
Norwalk	Stool
Rotavirus	Stool
Cytomegalovirus	Urine
Enteroviruses	Throat/Nasopharynx/Stool/Urine
Rubella	Urine
Adenoviruses	Stool
Cytomegalovirus	Urine
Coxsackie B	Throat/Nasopharynx/Stool
Lassa fever virus	Throat/Nasopharynx/Urine.

SPECIMEN: TRANSPORT AND STORAGE

Ideally, all specimens collected for detection of virus should be processed by the laboratory immediately. Specimens for viral isolation should not be allowed to be at room or higher temperature. Specimens should be placed in ice and transported to the laboratory at once, the specimen may also be refrigerated, not frozen, until processing occur. For storage up to 5 days, hold the specimen at 4°C. Storage for 6 or more days should be–20°C or preferably–70°C.

Specimens are collected using a swab. Most types of swab material, such as cotton, rayon and dacron are acceptable. Once collected the swab should be emulsified in viral transport medium. This medium contains protein such as serum, albumin or gelatin to stabilise virus and antimicrobials to prevent overgrowth of bacteria and fungi.

Examples are Amie's medium, Hanks balanced salt solution, Leibovitz-Emory medium and Eagles tissue culture medium.

Blood for culture is transported with anticoagulant and refrigerated at 4°C until processing and blood for viral serology should be transported to the laboratory as early as possible. Serum should be separated from the clot and stored for hours or days at 4°C or for weeks or months at–20°C or below.

SPECIMEN PROCESSING

Processing viral specimens should occur in a biological safety cabinet and they are processed as shown above.

Thus specimens include throat swabs, nasopharyngeal swab or aspirate, bronchial and bronchoalveolar washes, rectal swabs and stool specimens, urine, skin and mucous membrane lesions, sterile body fluids especially cerebro spinal, pericardial and pleural fluids, blood, bone marrow. Tissue specimens include lung, brain and gastrointestinal tract which are collected during surgical procedures and serum for antibody testing. Acute and convalescent serum specimens are needed to detect antibody to specific viruses.

General Principles

Specimens for viral culture should be processed immediately upon receipt in the laboratory. Each specimen for virus isolation should be accompanied by a requisition that provides the following information, in addition to patient identification and demographics: source of specimen, clinical history or viruses suspected, and date and time of specimen collection. Processing viral specimens should occur in a biological safety cabinet whenever possible. This protects specimens from contamination by the processing technologist.

Processing virology specimens is not complicated. Sterile fluid specimens can be inoculated directly. Viral transport medium or fluid specimens not in transport medium should be vortexed just before inoculation to break up virus-containing cells and resuspend the inoculum. Sterile glass beads added to the transport medium help break up cell clumps and release virus from cell aggregates. Grossly contaminated or potentially toxic specimens, such as minced or ground tissue, can be centrifuged ((1000/15 minutes) and the virus-containing supernatant used as inoculum. Each viral cell culture tube is inoculated with 0.25 ml of specimen. If insufficient specimen is available dilute with viral transport medium to increase volume. Blood for viral culture requires special processing to isolate leucocytes, which are then inoculated to cell culture tubes. Rapid shell vial cultures are used to detect many viruses.

Processing Based on Specimen Type

All lip and genital specimens should be cultured only for *herpes simplex virus.* Disease by other viruses at lip and external genital sites is unusual and detection of these agents, such as **VZV**, should be prompted by a special request. Urine specimens require a **CMV** detection test. This is done best using the shell vial assay. Stool specimens from infants should be tested for **rotavirus**. Stool or rectal swabs from adults, children, and infants should be tested for **enterovirus**. Stool for enterovirus should be tested in conjunction with throat and CSF specimens when possible. Respiratory specimens should be divided according to the patient's age and underlying medical condition. Immunocompromised patients require a comprehensive virus detection test consisting of cell culture and shell vial for CMV. Immunocompetent adults should have an **influenza virus** culture. Children younger than 10 years are susceptible to serious infection caused by influenza, parainfluenza, respiratory syncytial, and adenoviruses and need a full respiratory virus culture.

A rapid non-culture RSV detection test, such as **FA** staining or **enzyme immunoassay**, is appropriated in RSV bronchiolitis. Respiratory specimens from newborns with the possibility of congenital or perinatal viral disease should receive a comprehensive virus culture and shell vial for CMV. Blood for viral culture might contain CMV, VZV, adenovirus, or enteroviruses. Therefore, blood for viral culture should be processed for CMV only, unless other agents are mentioned by the requesting physician. All specimens from immunocompromised hosts and tissues or fluids from presumably sterile sites should be processed for comprehensive virus detection.

Processing Based on Requests for Specific Viruses

Serologic Tests

Arboviruses: Diagnosis of arbovirus encephalitis, such as Eastern, Western, Venezuelan, and St Louis, and that caused by California encephalitis, La Crosse and West Nile Viruses, requires detection of a rise in titre of antibody in acute and convalescent serum specimens. Detection of virus-specific **IgM in CSF** is available for most agents. Culture of arboviruses for diagnostic purposes is not practical. **Polymerase chain reaction** for some agents is available.

Cytomegalovirus: Cytomegalovirus can be detected in clinical specimens using conventional cell cultures, shell vial assay, or antigenemia immunoassay. CMV produces cytopathic effects (CPE) in diploid fibroblast cells in 3 to 28 days, averaging 7 days. Shell vial for CMV has a sensitivity equivalent to conventional cell culture. The **antigenemia** immunoassay use monoclonal antibody in an **indirect immunoperoxidase** or **indirect immunofluorescent** stain to detect CMV protein in peripheral blood leucocytes. Results are reported as number of positive **leucocytes** per total number of leucocytes in the smear.

Enteroviruses: Can be detected by conventional cell culture and PCR. Although most enteroviruses grow in primary monkey kidney cells, some strains grow faster in diploid fibroblast, buffalo green monkey kidney, or rhabdomyosarcoma cell lines. Presumptive diagnosis is based on CPE. Confirmation or definitive diagnosis is accomplished using commercially available FA stains. PCR is the test of choice for use with CSF to diagnose aseptic meningitis caused by enterovirus serotypes.

Epstein-Barr virus: Serology tests are used to help diagnose EBV associated diseases, including infectious mononucleosis. Isolation of EBV (in cultured B lymphocytes) is not routinely performed in clinical laboratories.

Hepatitis viruses: Disease or asymptomatic carriage caused by hepatitis A,B,C,D and E viruses is detected using serology, antigen detection, or PCR tests.

Herpes simplex virus: Herpes simplex virus grows rapidly in most cell lines. MRC-5 or mink lung fibroblast cell lines are recommened. Cultures should be examined daily and finalized if negative after 5 days of incubation.

Human immunodeficiency virus and other retroviruses: Human immunodeficiency virus 1 is detected by antibody, antigen, and PCR tests. HIV-1 ELISA antibody tests detect antibody to both HIV-1 and 2 confirmation of the ELISA screening test is accomplished with an HIV-1-specific Western blot test or with an ELISA test for HIV-2 followed by an HIV-2 specific Western blot test. HIV infection is identified and confirmed by positive HIV-1 ELISA and Western blot results.

Blood for transfusion is screened for antibody indicative of HIV-1 and 2 and HTLV-1 and 2 infection. HTLV-1 ELISA screening tests detect antibody to both HTLV-1 and 2. In addition HIV antigen (P^{24}) tests is performed to detect donors who may be recently infected and have not produced HIV antibody.

Influenza A and B viruses: Influenza A and B viruses are detected by using conventional cell culture, shell vial culture, membrane EIA (enzyme immunoassay), or direct staining of respiratory tract secretions using FA methods. Conventional cell culture using primary monkey kidney cells is superior. Fluorescent antibody staining is used for confirmation and typing of isolates as A or B.

Paediatric respiratory viruses: Influenza, parainfluenza, respiratory syncytial and adenoviruses should be sought in specimens from hospitalised infants and children younger than 10 years with suspected viral lower respiratory tract disease. All viruses can be detected by fluorescent staining of respiratory secretions or rapid cell culture (shell vial). If direct fluorescent staining is used, cell culture confirmation of all negatives is needed for children suspected to have viruses other than RSV. If conventional cell culture is used influenza and parainfluenza viruses are detected in PMK cells by CPE or haemadsorption. Fluorescent staining is used for confirmation and typing. Adenovirus and RSV are detected in HEp-2 cell culture and confirmed, if necessary, by fluorescence staining. Conventional ELISA is also very accurate and is recommended for large batches of specimens.

Gastroenteritis viruses: Electron microscopy (EM) can be used to identify viral agents known to cause gastroenteritis. Immunoassays for rotaviruses and enteric adenovirus types 40 and 41 are commercially available.

TORCH testing: TORCH is an acronym for *Toxoplasma*, rubella, cytomegalovirus, and herpes simplex virus. Testing for these agents, and other viral etiologies of infection in newborns, is appropriate during pregnancy, because transplacental infection followed by congenital defects can occur.

Varicella-zoster virus: Varicella-zoster virus (VZV) causes chickenpox (varicella) and shingles (zoster). VZV, a DNA containing virus, establishes latency in the dorsal nerve root ganglion. Months to years later, during periods of relative immune suppression, VZV reactivates to cause zoster. Zoster is a modified or limited form of varicella, localised to a specific dermatome, the cutaneous area served by the infected nerve ganglion. Virus is present in the vesicular fluid and in the cells at the base of the vesicle. Material for virus detection should be collected from newly formed vesicles.

Virus can be detected by staining cells from the base of the vesicles or by culturing cells and vesicular fluid. The **Tzack test**, which is a smear of cells from the base of the vesicle stained by the Giemsa, papanicolaou (Pap) or other suitable cytologic staining method, detects typical multinucleated giant cells and inclusions. The FA stain also can be used to detect VZV in Tzanck smears. Traditionally, diploid fibroblast cell culture has been used to detect VZV.

DIRECT DETECTION METHODS

Cytology and Histology

This involves the morphologic study of cells or tissue respectively. Viral inclusions are intracellular structures formed by aggregates of virus or viral components within an infected cell or abnormal accumulations of cellular materials resulting from viral induced metabolic disruption.

Pap or Giemsa's stained cytologic smears are examined for inclusions or syncytia. Cytology is most frequently used to detect infections with varicella-zoster and herpes simplex viruses A stained smear of cells from the base of a skin vesicle used to detect VZV or HSV inclusions is called a **Tzanck test.** Inclusions resulting from infection with CMV, adenovirus, parvovirus, papiloma virus, and molluscum contagiosum virus are detected by histologic examination of tissue stained with hematoxylin and eosin, Rabies virus inclusions in brain tissue are called Negri bodies.

Electron Microscopy

Electron microscopy is most helpful for the detection of viruses that do not grow readily in cell culture. EM allows visualisation of virus particles present in numbers too small for easy direct detection. Electron microscopy is still used by some clinical laboratories for the detection of viruses causing gastroenteritis (rotavirus, enteric adenoviruses, Norwalk agent virus, corona viruses, and caliciviruses) and encephalitis (herpes simplex virus, measles virus, and JC polyomavirus). Electron microscopy is no longer needed to detect common rotavirus types causing infection in humans because relatively simple and accurate enzyme immunoassay and latex agglutination antigen assay have been developed.

Viral Serology

Many serologic methods have been routinely used to detect antiviral antibody, prominent among these are complement fixation (CF) enzyme-linked immunosorbent assay (ELISA), indirect immunofluorescence, anticomplement immunofluorescence (ACIF), and western immunoblotting.

Immunodiagnosis: Commercially available viral antibody reagents have led to the development of fluorescent antibody, enzyme immunoassay, radio-immunoassay, latex agglutination, and immunoperoxidase tests that detect viral antigen in patient specimens.

Direct and indirect immuno FA methods are used. Direct immuno FA testing involves use of a labelled antiviral antibody; the label is usually fluorescein isothiocyanate (FITC) which is layered over specimen suspected of containing homologous virus. The indirect immuno FA procedure is a two step test in which unlabelled antiviral antibody is added to the slide first followed by a labelled (FITC) antiglobulin that binds to the first step antibody bound to virus in the specimen.

Enzyme immunoassay methods used most frequently in clinical virology are the solid phase enzyme-linked immunosorbent assay (solid-phase ELISA) and the membrane bound enzyme linked immunosorbent assay (Fig. 40.2) (membrane ELISA).

Anti-human IgM (or anti-IgG) antibody is used to detect specific IgM (or IgG) in the serum under test. Labelled antihuman antibody is used to detect the virus antibody; the label is an enzyme which reacts with a suitable substrate to produce a visible colour change. The enzyme substrates most often used are:

  i. Horseradish peroxidase and hydrogen peroxide: orthophenyldiamine
  ii. Alkaline phosphatase; paranitrophenyl phosphate.

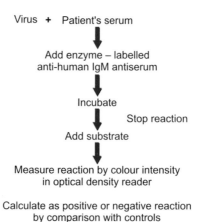

Virus + Patient's serum

Add enzyme – labelled
anti-human IgM antiserum

Incubate

Stop reaction

Add substrate

Measure reaction by colour intensity
in optical density reader

Calculate as positive or negative reaction
by comparison with controls

Fig. 40.2: Enzyme-linked immunoabsorbent assay (ELISA)

Radioimmunoassay (RIA)

Generally the most sensitive technique. Similar in principle to ELISA but the detecting antihuman antibody is tagged with an isotope-most often ^{125}I.

Antibody capture tests: Both ELISA and RIA tests can be made more sensitive and more specific by 'capturing' patient's IgM, reacting it with virus and then by adding labelled monoclonal antiviral antibody. This is illustrated in the following Figure 40.3.

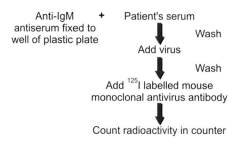

Fig. 40.3: Antibody capture

Complement Fixation Test

Virus antibody is detected by the fixation of added complement when the antibody combines with virus antigen. The fixation is rendered visible by later addition of sheep erythrocytes sensitised by addition of antierythrocyte antibody. If virus antibody is present complement is fixed and the sheep red cells do not haemolyse; if no virus antibody is present the complement lyses the sensitised erythrocytes as in Figure 40.4.

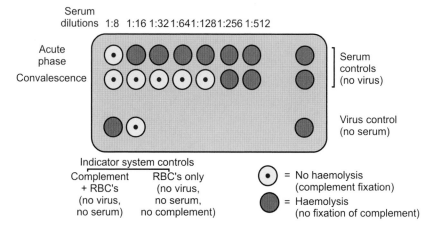

Fig. 40.4: Diagram for complement fixation test for viral antibody

Immunofluorescence

Virus-specific antibody is detected usually by the indirect or sandwich technique. Fluorescence is detected by examination in a microscope under ultraviolet light and indicates the presence of virus antibody. Sometimes virus antibody is detected by addition of complement to the reaction and then detecting its fixation by fluorescein-labelled anticomplement antibody.

Haemagglutination-inhibition Test

Many viruses haemagglutinate erythrocytes but virus antibody blocks this. Antibody can be detected in a patient's serum by inhibition of virus haemagglutination as in Figure 40.5.

Radial Immune Haemolysis

A useful qualitative test for antibody detection-but not titration with haemagglutinating viruses. Widely used as a screen test for immunity to rubella. Virus and erythrocytes are mixed in an agar gel in a plate with added complement. Patient's sera are added to wells cut in the agar; if antibody is present, zones of haemolysis appear round the wells on incubation.

Neutralisation

Antibody prevents virus infection of cells. Antibody can be detected by neutralisation of virus cytopathic effect (CPE) in tissue culture.

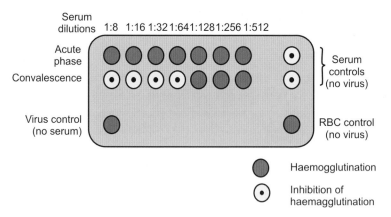

Fig. 40.5: Diagram for haemagglutination-inhibition test for viral antibody

VIRUS ISOLATION

Tissue Culture

The main types of tissue culture are:

1. *Primary culture:* These are laborious to prepare and short-lived but generally susceptible to a wide range of viruses. There is little cell division and although one subculture can be done, the cells disc in about two to three weeks, e.g. monkey kidney.
2. *Semi-continuous cell strains:* These are established from human embryo lung: easy to maintain and can be subcultured for about 30 to 40 passages before the cells die off. Susceptible to a wide ranges of viruses.
3. *Continuous cell lines:* It can be subcultured indefinitely and are therefore easy to maintain. Generally susceptible to fewer viruses than the other types of cell culture. HeLa (derived from human cervical cancer) is the most widely known.

Virus growth is recognised by:

1. *CPE (cytopathic effect):* The virus kills the cells which round up and fall off the glass. Some viruses cause cell fusion and their growth is recognized by the appearance of syncytia.
2. *Haemadsorption:* Added erythrocytes adhere to the surface of infected cells with haemagglutinating virus. Sometimes virus can be detected by haemagglutination in the medium.
3. *Immunofluorescence:* Infected cells are detected by fluorescence.

CONVENTIONAL CELL CULTURE

Viruses are strict intracellular parasites, requiring a living cell for multiplication and spread. To detect virus using living cells, suitable host cells, cell culture media and techniques in cell culture maintenance are necessary.

Host cells, referred to as cell cultures, originate as a few cells and grow into a monolayer on the sides of glass or plastic test tubes. Cells are kept moist and supplied with nutrients by keeping them continuously immersed in cell culture medium.

Once inoculated with specimen, cell cultures are incubated for 1 to 4 weeks, depending on the viruses suspected. Periodically the cells are inspected microscopically for the presence of virus, indicated by areas of dead or dying cells called cytopathic effect (CPE).

Two kinds of media, growth medium and maintenance medium are used for cell culture. Both are prepared with Eagle's minimum essential medium (EMEM) in Earle's balanced salt solution (EBSS) and include antimicrobials to prevent bacterial contamination. Usual antimicrobials to prevent bacterial contamination. Usual antimicrobials added are vancomycin (10 μg/ml), gentamicin (20 μg/ml), and amphotericin (2.5 μg/ml). Growth medium is a serum-rich (10% fetal, newborn, or agammaglobulinemic calf serum) nutrient medium designed to support rapid cell growth. This medium is used for initiating growth of cells. Maintenance medium is similar to growth medium but contains less serum and is used to keep cells in a steady state of metabolism.

Several kinds of cell cultures are routinely used for isolation of viruses. A cell culture becomes a cell line once it has been passed or subcultured *in vitro*. Cell lines are classified as primary, lowpassage or continuous.

Blind passage refers to passing cells and fluid to a secondary cell culture tube. Blind passage is used to detect viruses that may not produce CPE in the initial culture tube. Passage in both instances, is performed by scraping the monolayer by vortexing with sterile glass beads added to the culture tube.

Shell Vial Culture

The shell vial culture is a rapid modification of conventional cell culture. Virus is detected more quickly using the shell vial technique because the infected cells monolayer is stained for viral antigen found soon after infection, before the development of CPE. Viruses that normaly take days to weeks to produce CPE can be detected with in one to two days. A shell vial culture tube, a 15 × 45 nm vial, is prepared by adding a round cover slip to a bottom, covering this with growth medium and adding appropriate cells. During incubation a cell monolayer forms on top of the cover slip. Shell vial should be used 5 to 9 days after cells have been inoculated. Specimens are inoculated on to the shell vial cell monolayer by low speed centrifugation. Cover slip is stained using virus-specific immunofluorescent conjugates. Typical fluorescing inclusion confirm the presence of virus.

IDENTIFICATION OF VIRUSES DETECTED IN SHELL CULTURE

Viruses are most often detected in cell culture by the recognition of CPE. Virus infected cells change their glass surface while dying. Viruses have distinct CPEs, just as colonies of bacteria on agar plates have unique morphologies. CPE may be quantitated as indicated.

Some viruses, such as influenza, parainfluenza, and mumps, which produce little or no CPE, can be detected by haemadsorption, since infected cells contain viral haemadsorbing glycoproteins in their outer membranes. Interpretation of laboratory test results must be based on knowledge of the normal viral flora in the site sampled, the clinical findings, and the epidemiology of viruses.

Molecular Detection Using Nucleic Acid Probes and Polymerase Chain Reaction Assays

Nucleic acid probes are short segments of DNA that are designed to hybridise with complementary viral DNA or RNA segments. The probe is labelled with a fluorescent, chromogenic, or radioactive tag that allows detection if hybridisation occurs. The probe reaction can occur *in situ*, such as in a tissue thin section, in liquid, or on a reaction vessel surface or membrane. A DNA probe test is used to detect papillomavirus DNA in a smear of cervical cells.

Influenza and Respiratory Viruses

INFLUENZA

Influenza is an acute infectious disease of the respiratory tract which occurs in sporadic epidemic and pandemic forms.

Structure of the Virus

The influenza virus is an orthomyxovirus (Myxo-affinity for mucin) spherical with a diameter of 80 to 120 nm. The virus core consists of ribonucleoprotein in helical symmetry. The RNA genome is segmented and exists as eight pieces. The nucleo capsid is surrounded by an envelope and an inner membrane protein layer and an outer lipid layer projecting from the envelope are two types of spikes **haemagglutinin** spikes which are triangular and the mushroom shaped **neuraminidase** peplomers.

Clinical Features

Route of infection: Inhalation of respiratory secretions from an infected person.

Incubation Period: From 1 to 4 days.

Signs and symptoms: Fever, malaise, headaches, generalised aches, sometimes with nasal discharge and sneezing; cough is common and there may be sore throat and hoarseness.

Duration: Symptoms usually last for about 4 days but tiredness and weakness often persists for longer.

Complications

1. *Primary influenzal pneumonia:* Here severe respiratory distress and symptoms of hypoxia, dyspnoea and cyanosis, occur. Circulatory collapse follows and the patient almost always dies.

2. *Secondary bacterial pneumonia,* usually develops later in the course of influenza and is due to secondary invasion of the lungs by bacteria such as *Staphyloccus aureus, Haemophilus influenzae* or pneumococci. The signs and symptoms are like bacterial pneumonia.

Types of Virus

There are three influenza viruses, A, B and C which can be differentiated by complement fixation test:
1. The principal cause of epidemic influenza
2. Usually associated with a milder disease
3. Of doubtful pathogenicity for man.
   Influenza A viruses are also found in animals notably birds, pigs and horses.

Culture: This virus grow in monkey kidney tissue culture without CPE but with haemadsorption. Grows in amniotic cavity of the chick embryo and after passage or subculture-in the allantoic cavity also.

HAEMAGGLUTINATION BY INFLUENZA VIRUSES

Haemagglutination is due to adsorption of influenza virus particles to specific receptors on the erythrocyte surface.

Virus haemagglutinin is contained in the envelope round the virus particle; the haemagglutinin has a combining site which is antigenic and has an affinity for neuraminic acid.

Neuraminidase: Influenza virus particles also contain an enzyme which destroys the neuraminic receptors on erythrocytes. After viral haemagglutination, in the mixture of virus and erythrocytes is kept at 37°C, the neuraminidase becomes active and causes the virus to elute from the erythrocytes; as a result the haemagglutination is reversed and the erythrocytes disperse again.

Haemagglutination-inhibition: Treatment of the virus with specific antibody prevents haemagglutination.

Diagnosis

Serology: Complement fixation test: with the 'S' or soluble antigen.

Isolation

Specimen: Nasopharyngeal aspirates are best otherwise mouth washings or throat swabs.

Inoculation in Monkey Kidney Tissue Cultures

   i Observe for haemagglutination or haemadsorption of human group O erythrocytes.

  ii. Haemagglutination-inhibition with specific antisera.

Inoculation in Amniotic Cavity of Chick Embryo

a. Observe for haemagglutination of fowl erythrocytes

b. Inhibition of haemagglutination with standard antisera.

Direct Demonstration

Specimen; nasopharyngeal aspirate

Detection of viral antigen by indirect immunofluorescence.

Other Respiratory Tract Infections

Most respiratory viruses infect both upper and lower respiratory tracts and spreads rapidly by inhalation of respiratory secretions. Viruses which affect the respiratory tract are as follows (Table 41.1):

Table 41.1: Viruses and respiratory tract infections	
Virus	*Infections*
Parainfluenza viruses	Croup; colds, lower respiratory infections in children
Respiratory syncytial virus	Bronchiolitis and pneumonia in infants, colds in older children
Rhinoviruses	Colds
Adenoviruses	Pharyngitis and conjunctivitis
Coronaviruses	Colds
Coxsackieviruses	Colds
Echoviruses	Colds

Parainfluenza Viruses

The family Paramyxoviridae contains important pathogens of infants and children, namely parainfluenza viruses and respiratory syncytial virus.

    Paramyxoviruses resemble orthomyxoviruses in morphology but are larger and more pleomorphic. They are roughly spherical in shape and range in size from 100 to 300 nm. The helical nucleocapsid is surrounded by a lipid envelope which has the matrix (M) protein at its base and two types of transmembrane glycoprotein spikes at the surface. The longer spike is the haemagglutinin (H) which may also possess neuraminidase (N) activity and is hence known as H or HN Protein.

Clinical Features of Parainfluenza Viruses

Common cold: Coryza, sore throat, hoarseness and cough and sometimes fever.

Croup or acute laryngo-tracheobronchitis is characterised by hoarseness, cough and inspiratory stridor in infants, the disease may be severe with respiratory distress, marked stridor and cyanosis.

Bronchiolitis and pneumonia in young children are also some times caused by parainfluenza viruses.

This infection is most common in children under 5 years old and the more severe respiratory tract infections are seen mainly in pre-school children.

Serotypes: There are four parainfluenza viruses-types 1, 2, 3 and 4.

Immunity: It is not long-lasting and reinfections are common.

Specific Features

   i. Haemagglutinate human group O erythrocytes
  ii. Grows in monkey kidney tissue cultures with haemadsorption.

Diagnosis

Specimen: Mouth washings, throat swabs.

a. Inoculated in monkey kidney tissue culture and haemadsorption with human O erythrocytes or immunofluorescence.
b. Direct demonstration of virus in nasopharyngeal aspirates by immuno-fluorescence.
c. Serology is of limited value.

Respiratory Syncytial Virus (RSV)

RSV is pleomorphic and ranges in size from 150 to 300 nm (Fig. 41.1). The viral envelope has two glycoproteins the G protein by which the virus attaches

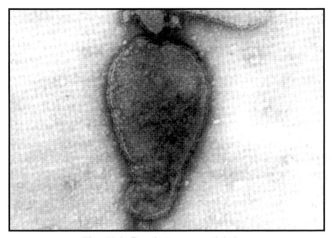

Fig. 41.1: Respiratory syncytial virus

to cell surfaces and the fusion (F) protein which brings about fusion between viral and host cell membranes.

RSV differs from other paramyxoviruses in not possessing haemagglutinin activity. It does not have neuraminidase or haemolytic properties.

RSV does not grow in eggs but can be propagated on heteroploid human cell cultures, such as HeLa and Hep-2.

Respiratory Syncytial Virus

Respiratory syncytial virus causes common colds but its importance lies in its tendency to invade the lower respiratory tract in infants under one-year-old causing bronchiolitis or pneumonia.

Clinical Features

Common cold is the most common manifestation of infection: usually seen in children, especially in those under 5 years of age but sometimes infect the elderly. Bronchiolitis is a common form of disease in infants. The infection usually starts with nasal obstruction and discharge followed by fever, cough, rapid breathing, expiratory wheezes and signs of respiratory distress.

Pneumonia is mainly seen in small infants.

Diagnosis

Direct demonstration of virus in nasopharyngeal aspirates by immuno-fluorescence using monoclonal antibody.

Isolation

Specimens: Mouth washings, nasal secretions (not frozen during delivery because the virus is inactivated by freezing).

Inoculated into HeLa cells (Bristol strain), Hep-2 cells and observe for characteristic CPE of syncytia of multinucleated giant cells.

Serology: Complement fixation test.

Rhinoviruses

Rhinoviruses resemble other picorna viruses in size and structure. These are RNA viruses, single-stranded, icosahedral particles 22 to 30 nm (Fig. 41.2). Grows in tissue cultures at 33°C instead of the usual 37°C. There are two groups of viruses.

1. 'H' viruses which grow only in human embryocells where they produce a CPE.

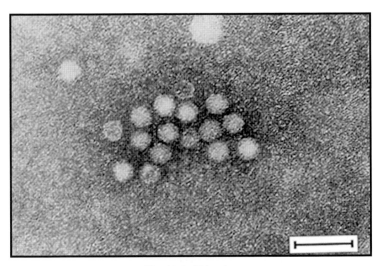

Fig. 41.2: Rhinovirus

2. 'M' rhinoviruses grow in monkey kidney as well as human embryo cells—also with CPE.

Clinical Features

Rhinoviruses are the major cause of common colds:

Nasal discharge with nasal obstruction, sneezing, sore throat and cough; about half the patients are mildly febrile; hoarseness and headache are common especially in adults. Rhinoviruses spread by contact (i.e. from hand to nose as well as by inhalation of respiratory secretions.

Infections are most frequent in pre-school children; thereafter, the attack rate falls but infections are common even amongst adults.

Diagnosis

Isolation

Specimens: Nasal secretions, mouth washings.

Inoculate human embryo lung and monkey kidney cell cultures and observe: for CPE.

Adenoviruses

Adenoviruses are double standard DNA viruses that causes various syndromes which are described in Table 41.2.

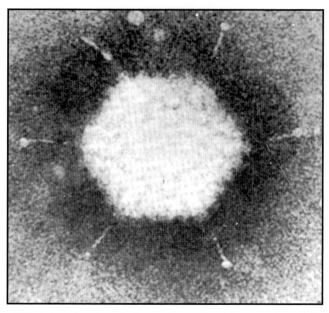

Fig. 41.3: Adenovirus. Icosahedron particle with cubic symmetry and fibres which project from the vertices. × 200 000

SI No	Syndrome	Adenovirus types
	Table 41.2: Syndromes caused by adenoviruses	
1.	Epidemic infection Pharyngo-conjunctival fever, acute respiratory disease	3, 4, 7,14, 21
2.	Endemic infection Pharyngitis, follicular conjunctivitis	1, 2, 3, 5, 6, 7
3.	Epidemic kerato-conjunctivitis or 'shipyard eye'	8, 19, 37

41 serological types have been isolated. There are icosahedron-shaped with cubic symmetry and with fibers topped with knobs projecting from the vertices. Most of these viruses haemagglutinate and grow slowly in tissue cultures-human embryonic cells or HeLa cells with CPE of clusters where the cells are rounded and ballooned (Fig. 41.3).

Clinical Features

Clinically the main symptoms of adenovirus respiratory infection are pharyngitis and conjunctivitis.

1. *Epidemic infection:* Common in recruit camps and also seen in children's institutions due to crowding together of susceptible young hosts.
2. *Endemic infection:* Adenovirus infections are endemic but at a low level in the general population. Types 1, 2, 5, and 6 are associated with endemic

infection, but cases of infection due to types 3 and 7 are common in the community and tend to be found in cluster.

3. *Epidemic keratoconjunctivitis*: A form of eye infection which is spread mainly by contaminated instruments at eye clinics and surgeries.

Faecal adenoviruses: Adenoviruses are often found in the intestine-sometimes in association with respiratory infection. They can cause viral gastroenteritis and this is particularly associated with types 40 and 41 which are 'fastidious' and do not grow in routine cell cultures.

Other Syndromes

Alimentary tract: Faecal adenoviruses may play a role in mesenteric adenitis and possibly intussusception in children.

Bone marrow transplation: Adenovirus infection-mostly enteric is common in recipients of these transplants.

Persistent infection: Adenoviruses have a tendency to persist for long periods in tissues such as the tonsils, adenoids and less often, kidneys.

Oncogenic properties: Several adenoviruses cause cancer on injection into hamsters; the most highly oncogenic are types 12, 18 and 31. However, adenoviruses do not cause tumours in man.

Diagnosis

Isolation

Specimens: Mouth washings, throat swabs, faces.

Inoculated into human embryonic cell cultures or HeLa cells and observe: for characteristic CPE or large rounded cells arranged like 'bunches or grapes'.

Serology: Complement fixation test detects antibody to adenovirus group antigen but not the serotype of the adenovirus responsible.

OTHER VIRUSES CAUSING COMMON COLDS

Corona Viruses

Medium Sized (80 to 100 nm) RNA single stranded, positive sense viruses; characteristically enveloped particles surrounded by a fringe of club-shaped projections; haemagglutinate; can only be isolated in organ cultures of human embryo trachea although some strains, notably 229E, have been adapted to growth in the L 132 line of human embryo lung cells with CPE. There are at least 3 antigenic types although with some antigenic cross-reactions or sharing between the types.

Enteroviruses

Some enteroviruses cause respiratory infections; the main types associated with respiratory disease are coxsackievirus A21 (formerly called Coe virus), B3 and echovirus types 11 and 20.

Chapter
42

Neurological Disease due to Viruses

Neurological diseases caused by viruses are clinically classified into two categories namely acute and chronic.

Acute neurological diseases and their causative agents are as follows:

 i. *Encephalitis:* A destructive lesion in grey matter of the brain resulting in neuronal damage. The viruses involved are herpes simplex, arboviruses, and rabies.

 ii. *Paralysis (Poliomyelitis):* Destructive lesions of lower motor neurons with Meningitis resulting in the damage of anterior horn cells of spinal cord. The causative agents belong to the group enteroviruses especially polio-viruses.

 iii. *Aseptic meningitis:* Inflammation of meninges resulting in the presence of cells in CSF (Usually lymphocytes). The causative agents are entero-viruses, mumps, lymphocytic, choriomeningitic viruses.

 iv. *Post-infectious encephalomyelitis:* This causes perivascular infiltration in the brain resulting in microglial proliferation and demyelination. The Viruses causing these complications are measles, rubella, varicella-zoster and vaccinia.

CHRONIC VIRAL NEUROLOGICAL DISEASES

Viruses cause several chronic neurological diseases which are as follows:

1. *Subacute sclerosing panencephalitis:* This involves the brain resulting in neuronal degeneration, intranuclear inclusions within the cells. The causative agents are measles, rubella (after congenital infection).

2. *Progressive multifocal neuro-encephalopathy:* This again damages the Brain by causing multiple foci of degeneration. This is caused by JC Virus.

3. *Creutzfeldt-Jakob disease:* Here spongiform degeneration of the brain and spinal cold occurs. These are transmissible by filter passing agents.

4. *Kuru:* In this case the spongiform degeneration occur in the brain especially in cerebellum and are transmissible by filter passing agent.

The main features of the chronic virus neurological disease are as follows:

i. All the diseases are very rare.

ii. Signs and symptoms are numerous and varied, neurological and often affect intellectual capacity as well as both motor and sensory function.

iii. Duration: The diseases may last for months or even years and are always fatal.

Chapter

43

Enterovirus Infections

Enteroviruses are a large family of viruses, of which the primary site of infection is the gut. These diseases are the result of spread of the viruses to other sites of the body-particularly the CNS.

Enteroviruses of medical importance include:

1. Poliovirus types   1-3
2. Coxsackie virus A types   1-24
3. Coxsackie virus B types   1-6
4. ECHO virus types   1-34
5. Enterovirus types   68-71

POLIOVIRUS

The virion is a spherical particle about 27 nm in diameter (Fig. 43.1). The genome is a single strand of positive sense RNA. The virus grows readily in

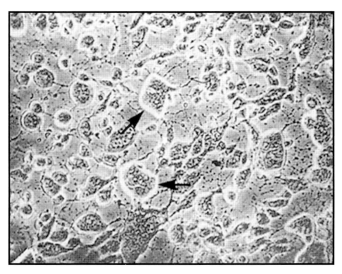

Fig. 43.1: Poliovirus

tissue cultures of primate origin. Primary monkey kidney cultures are used for diagnostic cultures and vaccine production. The infected cells round up and become refractile and pyknotic. Eosinophilic intranuclear inclusion bodies may be demonstrated in stained preparations.

Clinical Features

The virus is transmitted by the faecal-oral route through ingestion. The virus multiplies in the alimentary canal and spreads to the regional lymph nodes and enters the blood. After further multiplication, the virus is carried to the spinal cord and brain.

Illnesses of Poliovirus

This consists of fever, headache, sore throat and malaise lasting 1 to 5 days. This is called minor illness. If the infection progresses, the minor illness is followed by the major illness. The fever is biphasic along with headache, stiff neck and features of meningitis. In some cases flaccid paralysis develops and may lead on to residual paralysis.

Laboratory Diagnosis

1. *Virus isolation in tissue culture:* Virus can be isolated from the following specimens, including blood, CSF, throat swab and faeces. Primary monkey kidney cells are usually employed and the growth is indicated by typical cytopathic effects in 2 to 3 days.
2. *Serodiagnosis:* Antibody rise can be demonstrated in paired sera by neutralisation or complement fixation tests (Fig. 43.2).

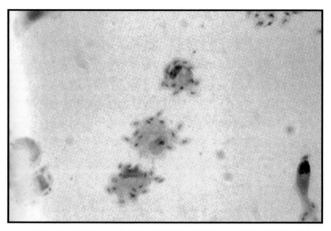

Fig. 43.2: Histopathological appearance of polio infection

Prophylaxis

Salk's killed polio vaccines and Sabin's live polio vaccines are available.

COXSACKIEVIRUSES A AND B

Coxsackieviruses are typical enteroviruses and are classified into two groups A and B. The characteristic feature of this group is its ability to infect suckling but not adult mice.

Clinical Features

Coxsackieviruses produce a variety of clinical syndromes in humans. They are as follows:

1. Herpangina (Vesicular pharyngitis) is a common clinical feature of coxsackie group A infection in children. The features are severe febrile pharyngitis with headache, vomiting and pain in the abdomen.
2. Aseptic meningitis may be caused by A and B group viruses.
3. Hand-foot-and-mouth-disease (HFMD)–It is caused by coxsackie A and B. It starts as an exanthematous fever and results in complications like aseptic meningitis, encephalitis, flaccid paralysis, pulmonary haemorrhage.
4. Minor respiratory infections resembling common cold is caused by coxsackie A.
5. Epidemic pleurodynia is a febrile disease with pain in the chest and abdomen. This is caused by group B viruses.
6. Myocarditis and pericarditis in the new born and juvenile diabetes are caused by coxsackie B virus. Orchitis due to coxsackie virus has also been reported.

Laboratory Diagnosis

Virus isolation from the lesions or from faeces may be made by inoculation into suckling mice. Serodiagnosis is not practicable.

ECHO VIRUS

ECHO viruses resemble other enteroviruses in their properties. By neutralisation tests, they have been classified into 34 serotypes.

Clinical Features

In general, the clinical features resemble those produced by coxsackie viruses. Fever with rash and aseptic meningitis can be produced by several serotypes.

ECHO viruses have been isolated from respiratory disease in children and gastroenteritis.

Laboratory Diagnosis

Faeces, throat swabs or CSF may be inoculated into monkey kidney tissue cultures and virus growth detected by cytopathic changes. Serological diagnosis is not practicable.

NEW ENTEROVIRUS TYPES

Type 68 was isolated from pharyngeal secretions of children with pneumonia and bronchitis. Type 70 causes acute haemorrhagic conjunctivitis. Enterovirus-71, originally isolated from cases of meningitis and encephalitis, causes many other syndromes, including HFMD.

Viral Gastroenteritis

Viruses are an important cause of acute diarrhoea-most often, but not exclusively in young children. Several viruses are responsible. They are:

Virus	Nucleic acid	Particle
Rotavirus	DS RNA in 11 segments	70 nm double-shelled with wheel-like surface structure
Astrovirus	SS RNA	28 nm, 6 pointed star surface structure
Adenovirus	DS RNA	74 nm, classical icosahedron with rounded capsomeres
Calcivirus	SS RNA	33 nm, 6-pointed star surface structure with central 'hole'
Norwalk	SS RNA	35 nm, indistinct surface structure
Small round viruses (SRVs)	Not known	22-25 nm, small, round, featureless
Small round structured viruses (SRVs)	Not known	35 nm, ill-defined surface structure

Clinical Features

The symptoms are an acute onset of diarrhoea, often with vomiting and sometimes with fever; abdominal cramps, dehydration is a common complication and requires treatment with rehydration fluids.

VIRUSES

Rotavirus

This is the first virus to be associated with acute non-bacterial gastroenteritis mainly affects infants but has been reported in adults during out breaks.

This virus belongs to Reovirus family consisting of RNA, double-stranded in 11 fragments (Fig. 44.1). Five groups A-E have been identified. Rotaviruses grow in tissue culture.

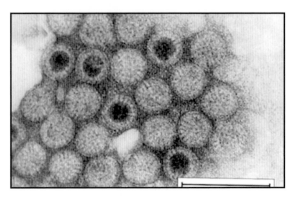

Fig. 44.1: Rotavirus

Clinical Features

The virus enters via faecal-oral route from case-to-case and causes gastroenteritis. The virus can be demonstrated in the stools of healthy controls. Infection in the neonate is often not accompanied with diarrhoea. This virus also causes respiratory infection, where respiratory secretions may also be a source of spread.

Diagnosis

Demonstration of virus in the stools by
1. Electron microscopy
2. Serology-ELISA, latex agglutination as in (Figs 44.2A and B)

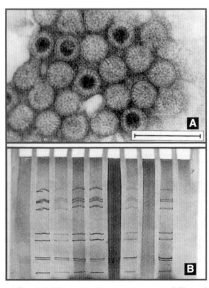

Figs 44.2A and B : (A) Microscopic appearance of Rotavirus (B) Serologic investigation (gel electrophoresis) of rotavirus

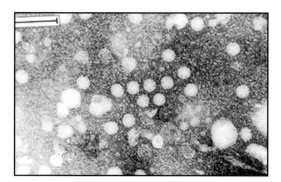

Fig. 44.3: Astrovirus

ASTROVIRUSES

These are single stranded RNA viruses consisting of 5 serotypes. Astroviruses are clearly associated with diarrhoea (Fig. 44.3).

Adenovirus

Diarrhoea causing adenoviruses are sometimes called group F adenovirus (Fig. 44.4). These are 'fastidious' and do not grow in routine cell cultures but are serologically distinct from respiratory strains.

Diagnosis: Electron microscopy of stools.

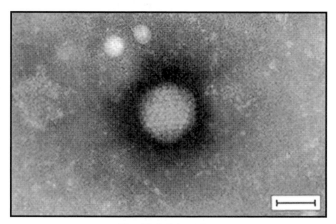

Fig. 44.4: Adenovirus

Calcivirus

Calciviruses are found more often in stools of children with diarrhoea. They are found in outbreaks of vomiting and diarrhoea. Vomiting tends to be a

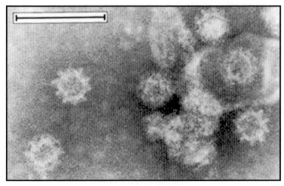

Fig. 44.5: Calcivirus

prominent symptom and calciviruses have been found as a major cause of outbreaks of vomiting disease.

Calciviruses are single stranded, positive sense RNA (Fig. 44.5). These viral particles are difficult to differentiate from those of the other diarrhoea associated viruses but serologically, there is evidence of three serotypes.

Diagnosis: Electron microscopy of stools.

Norwalk Virus

These are single stranded, positive sense RNA viruses (Fig. 44.6) which have been repeatedly reported as a cause of out-breaks. Vomiting is a prominent symp-tom in norwalk virus infection.

Diagnosis: Electron microscopy of stools.

Small Round Viruses: Structured (SRSVS) and Non-structured (SRVS)

These viruses are traced during outbreaks. Little is known about these viruses. SRVs are smaller than SRSVS of which the structure detectable on the surface of the particles is ill defined. There is some indica-tion that SRSVs may be a different serotype in the same virus family group as calcivirus and norwalk virus.

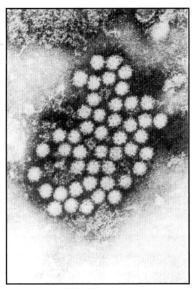

Fig. 44.6: Norwalk Virus

Chapter

45

Arthropod Borne Viral Diseases

The arthropod borne viruses are a classification of viruses that are transmitted to man by biting insects. Such viruses multiply in the insect vector and are injected into man when the insect feeds on the host. The arboviruses are extremely numerous and include many unrelated viruses.

CLASSIFICATION OF VIRUSES

Flaviviruses

These are small (40-50 nm), enveloped viruses with a single stranded, and positive-sense RNA genome (of approximately 10 kb). The symmetry of their capsid is undefined. Many have the mosquito as the vector and cause viral haemorrhagic fever (yellow fever and dengue viruses) and meningo-encephalitis (Japanese encephalitis and St Louis viruses). Diagnosis is by virus isolation (which requires special facilities) or the detection of an antibody response.

Togaviruses

The alphaviruses are transmitted principally via mosquitoes and cause encephalitis, (e.g. Eastern equine encephalitis virus) or a febrile exanthem with polyarthritis (e.g. Chikungunya). All have a single stranded, linear, positive-sense RNA genome (of about 12 kb).Viruses replicate within the cytoplasm and release by budding. Diagnosis is by serology and/or virus isolation.

BUNYA VIRIDAE

This is a large family of viruses some of which are transmitted by insects (bunyamvera, nairo and phleboviruses). Some infect plants (tospovirus) and some are zoonotic (hantaviruses.) All however, are spherical (95 nm) enveloped viruses with a cuboidal nucleocapsid. The RNA genome is linear in three segments and is mostly negative sense, although some are ambisense. The bunyamwera viruses are found worldwide, are transmitted by mosquitoes and may cause encephalitis (e.g. La Crosse) fever with myalgia (e.g. Guama) or undifferentiated febrile illness (e.g. Tahnya). There are approximately 45 phleboviruses but not all cause disease in man. Sandfly fever is transmitted by phlebotomine flies and is a febrile illness with headache, photophobia and joint pains. Nairoviruses are transmitted by ticks, the most important disease being Congo-Crimean haemorrhagic fever. It may also be transmitted person-to-person. The recently described hantavirus pulmonary syndrome is the result of Sin Nombre and numerous other viruses and cases have occurred with high mortality throughout North, Central and South America. The diagnosis of each of the above infections is by serology, antigen detection, culture or RT-PCR. They require special containment facilities.

Yellow Fever

Yellow fever is one of the most important haemorrhagic fevers, but with liver involvement and jaundice as characteristic features.

Clinically the most striking feature of yellow fever is jaundice due to viral involvement of the liver causing hepatitis; haemorrhages are often seen and a toxic nephrosis with proteinuria is a common feature.

Diagnosis

Serology: By haemagglutination-inhibition, neutralisation or complement fixation tests.

KYASANUR FOREST FEVER

Dengue

A major health problem in south-east Asia, India, the pacific Islands, and the Caribbean; infection is widespread in these areas; monkeys are probably the main reservoir of infection and the main vector is *Aedes aegypti.*

Clinically, dengue typically presents as a severe febrile disease with pain in the limbs and rash..

Dengue haemorrhagic shock syndrome: A serious complication of dengue in young children. In this syndrome, an attack of dengue progresses to a more severe disease characterised by haemorrhages and shock; seen in children who have experienced a previous attack of dengue due to a different sub-type of virus. On re-infection with the second virus, immune complexes are formed due to production of excess antibody; the immune complexes with complement activation are responsible for the haemorrhagic shock syndrome.

Chikungunya

The cause of febrile disease-sometimes in widespread epidemics-in Africa and Asia. In Asia the disease has had haemorrhagic manifestations.

Clinically, characterised by sudden fever with severe pain in the joints. Residual joint pains may persist after recovery from the acute disease.

O'nyong-nyong: Due to a similar virus is a febrile disease with joint pain (the name means 'break-bone fever') seen in Africa in small outbreaks but there was a large epidemic in 1959-1960.

Chapter

46

Herpes Viruses Diseases

Herpes viruses are classified as alpha herpes viruses. These are roughly spherical particle with cubic symmetry, medium size 100 nm with 162 projecting hollow cored capsomeres. This consists of double stranded DNA. There are two types of viruses. Type 1 and type 2 and grow in various tissue cultures with characteristic cytopathic effects (CPE) with ballooning and rounding of cell. It also grows on chorioallontoic membrane with production of tiny white pocks

There are six recognised human herpes viruses:
1. Herpes simplex virus type 1
2. Herpes simplex virus type 2
3. Varicella-zoster virus
4. Cytomegalovirus
5. Epstein-Barr virus
6. Human herpes virus 6.

HERPES SIMPLEX VIRUS

Herpes simplex virus is unusual among viruses in causing a wide variety of clinical syndromes; the basic lesions are vesicles but these can take many different forms.

There are two types of herpes simplex virus—type 1 and type 2. The two viruses are biologically and serologically related but can be differentiated.

Type 1: The commonest, causes oro-facial lesions but also a proportion of cases of genital herpes.

Type 2: The main cause of genital herpes.

Clinical Features

Primary Infections

Virtually everyone becomes infected with the virus but most primary infections are symptomless. The main clinical manifestations of primary infection is accompanied by the following symptoms:

1. *Gingivostomatitis:* Vesicles inside the mouth on the buccal mucosa and on the gums; these ulcerate and become coated with a grayish slough.

2. *Herpetic whitlow:* Due to implantation of the virus into the fingers; the lesion produced is very similar to a staphylococcal whitlow but the exudates is serous rather than purulent.

3. *Conjunctivitis and keratitis:* Primary herpes can involve the eye-both conjunctiva and cornea; the eyelids are generally swollen and there are often vesicles and ulcers on them.

4. *Kaposi's varicelliform eruption:* It is a superinfection of eczematous skin. Mainly seen in young children, and sometimes a serious disease with a significant fatality rate.

5. *Acute necrotising encephalitis:* Herpes encephalitis is very rare but extremely severe disease. Clinically, it presents with the sudden onset of fever, mental confusion and headache. The main site of infection is the temporal lobe where the disease causes necrosis.

6. Genital herpes a vesicular eruption of the genital area most often due to herpes simplex virus type 2. Genital herpes is usually sexually transmitted.

7. Neonatal infection with severe generalised infection in neonates is usually acquired from a primary genital infection in the mother. Affected infants have jaundice, hepatosplenomegaly, thrombocytopenia and large vesicular lesions on the skin. There is a high case fatality rate. Usually due to herpes simplex virus type 2.

8. Generalised infection in adults is a rare manifestation of primary infections with type 1 virus with disseminated vesicular skin lesions and virus in viscera and other body organs and tissues. Herpes hepatitis has also been described.

Diagnosis

Isolation

Specimens vesicle fluid, skin swab, saliva, conjunctival fluid, corneal scrapings, brain biopsy.

Inoculate into cell cultures, e.g. BHK 21 (a hamster kidney cell line), human embryo lung cells and observe for CPE of rounded cells.

Serology

Complement fixation test: Direct demonstration of virus or virus antigen in vesicle or other fluids or tissues by electron microscopy or immunofluorescence.

Treatment

Acyclovir: Treatment of herpes simplex has been revolutionised by the introduction of this non-toxic drug which has a specific inhibitory action on herpes simplex virus replication.

Administered intravenously, orally or topically.

Indications: Herpes encephalitis, severe or generalized herpes (given systemically). Genital herpes (systemically or orally). Dendritic ulcers, cold sores, possibly genital herpes topically. Prophylaxis in immunocompromised patients.

Idoxuridine (0.1% solution) still used topically in the treatment of herpes keratitis.

VARICELLA-ZOSTER VIRUS

These are classified as an alpha herpes virus and in morphologically identical to that of herpes simplex. These are double stranded DNA viruses and grows in tissue cultures of human cells, e.g. human embryo lung with focal CPE.

Varicella (chickenpox) and zoster (shingles)-but also sometimes called 'herpes-zoster' are different disease due to the same virus

Varicella is the primary illness.

Zoster is a reaction of infections.

Clinical Features

Varicella—one of the common childhood fevers. A mild febrile illness with a characteristic vesicular rash. Vesicles appear in successive waves so that lesions of different age are present together. The vesicles develop into pustules.

Complications are rare postinfectious encephalomyelitis, haemorrhagic (fulminating) varicella. In adults, pneumonia is a relatively common and serious complication and may be followed by permanent pulmonary calcification.

Perinatal or neonatal varicella: If the mother contracts varicella more than 5 days before delivery, the disease in the child is usually mild. When maternal varicella is contracted within 5 days of delivery, there is not time for maternal antibody to be produced and cross the placenta, and the child is liable to develop severe disease. Varicella may also be an unusually severe disease, with pneumonia, in pregnant women.

Zoster

A reactivation of virus latent in dorsal root or cranial nerve ganglia following-and usually many years after childhood varicella. Virus travels down sensory nerves to produce painful vesicles in the area of skin (dermatome) enervated from the affected ganglion.

The virus is present in the skin vesicles and in the ganglia involved (where there are cytopathic changes of cell destruction and marked inflammatory infiltration). Adults are affected much more often than children. Ganglia dorsal root ganglia and therefore the thoracic nerves supplying the chest wall are most often affected. There is a segmental rash which extends from the middle of the back in a horizontal strip round the side of the chest wall.

Cranial zoster: When the ophthalmic nerve of the trigeminal ganglion is affected. This may cause a sharply demarcated area of lesions down one side of the forehead and scalp.

Ramsay-Hunt syndrome is a rare form of zoster. The eruption is on the tympanic membrane and the external auditory canal and there is often a facial nerve palsy.

Neurological signs are sometimes seen, e.g. paralysis.

Diagnosis

Serology

Complement fixation test useful for both varicella and zoster. Unlike reactivations of herpes simplex, zoster usually causes a rise in antibody titre.

Direct demonstration of typical herpes virus particles in vesicle fluid by electron microscopy is a quick method of confirming a clinical diagnosis.

Isolation

The virus is slow-growing, markedly cell-associated and difficult to passage. The CPE produced is focal. Immunofluorescence can be used to confirm the identity of the virus.

Treatment

Severe varicella or zoster responds well to acyclovir.

CYTOMEGALOVIRUS

These are classified as a beta herpes virus made up of double stranded DNA. This grows slowly in cultures of human embryo lung cells with characteristic focal CPE and intranuclear 'owl's 'eye' inclusions.

Clinical Features

There are two types of disease due to cytomegalovirus.

Congenital

A more difficult problem than congenital rubella because:

a. Maternal infections is almost always symptomless.
b. The fetus can be damaged by infections in any of the three trimesters of pregnancy.
c. Fetal infection can follow reactivation as well as primary maternal infection.

The majority of congenitally infected neonates show no signs or symptoms and diagnosis is made by virological tests. Many of the children develop normally but some show neurological sequelae late in life, principally,

   i. Deafness
  ii. Mental retardation.

about a fifth of infected children do show clinical signs of infection and in some cases this take the form of the severe generalised or cytomegalic inclusion-disease.

Severe generalised infection (cytomegalic inclusion disease): The affected infants have jaundice, hepatosplenomegaly, blood dyscrasias such as thrombocytopenia and haemolytic anaemia. The brain is almost always involved and some infants have microcephaly motor disorders are common. Surviving infants are usually deaf and mentally retarded.

Postnatal

Hepatitis: In young children, cytomegalovirus causes hepatitis with enlargement of the liver and disturbance of liver function tests. Jaundice may or may not be present.

Infectious mononucleosis: In adults and in older children, infection may take the form of an illness-like infectious mononucleosis but with a negative. Paul-Bunnell reaction and no lymphadenopathy or pharyngitis. There is fever, hepatitis and lymphocytosis with atypical lymphocytes in the peripheral blood.

Diagnosis

Isolation

Specimens: Urine, throat swab.

Inoculate: Human embryo lung cell cultures.

Observe: For characteristic CPE of foci of swollen cells, this may take from 2 to 3 weeks to appear.

Serology

Immunofluorescence, ELISA tests for IgM. Complement fixation test.

Demonstration

Of typical intranuclear 'owl's eye' inclusions in cells of urinary sediment or other tissues.

Treatment

    i. Ganciclovir
  ii. Foscarnet
      Both are being used for severe cytomegalovirus infection.

EBSTEIN-BARR VIRUS IS CLASSIFIED AS GAMMA HERPES VIRUS

Epstein-Barr virus is named after the virologists who first observed it when examining cultures of lymphoblasts from Burkitt's lymphoma in the electron microscope. It is double-stranded DNA Virus. Grows in suspension cultures of human lymphoblasts.

Most infections are symptomless especially if acquired during childhood; if infection is delayed until adult life there is greater likelihood of disease; this takes the form of infectious mononucleosis or glandular fever.

Human cancer: EB virus has a strong association, and is almost certainly a co-factor in causing Burkitt's lymphoma and nasopharyngeal carcinoma.

INFECTIOUS MONONUCLEOSIS (GLANDULAR FEVER)

Clinical Features

Signs and symptoms: Low-grade fever with generalised lymphadenopathy and sore throat due to exudative tonsillitis; malaise, anorexia and tiredness. splenomegaly is common and most cases have abnormal liver function tests;

Mononucleosis: Or more correctly a relative and absolute lymphocytosis is a diagnostic feature.

Paul-Bunnell test: Infectious mononucleosis is characteristically associated with the appearance in the blood of heterophil antibodies to sheep erythrocytes. this antibody can be removed by absorption with ox erythrocytes but not with absorption with guinea-pig kidney. The differential absorption and the haemagglutination test with sheep erythrocytes constitute the Paul-Bannell test which is diagnostic of infectious mononucleosis; development of other non-specific antibodies are also features of the diseases.

EB virus antibody is produced during infection but antibody is usually present before symptoms develop; the detection of EB virus-specific IgM is a useful confirmatory diagnostic test.

EB Virus and Burkitt's Lymphoma

A highly malignant tumour which is common in African children. Primarily a tumour of lymphoid tissue but the earliest manifestations are often large tumours of the jaw and, in girls, some times of the ovaries; it spreads rapidly with widespread metastases.

EB virus is certainly associated with Burkitt's lymphoma; the virus is found in cell cultures established from Burkitt's lymphoma and EB virus DNA is present although not integrated in the lymphoblasts of the tumour. Patients with Burkitt's lymphoma uniformly have antibody to the virus.

Diagnosis

Serology

1. Paul-Bunnell test
2. Demonstration of EB virus-specific IgM by immunofluorescence; the antibody tested is that directed against the viral capsid antigen.
   Haematology: Demonstration of atypical lymphocytes in the peripheral blood.

HUMAN HERPES VIRUS-6

Recently discovered, this virus (HHV-6) (Fig. 46.1) was found as a latent infection of lymphoid tissue. Most infections appear to be symptomless.

Fig. 46.1: HHV-6

  i. Exanthem subitum (also known as Roseola infantum). This mild facial rash in small babies seems to be associated with HHV-6 infection.
 ii. Mononucleosis with cervical lymphadenopathy: It has been described in a few adults undergoing primary infection.

Chapter
47

Childhood Fever

Mumps, measles, rubella are the common childhood fevers.

MUMPS

This belongs to the paramyxovirus group, consisting of single stranded RNA virus. This haemagglutinates fowl erythrocytes and grows in amniotic cavity of chick embryo and in monkey kidney and other tissue cultures with haemadsorption.

Clinical Features

Clinically Classical mumps is a febrile illness with inflammation of salivary glands causing characteristic swelling of parotid and submaxillary glands.

Aseptic meningitis: (less often meningoencephalitis) is a frequent neurological complication of mumps; occasionally there is muscular weakness or paralysis.

Other complications: Orchitis, pancreatitis and rarely oophoritis and thyroiditis are seen in association with mumps.

Mumps is a generalised infection by a virus with a predilection for the CNS and for glandular tissue.

Diagnosis

Serology

Complement fixation test: Two antigens are used:
1. 'S' or soluble antigen (the nucleoprotein core of the virus particle)
2. 'V' or viral antigen (found on the surface of the virus particle)

Antibody to 'S' antigen tends to diminish sooner than antibody to 'V' antigen; it can therefore be a useful indicator of recent infection. 'V' antibody usually persists for years.

Isolation

Isolation (mainly used for diagnosis of mumps meningitis)
Specimens; CSF, *possibly throat washings.*
Inoculate into monkey kidney tissue cultures and *observe* for haemadsorption of fowl erythrocytes.
Identify virus by inhibition of haemadsorption or haemagglutination with standard antiserum.

MEASLES

This again is a paramyxovirus consisting of a single stranded RNA virus. It haemagglutinates and haemolyses monkey erythrocytes and grows in human embryo and primary monkey kidney cells with syncytial CPE of multinucleated giant cells.

Measles is the most common of childhood fevers; in uncomplicated cases it is a mild disease but complications are relatively frequent.

Prodromal symptom are respiratory, e.g. nasal discharge and suffusion of the eyes.

The main illness of measles follows; fever which may be high with a maculopapular rash lasting from two to five days; the rash is an enanthem (as-well-as an exanthem) and characteristic spots (Koplik's spots) appear in the buccal mucosa inside the cheek and mouth.

Complications

1. Respiratory
2. Neurological
1. **Respiratory infections** are the most common and are seen in about 4 per cent of patients; these include bronchitis, bronchiolitis, croup and bronchopneumonia; otitis media is also seen in about 2.5 per cent of cases.
   Giant cell pneumonia rare complication, seen in children immunodeficient or with chronic debilitating disease; due to direct invasion of the lungs by measles virus and usually fatal; there are numerous multinucleated giant cells in the lungs at postmortem.
2. **Neurological complications:** Two types of encephalitis are seen:
   i. *Encephalitis or post infectious encephalomyelitis:* A serious condition which follows measles in about one in every 1000 cases. Encephalitis commonly presents with drowsiness, vomiting, headache and convulsions.
   ii. *Subacute sclerosing panencephalitis:* A rare but severe, chronic, neurological disease seen in children and young adults. The presenting symptoms

are of personality and behavioural changes with intellectual impairment; the disease progresses to convulsions, myoclonic movements and increasing neurological deterioration leading to coma and death. Affected children have high titres of measles antibody in their serum and both IgM and IgG measles-specific antibody in the CSF.

At postmortem, there are numerous intranuclear inclusions throughout the brain; measles virus has been grown from brain tissue.

Diagnosis

Serology

Complement fixation test; detection of measles IgM by immunofluorescence. In subacute sclerosing panencephalitis, the diagnosis can be confirmed by demonstration of measles antibody in the CSF.

Direct demonstration of viral antigen by immunofluorescence in nasopharyngeal aspirates.

Isolation

Specimens like throat washings, blood, urinary sediment, etc. are *Cultured on* primary human embryo kidney cells

Observation: Characteristic CPE.

RUBELLA

This is a non-arthropod-borne togavirus composed of single stranded RNA virus. This haemagglutinates bird erythrocytes and grows in a rabbit kidney cell line RK 13 with production of CPE and in other tissue cultures but without CPE.

Rubella is a mild childhood fever but if infection is contracted in early pregnancy the virus can cause severe congenital abnormalities and disease in the fetus.

Clinical Features

A mild febrile illness with a macular rash which spreads down from the face and behind the ears; there is usually pharyngitis and enlargement of the cervical and especially the posterior cervical lymph glands.

Virus is present in both blood and pharyngeal secretions and is excreted during the incubation period for up to seven days before the appearance of the rash.

Infection is symptomless in a proportion of cases.

Complications are rare: Post infectious encephalomyelitis, thrombocytopenic purpura and arthralgia or painful joints.

Congenital Infection

The main defects are a triad of:
- Cataract
- Nerve deafness
- Cardiac abnormalities (e.g. Patent ductus arteriosus, ventricular septal defect, pulmonary artery stenosis, Fallot's tetralogy).

However, affected infants have various other disorders due to generalised infection which together with the defects, are known as the rubella syndrome these are
- Hepatosplenomegaly
- Thrombocytopenic purpura
- Low birth weight
- Mental retardation
- Jaundice
- Anaemia
- Lesions in the metaphyses of the long bones.

The incidence of defects after maternal rubella in the first three months of pregnancy has varied from 10 to 54 per cent in different studies. Maternal rubella at this time also is associated with a higher proportion of abortions and stillbirths. The severity and multiplicity of defects are increased when infection is in the earliest weeks of pregnancy.

Subacute sclerosing panencephalitis has been reported as a rare, late, complication of congential rubella.

Infants with the rubella syndrome have IgM antibody to rubella virus and therefore are immunologically competent (the maternal antibody which cross the placenta is IgG antibody.

Diagnosis

Laboratory diagnosis is now widely used for confirmation of the diagnosis of rubella.

Serology

IgM antibody: Recent infection with rubella virus is best diagnosed by the demonstration of IgM rubella antibody in a single sample of blood; detected by ELISA or by immunofluorescence.

Haemagglutination inhibition test: Quite a sensitive technique for detecting rubella antibody, active rubella in pregnancy can be diagnosed by demonstration of a rising titre of IgG. Specimens need only be 3 days apart.

Isolation

Most often used for the diagnosis of congenital rubella in the fetus.

Inoculate: RK 13 (rabbit kidney) or SIRC (rabbit cornea) cell lines.

Observe: For CPE. (Cytopathological examination).

48

Viral Hepatitis

Hepatitis is a common infection with many different viruses. Namely yellow fever, cytomegalovirus and Epstein-Barr virus infection and congenital rubella. Apart from these, are specific viruses which causes 'Viral hepatitis'. These are classified as hepatitis viruses and are listed as A,B,C,D,E and G viruses.

Clinical Features

In general all of these viruses produce Jaundice along with marked anorexia, nausea and malaise.

HEPATITIS A

This belongs to the picorna virus group and is made up single standed RNA. There are small spherical particles, 27 nm in size and grows in tissue culture.

Type A hepatitis is also known as infectious hepatitis and usually affects children aged 5 to 15 years. This infection is transmitted by the faeco-oral route and it multiplies in the gut. Later these spreads to the liver and affects the hepatocytes.

Diagnosis

1. Serology: Detection of virus specific IgM by RIA or ELISA.
2. Demonstration of virus in stools by electron microscopy.

HEPATITIS B

Hepatitis B belongs to the family hepadna virus, consisting of double stranded DNA under electron microscopy, this consist of Dane particles which is the virion of hepatitis B and tubular structures which are aggregates of the virus coat protein. The Dane particle is spherical about 22 nm is size and consists of the DNA polymerase. (Fig. 48.1).

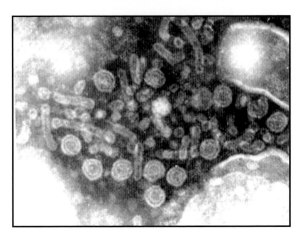

Fig. 48.1: Hepatitis B

Clinical Features

Hepatitis B is a major cause of post-transfusion hepatitis and blood is highly infectious. Drug abusers are at high risk as the infection is transmitted by sharing of unsterilised syringes. Hepatitis B virus is also transmitted sexually especially male homosexuals. Carriers of hepatitis B virus can also spread Hepatitis. Tatooing and Acupuncture has been the source of outbreaks. The virus is also transmitted transparentally *in utero* or during delivery, the infected children become chronic carriers of hepatitis B antigen.

As a result of this, chronic hepatitis occurs where the liver is affected resulting in cirrhosis and progressive liver failure. This chronic persistent hepatitis is beningn and sometimes results in hepatocellular carcinoma which is otherwise known as primary liver cell cancer.

Diagnosis

Serology

Detection of hepatitis B surface antigen (HbsAg) by RIA, ELISA test.

Anti HBc-(IgM) by ELISA-distinguishes active infection from long-term carriage in HBsAg-positive people e antigen (HBeAg) by RIA or ELISA—a marker of infectiousness. Tests of immunity detection of anti-HBs by RIA or ELISA.

Vaccine

Contains: **HBsAg** *now prepared by genetic engineering.*

Administered: Intramuscularly in three doses at intervals of one and then six months.

Babies born to HBsAg-positive mothers should be immunised within a few hours of birth; vaccine should be given simultaneously but at a different site, with hepatitis B specific immunoglobulin.

Passsive immunisation: Injection of hepatitis B specific immunoglobulin gives partial but significant protection against the disease.

HEPATITIS C

Hepatitis C comes under the family flaviviruses. This is a single stranded RNA virus and does not grow in cell culture.

Clinical Features

Post-transfusion hepatitis: *Milder form of hepatitis than hepatitis B.*

The blood donors, drug-abusers and haemophilics are at higher risk of this infection, resulting in chronic active hepatitis which may constitute hepatocellular carcinoma.

Diagnosis is by serological estimation by ELISA and RIA.

HEPATITIS D

This virus is also known as delta agent and is defective. It can replicate only in the presence of an appropriate helper virus hepatits B. Hence, hepatitis D virus is found only in patients infected with hepatitis B. Hepatitis D virus has single-stranded RNA and small particle coated with HBs Ag.

Clinical Features

The viruses affects the liver and hence increases the severity of the clinical features of hepatitis B.

Diagnosis

ELISA test for antigen and antibody are available.

HEPATITIS E

HEV is a spherical non enveloped virus, with a single stranded RNA genome. It has been classified in the genus herpes virus under the family calciviridae. This is transmitted by faeco-oral route and causes several outbreaks. The disease is generally mild. Carrier state has been observed. This is diagnosed by the detection of IgG and IgM antibodies by ELISA.

Chapter

49 *Chronic Neurological Diseases due to Viruses*

Viruses can cause chronic disease of the central nervous system resulting in 'slow virus diseases'. These disease have a long incubation period and the symptoms slowly develop. Some are chronic infections with conventional viruses but others are due to unconventional agents.

Viruses	Diseases
Measles	Subacute sclerosing panencephalitis
Rubella	Subacute sclerosing panencephalitis (follows congenital infection)
JC (human polyoma virus)	Progressive multifocal Leucoencephalopathy

Progressive Multifocal Leucoencephalopathy

This is caused by a papova virus known as JC, one of two human polyoma viruses. The other is BK virus and is not neurotropic. This is a papova virus consisting of double stranded DNA. Under electron microscopy, small icosahedral particles with capsomeres are seen. This virus grows in human fetal glial tissue cultures and haemagglutinates human and guinea-pig erythrocytes.

Clinical features: Progressive multifocal leucoencephalopathy is due to reactivation of JC viruses latent in the brain-resulting in hemiparesis, dementia dysphasia, impaired vision and haemianaesthesia. Both JC and BK viruses are oncogenic *in vitro*.

NEUROLOGICAL VIRUS DISEASES DUE TO UNCONVENTIONAL AGENTS

There are three virus diseases, all of which involve the CNS. These are clearly not typical viruses and have never been seen in the electron microscope.

Three diseases of this type are listed in Table 50.1

Table 50.1: Differentiating features of kuru, serapic and Creutzfeldt–Jackobdisease			
Disease	*Host species*	*Pathological features*	*Disease syndromes*
Creutzfeldt-Jakob disease	Man	Subacute degeneration of brain and spinal cord with status spongiosus of cortex	Presenile dementia; ataxia, spasticity involuntary movements
Kuru	Man	Subacute cerebellar degeneration; status spongiosus	Postural instability; ataxia, tremor
Scrapie	Sheep	Subacute cerebellar degeneration	Ataxia, tremor, constant rubbing; susceptibility to infection is genetically determined.

Scrapie

Scrapie agent is a prion a new class of agents which are infectious proteins. These protein are small in size and no nucleic acid has been demonstrated even by electron microscopy. Scrapie is transmitted by intracerebral or subcutaneous inoculation using the brains of infected sheep.

Clinical Features

This affects the brain at the cerebellar level and causes in co-ordination, ataxia, tremor and scratching or rubbing sensation due to sensory neurological disturbance. These symptoms progress to paralysis and death.

Creutzfeldt-Jakob Disease (CJD)

This is a rare progressive neurological disease transmitted accidentally via corneal graft and injection of pituitary growth hormone, presumably from the donors.

Clinical Features

The disease starts with tiredness, apathy and vague neurological symptoms. The patients develop ataxia, dysarthria and progressive spasticity of the limbs, associated with dementia. The disease progresses steadily until death.

Kuru

Kuru means trembling with cold and fever. Cannibalism has been responsible for the spread of Kuru. The women and children eat the viscera and brains of relatives including those who have died of Kuru. Spread may have been through contact of infected tissues with abrasions on skin or ingestion.

Clinical features: The disease starts with unsteadiness in walking, ataxia and tremor. Speech becomes slurred and tremulous and later the patient becomes sedentary. The patient becomes progressively more paralysed and emaciated until death which is due to bulbar depression.

Warts are one of the commonest virus infections. The main types of wart and the human parts with which they are associated are as follows:

Wart	Papillomavirus type
Plantar	1, 4
Hand	2
Flat; juvenile	3, 10
Condylomata acuminata	6, 11, 16
Carcinoma of cervix	16, 18
Laryngeal	6, 11,30
Warts and macules in epidermodysplasia verruciformis	5, 8 (and other types not found in other warts)
Butcher's warts	7

Clinical Features

Warts are benign tumours of the skin, most commonly on hands and feet, warts also infect the genitalia and anus where they may be large and known as condylomata acuminata. Warts are also found in the larynx and oral cavity.

Papilloma viruses infect the immunocompromised women causing warts which become malignant.

Laryngeal papilloma occurs during birth from maternal condylomata acuminata. Butcher's warts occur in slaughter men and butchers. Papilloma virus also causes malignancies like cervical cancer which is due to sexually transmitted agent and squamous cell carcinomas of skin.

Diagnosis

The virus DNA can be demonstrated in tissue by radioactive DNA probe.

Retroviruses

Human retroviruses are grouped under oncornavirus which consists of RNA as the genome structure. The viruses are as follows:

Virus	Disease
Human T-cell lymphotropic virus	T-cell leukaemia/lymphoma: tropical spastic
Type 1 (HTLV-1)	paraparesis
Type 2 (THLV-2)	Unknown
Human immunodeficiency virus	
Type 1 (HIV-1)	Acquired immunodeficiency syndrome (AIDS)
Type 2 (HIV-2)	AIDS

HTLV-1 is an infection, which causes adult T-cell leukaemia and lymphoma malignancy of T4-lymphocytes, with lymphadenopathy and hepatospleno-megaly

Clinical Features

The disease often presents with lymphomatous infiltration of the skin along with tropical spastic paraparesis—a non-demyelinating spastic paralysis.

Diagnosis

Electron microscopy reveals C type particles which contain reverse transcriptase. Culture in human T-lymphocytes shows cell transformation without cytopathic effect.

Serologically RIA, ELISA and immunofluorescence tests are used in the diagnosis.

HTLV-2: Similar to HTLV-1. Its pathogenesis is unknown but it has been isolated from patients with hairycell leukaemia.

HIV

There are two types of virus—HIV-1, the main cause of AIDS pandemic and HIV-2 causing a significant spread.

Clinical Features

It starts with fever, sore throat night sweats, malaise, hymphadenopathy, diarrhoea and mouth and genital ulcers. Then there occurs a depression in T4 lymphocyte count. Persistent generalised lymphoadenopathy with reappearance of symptoms especially enlarged lymph nodes. Opportunistic infections may begin to appear-most often with diarrhoea and oral candidiasis. Full blown AIDS is characterised by immunodeficiency.

AIDS is a syndrome with following features.

1. Constitutional disease: Fever, diarrhoea, weight loss.
2. Neurological disease: Dementia, sometimes myelopathy or peripheral neuropathy.
3. Secondary infectious diseases: Opportunistic infections as in Table 51.1.
4. Secondary cancers.
5. Other conditions: Aids has an extraordinarily wide variety of associated signs and symptoms. Two diseases are particularly characteristic.
   a. Kaposi's sarcoma.
   b. Hairy leukoplakia of the tongue.

Table 51.1: Infections associated with AIDS	
Infectious agents	*Diseases*
Parasites	Pneumonia
Pneumocystis carinii	Diarrhoea
Cryptosporidium	Diarrhoea
Isospora belli	Diarrhoea
Strongyloides stercoralis	Encephalitis
Toxoplasma gondii	Chorioretinitis
Viruses	
Herpes simples	Oral ulceration
Cytomegalovirus	Pneumonia
	Retinitis
JC virus	Progressive multifocal
	Leucoencephalopathy
Bacteria	
Mycobacterium avium-intracellulare	Respiratory disease
	Disseminated infection
Mycobacterium tuberculosis	Tuberculosis
Salmonella spp	Diarrhoea
Fungi	
Candida albicans	Oral ulceration, oesophagitis
Cryptococcus neoformans	Pneumonia, meningitis

Diagnosis

Serology: ELISA test if positive, confirm with western blot to analyse antibodies against virus structural proteins. HIV-1 and HIV-2 infections can be distinguished by serological test.

Demonstration of virus products
1. Antigen: Detected by ELISA in blood, other body fluids
2. Reverse transcriptase: Assay of reverse transcriptase. Assay can be used as a measure of virion-associated enzyme activity and virus titre.

Treatment

Antiviral drugs and appropriate therapy for the opportunistic infection of AIDS.

Chapter

52

Rabies, Non-arthropod Borne Haemorrhagic Fevers, Arenavirus Infections

RABIES

Rabies is a lethal form of encephalitis due to a virus which affects a wide variety of animal species; rabies is transmitted to man via the bite of an infected animal which is usually-but not always a dog.

Clinical Features

Virus spread from the wound to the CNS is via the nerves.

Symptoms; initially: Paraesthesia in the wound is an early symptom. Thereafter there are two forms of rabies:

i. *Furious:* The more common of the two, in which the symptoms are excitement, with tremor, muscular contractions and convulsions; typically spasm of the muscle of swallowing (hence the older name for the disease of 'hydrophobia' or fear of water) and increased sensitivity of the sensory nervous system.

ii. *Dumb:* Dumle or paralytic rabies in which the symptoms are of ascending paralysis, eventually involving the muscles of swallowing, speech and respiration.

Rabies is a natural infection of dogs, cats, bats and carnivorous wild animals such as foxes, wolves, skunks; infection is also found in rodents and cattle. Human exposure to rabies is generally most common from dogs. In under developed countries, infection in urban dogs poses a major problem and a risk to man. Virus is present in the saliva of infected dogs—sometimes for up to four days before the onset of symptoms of the disease; dogs and cats which remain healthy for ten days after biting can be regarded as being free of virus at the time of biting.

Aerosol infection: has been recorded-as a result of laboratory accident.

Corneal transplant: Cases of rabies in recipients of corneas from donors of undiagnosed rabies have been reported.

VIROLOGY

A member of the *Lyssavirus* genus within the rhabdovirus family: rabies is the most neurotropic of the antigenically related group of lyssaviruses which naturally infect many different animal species.

RNA virus-negative sense, single-stranded RNA.

Bullet-shaped, enveloped particles containing helically-coiled nucleoprotein; length 180 nm, diameter 70 to 80 nm.

Haemagglutinates goose erythrocytes.

Grows in hamster and chick embryo cell tissue cultures with eosinophilic inclusions but usually without CPE.

Pathogenic for mice and other laboratory animals.

Diagnosis

Direct Demonstration of Virus

Specimens: Hair-bearing skin (e.g. back of neck), corneal impression smears, brain tissue.
Examine: For presence of rabies virus antigen by immunofluorescence.
  Negri bodies: A less sensitive method of diagnosis: examine brain smears of the Ammon's horn of the hippocampus stained with seller's stain for red intracytoplasmic intrusions (Negri bodies).

Isolation

Specimens like brain tissue, saliva, CSF, urine.
Inoculated are mice intra-cerebrally.
Observation paralysis, convulsions; postmortem for immunofluorescence with rabies antiserum and Negri bodies in brain cells.

Vaccination

Human Diploid Cell Vaccine (HDCV)

Now the vaccine of choice but, unfortunately, its cost limits its use to developed countries, which are not the countries that have the major problem of human rabies. Which contains: inactivated virus and prepared in WI 38 or MRC-5 diploid human embryo lung cells.

Administered intramuscularly or subcutaneously into the deltoid area of the upper arm in 6 doses spaced at 0, 3, 7, 14, 30 and 90 days.

Simple Vaccine

The successor to the original Pasteur vaccine and still in use in many developing countries despite the risk of severe side effects. This contains virus inactivated by phenol and prepared from infected brain tissue from rabbits, sheep or goats. This is administered subcutaneously in 21 daily injections with later booster doses.

Pre-exposure Vaccination

Veterinary surgeons, animal handlers, laboratory workers or others at high risk from rabies should be given two doses of diploid cell vaccine one month apart with a booster dose one year later; two booster doses should be given if they are exposed to infection.

Rabies antibody levels should be checked 3 to 4 weeks after the primary course.

Chapter

53

Antiviral Therapy

A PRINCIPAL ANTIVIRAL AGENTS IN CURRENT USE

Amantadine and Rimantadine

They are available only as oral preparations. These drugs are used for the treatment of influenza. They are also used as prophylactic drugs during influenza outbreaks.

Oseltamivir and Zanamivir

They are used to reduce influenza symptoms and shortens the course of illness by 1 to 15 days.

Idoxuridine and Trifluorothymidine

Idoxuridine can be used topically as effective treatment of herpetic infection of the cornea (Keratitis). Trifluorothymidine is effective in treating herpetic corned infections.

Acyclovir

It is effective in the treatment of primary (HSV) herpes simplex viral mucocutaneous infections or for severe recurrences in immunocompromised patients. This is also useful in neonatal infectious herpes encephalitis and is also recommended for varicella. Zoster viral infections in immunocompromised patients and varicella in older children or adults. In patents with severe genital herpes, the oral form is effective in preventing recurrences. Acyclovir is minimally effective in the treatment of recurrent genital or labial herpes in otherwise healthy individuals.

Valacyclovir, Famciclovir and Penciclovir

Valacyclovir is used in herpes simplex and varicella zoster infections in immunocompetent adult patients. Famciclovir is currently approved for treatment of herpes simplex virus and varicella zoster virus infections. Penciclovir is used for topical treatment of recurrent herpes labialis.

Mechanism of Action	Antiviral Agent	Viral Spectrum
Inhibition of viral uncoating penetration	Amantadine	Flu A
	Rimantadine	Flu A
Neuraminidase inhibition	Oseltamivir	Flu A, Flu B
	Zanamivir	Flu A, Flu B
Inhibition of viral DNA polymerase	Acyclovir	HSV, VZV
	Famciclovir	HSV, VZV
	Penciclovir	HSV
	Valacyclovir	HSV, VZV
	Ganciclovir	CMV, HSV, VZV
	Foscarnet	CMV resistant HSV
	Cidofovir	CMV
	Trifluridine	HSAV, VZV
Inhibition of viral reverse transcriptase	Zidovudine	HIV
	Dideoxyinozine	HIV
	Dideoxycytidine	HIV
	Stavudine	HIV
	Lamivudine	HIV, HBV
	Nevirapine	HIV
	Delavirdine	HIV
	Efavirenz	HIV
Inhibition of viral protease	Saquinavir	HIV
	Indinavir	HIV
	Ritonavir	HIV
	Nelfinavir	HIV
	Lopinavir	HIV
Inhibition of viral protein synthesis	Interferon-a	HBV, HCV, HPV
Antisense inhibition of viral mRNA synthesis	Fomiviresen	CMV
Inhibition of viral RNA Polymerase	Ribavirin	RSV, HCG, Lassa fever

Ganciclovir

Ganciclovir is available in oral as well as in injectable form. Administration of ganciclovir is indicated for the treatment of active cytomegalovirus infection in immunocompromised patients, other herpes viruses particularly HSV-1, HSV-2 and VZV are also susceptible. AIDS patients with concurrent illness of herpes viruses may be benefited with this drug.

Ribavirin

This drug is beneficial of given early by aerosol to infants who are infected with respiratory syncytial virus. Oral and intravenous forms have been used for patients with Lassa fever. The oral form has limited activity against hepatitis C as monotherapy.

Azidothymidine (AZT)

AZT was the first useful treatment for HIV infection but now is recommended for used only in combination with lamivudine.

Stavudine and Lamivudine

Stavudine inhibits HIV replication and lamivudine is a safe and well-tolerated agent used in combination with AZT. When combined with interferon-α, lamivudine is also useful for treating hepatitis B.

Other Antiviral Agents

Interferons

Interferon-α is given for 6 to 12 months to treat chronic hepatitis C disease and combination with ribavirin usually produces improved results. Topical interferon application is beneficial in the treatment of human papilloma virus infections.

Fomivirsen

This drug is approved for the local (intravitreal) therapy of cytomegalovirus retinitis in patients who have failed other therapies.

SECTION

5

PARASITOLOGY

Chapter
54

Intestinal Protozoa

Protozoa are unicellular parasites and they exist in both animate and inanimate environment. Some are even commensals of man. Only a small number of protozoa are pathogenic for man.

The intestinal protozoa are numerous and the medically important ones are discussed here.

ENTAMOEBA HISTOLYTICA

This exists both in trophozoite form and cyst form. This protozoa is spread to man by ingestion of food or water contaminated by cysts. It causes amoebic dysentery and can invade beyond the large intestine to cause liver abscesses.

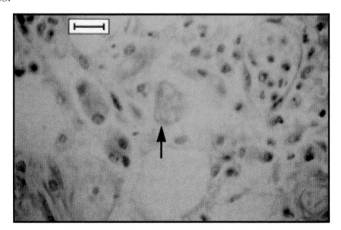

Fig. 54.1: *Entamoeba histolytica* trophozoites

Trophozoite and cyst forms: It measures 10 to 60 μm (average 20 to 30 μm) in diameter. The cytoplasm of the trophozoite can be divided into a clear outer ectoplasm and an inner finely granular endoplasm in which red blood cells, leucocytes and tissue debris are found within the food vacuoles. Trophozoites are motile (Fig. 54.1).

Cyst: It is spherical, 10 to 15 mm in diameter. It is surrounded by a thick chitinous wall which makes it highly resistant to the gastric acid. It starts as a uninucleate body but later the nucleus divides to form two and then four nuclei. Cysts are present only in the lumen of the colon and in formed faeces. Stools may contain cysts with 1 to 4 nuclei depending on their degree of maturation.

Clinical Features

Intestinal amoebiasis: After an incubation period of 1 to 4 weeks, the amoebae invade the colonic mucosa, producing characteristic ulcerative lesions and a profuse bloody diarrhoea (amoebic dysentery). Ulcers are discrete and vary in size from pin-head size to more than 2.5 cm in diameter. They may be deep or superficial.

E. histolytica may also cause amoebic appendicitis and amoebomas

Extraintestinal amoebiasis: Trophozoites of E. histolytica are carried as emboli by the radicles of the portal vein from the base of the amoebic ulcer in the large intestine. They multiply in the liver. Amoebic liver abscess varies greatly in size. It may occur in any part of the liver but it is generally confined to postero-superior surface of the right lobe.

Pus of liver abscess: The centre of an amoebic liver abscess contains a viscous red-brown (anchovy sauce appearance) or grey-yellow fluid consisting of cytolysed liver cells, red blood cells and leucocytes.

LABORATORY DIAGNOSIS (FIGS 54.2 TO 54.6)

Intestinal Amoebiasis

Stool Examination

In acute amoebiasis, stool or colonic scrapings from ulcerated areas are examined by macroscopic and microscopic examination. For microscopic examination, stool is picked up with a matchstick or a platinum loop and emulsified in a drop of normal saline on a clean glass slide. A clean coverslip is placed over it and examined under microscope, first under low power and then under high power. This method is specially useful for the demonstration of the actively motile trophozoites of E. histolytica. For the demonstration of cysts or dead trophozoites, stained preparation may be required for the study of the nuclear character. For this purpose iodine stained preparation is commonly employed. Stool is emulsified in a drop of five times diluted solution

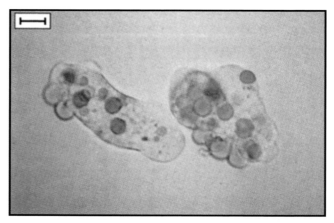

Fig. 54.2: *Entamoeba histolytica* found at biopsy

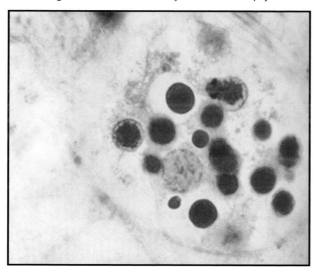

Fig. 54.3: *Entamoeba histolytica* trophozoite containing ingested red blood cells

of Lugol's iodine, covered with a clean coverslip and examined under microscope. Stool culture is useful especially in the cases of chronic and asymptomatic intestinal infections, excreting less number of cysts in the faeces. Biopsy specimens should also be submitted for histopatholgic studies. DNA Probe has been used recently to identify *E. histolytica* in the stool specimen and specific sequences can be amplified by polymerase chain reaction (PCR).

Blood examination: It shows moderate leucocytosis.

Serological tests: Antibodies appear and serological test become positive. These tests include indirect haemagglutination (IHA) indirect fluorescent antibody (IFA) test and enzyme-linked immunosorbent assay (ELISA).

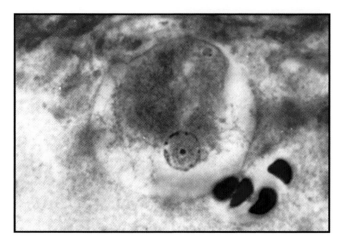

Fig. 54.4: *Entamoeba histolytica/Entamoeba dispar* trophozoite;
no ingested red blood cells are present

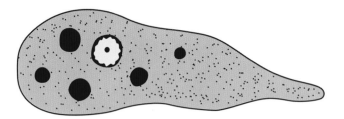

Fig. 54.5: Trophozoite of *Entamoeba histolytica*

Fig. 54.6: *Entamoeba dispar*

Amoebic Liver Abscess

1. *Diagnostic aspiration:* Trophozoites of E. histolytica may be demonstrated by microscopy of the pus aspirated by puncture of amoebic liver abscess.
2. *Liver biopsy:* Trophozoites of E. histolytica can be demonstrated in the specimens of liver biopsy.
3. *Stool examination:* Cysts of E. histolytica can be demonstrated in the stool.
4. *Serological:* Serological tests like IHA, IFA coagglutination test and ELISA are performed.
5. *Molecular methods:* DNA probes and PCR are the recent molecular methods for the detection of E. histolytica in stool and liver aspirates.

Treatment: Treatment of amoebiasis is by using antiamoebic drugs.

Prevention: They are as follow:

Amoebicides with luminal action

- Di-iodohydroxyquin
- Diloxanide furoate
- Paromomycin.

Amoebicides effective in the liver, intestinal wall and other tissues

- Emetine
- Dehydroemetine.

Amoebicides effective only in the liver

- Chloroquine.

Amoebicides effective in both tissues and the intestinal lumen

- Metronidazole
- Nitroimidazole.

Entamoeba coli: This parasite lives freely in the lumen of large intestine and is non pathogenic. It resembles E. histolytica in morphology and its life cycle, the cyst for alone is small and spherical, and the number of nuclei is 1 to 4, whereas in E. histolytica the cysts are large and has 1 to 8 nuclei.

Giardia Intestinalis (lamblia): This is a protozoan parasite with flagellae and is pear shaped in morphology it exists in two forms (i) trophozoite and (ii) cyst.

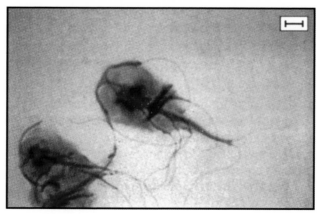

Fig. 54.7: *Giardia intestinalis* (lamblia) trophozoites

Trophozoite

It is pear-shaped with rounded anterior and pointed posterior end. The dorsal surface is convex while on the ventral surface it has a shallow and bears sucking disc. It is bilaterally symmetrical and has one pair of nuclei, one on each side of the midline, one pair of axostyles, one pair of parabasal bodies present on the axostyles, four pairs of flagella. The nuclei are rounded and possess a central karyosome. By rapid movement of the flagella, the trophozoites move from place to place and by applying their sucking discs to epithelial surfaces they become firmly attached (Figs 54.7 to 54.9).

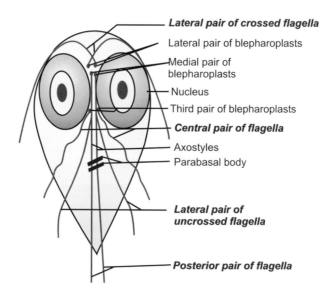

Fig. 54.8: Trophozoite of Girdia lambia

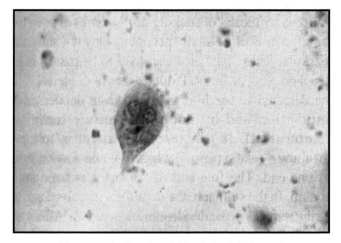

Fig. 54.9: Trophozoite of *Giardia lamblia* in stool

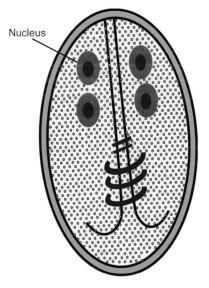

Fig. 54.10: Cyst of Giardia lamblia

Cyst: Mature cyst is oval in shape and measures 11 to 14 μm × 7 to 10 μm in size. It has two pairs of nuclei which may remain clustered at one end or lie in pairs at opposite ends. The remains of the flagella and margins of the sucking disc may be seen inside the cytoplasm of the cyst (Fig. 54.10).

Laboratory Diagnosis

Giardiasis can be diagnosed by identification of cysts of *G. lamblia* in the formed stools and the trophozoites of the parasite in diarrhoeal stools by normal saline

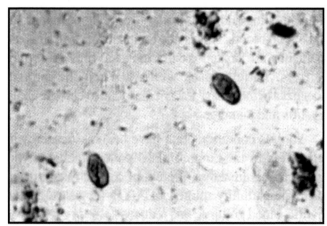

Fig. 54.11: Cysts of *Giardia lamblia* in stool

and iodine preparation. For the detection of *G. lamblia* in faecal specimens, a fluorescent method using monoclonal antibodies is extremely sensitive and specific. ELISA test has been developed for the detection of *Giardia* antigen in faeces. Anti-Giardia antibodies, in the patient serum, may be detected by ELISA and indirect fluorescent antibody tests (Fig. 54.11).

Treatment

Treatment of giardiasis is carried out with metronidazole, tinidazole and furazolidone. Tinidazole has proven more effective than metronidazole as a single dose. Furazolidone is often used for treating children.

Trichomonas Vaginalis

These are common flagellates which exists only in the trophozoite phase. The flagellates are pear-shaped bodies and measure 10 to 12 μm in length. The single ovoid nucleus is situated at the anterior end and a cleft like depression lies at its side known as cytostome (Fig. 54.12).

In a wet mount the trophozoite has a characteristic jerky motility. The normal habitat of the parasite is human vagina, prostate, and urinary tract of both males and females. Human trichomoniasis is a widely prevalent sexually transmitted disease. The organism is responsible for a mild vaginitis with discharge. Vaginal discharge contains a large number of parasites and leucocytes and is liquid, greenish or yellow. Male patients usually have mild or asymptomatic infections. They may develop itching and discomfort inside penile urethra especially during urination. The parasite is transmitted by sexual intercourse.

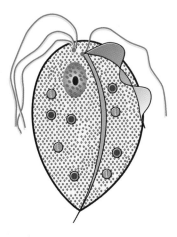

Fig. 54.12: *Trichomonas vaginalis*

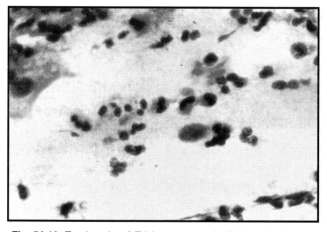

Fig. 54.13: Trophozoits of *Trichomonas vaginalis* in vaginal smear

The diagnosis can be made by demonstration of trophozoites of *T. vaginalis* in wet mounts of the sedimented urine, vaginal secretions or vaginal scrapings by bright-field, dark-field, or phase-contrast microscopy (Fig. 54.13). In males it may be found in urine or prostatic secretions.

Fixed smears may be stained with Papanicolaou, Giemsa, Leishmann and periodic acid-Schiff's stain and seen under light microscope. The parasites may also be detected by fluorescent microscopy by staining with fluorescein labeled monoclonal antibody. Trussell and Johnson's medium is a simple medium that gives good growth. It consists of proteose peptone, sodium chloride, sodium thioglycollate and normal human serum. Simplified trypticase serum medium is also suitable for the isolation of *T. vaginalis*. Polymerase chain reaction (PCR) for the diagnosis of trichomoniasis has also been developed.

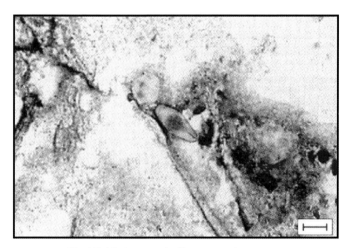

Fig. 54.14: *Isospora belli* oocyst and sporocyst

Treatment

Metronidazole is highly effective therapeutic agent. It is given orally 250 mg three times daily for seven days or 2 grams orally as a single dose.

Isospora Belli

This is usually an asymptomatic infection but may cause severe diarrhoeal disease in immunocompromised patients, particularly those with AIDS. It is spread faeco-orally the infective form being the oocyst, which contains two sporocysts. The oocyst is immature when excreted and matures in the stool to become infective (Fig. 54.14).

Diagnosis is by modified Ziehl-Neelsen or safranin methylene blue stains of faecal smears. Co-trimoxazole is the treatment of choice.

CRYPTOSPORIDIUM SPECIES

This is a coccidian parasite and is a major cause of diarrhoeal disease in children and life-threatening pathogen in the immunocompromised host. It is spread faeco-orally. *Cryptosporidium hominis* is the major human pathogen, but *C. parvum*, which has microscopically indistinguishable ooccysts, is the major zoonotic pathogen. Other zoonotic cryptosporidia include *C. canis, C. felis, C. meleagridis* and *C. muris*. The oocyst is the infective form which is fully infective. The oocyst is small (4-5 µm). The oocyst contains four sporozoites that attach to and penetrate the enterocytes. The develop to form trophozoites which are described as being intra-enterocytic. Diagnosis is by the examination of fecal smears by modified Ziehl-Neelsen, safrannin methylene blue or auramine

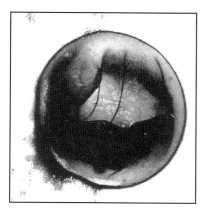

Fig. 54.15: *Cryptosporidium parvum* oocyst electron micrograph

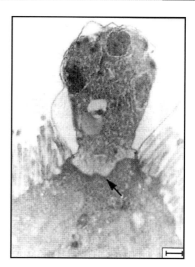

Fig. 54.16: *Cryptosporidium parvum* trophozoite found at biopsy

phenol stains. Immunofluorescent antigen test and enzyme-linked immunosorbent assay tests are also available for the detection of antigen or of serological response. The newly developed agent nitazoxanide appears to be effective therapeutically (Figs 54.15 and 54.16).

BALANTIDIUM COLI

Balantidium coli: Balantidium coli belongs to the phylum *Ciliophora*. It *inhabits* the large intestine of man, monkeys and pigs.

Morphology: B. *coli* is the largest protozoal parasite inhabiting the large intestine of man. It has a trophozoite and a cyst stage. The trophozoite is found in dysenteric stool. It is actively motile and is the invasive stage. On the other hand the cyst is found in chronic cases and carriers.

Trophozoite: It is an oval organism measuring 60 × 45 μm or more. The anterior end is somewhat pointed and has a groove (peristome) leading to a mouth (cytostome) terminating in a short funnel-shaped gullet (cytopharynx) extending up to anterior one-third of the body. The posterior end is broadly rounded and has an excretory opening known as cytopyge.

The body is covered with a delicate pellicle showing longitudinal striations. The cilia that line the mouth part are longer and are called adoral cilia. These are used for propelling food into the cytopharynx (Fig. 54.17).

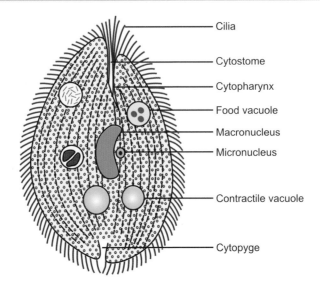

Fig. 54.17 : Morphological forms of *Balantidium coli* Trophozoite

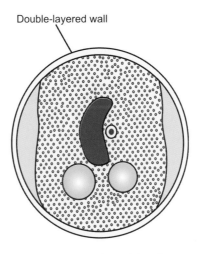

Fig. 54.18 : *Cyst of B. coli*

The cytoplasm of the trophozoite has:
1. *Two nuclei:* A macronucleus which is large and situated near the middle of the body. It may be kidney shaped, spherical, curved or elongate. The micronucleus is small round and lies in close proximity to the macronucleus.
2. *Two contractile vacuoles:* Which may lie side by side or one above the other.

3. *Numerous food vacuoles:* The food particles on being ingested become surrounded by a vacuolar membrane and digestion takes place inside the vacuoles.

Cyst: The cyst of *B. coli* is spherical or oval measuring 40 to 60 μm in diameter. It is surrounded by a thick and transparent double-layered wall. The macronucleus micronucleus and vacuoles are present in the cyst also (Fig. 54.18).

Laboratory Diagnosis

1. *Stool examination:* Diagnosis is based on faecal examination, which reveals mainly trophozoites in acutely infected patients and cysts in chronic cases and carries. It is generally easy to recognize *B. coli* in stool specimens because of its large size (60 × 45 μm or more) an outer membrane covered with short cilia, and its large kidney-shaped macronucleus. When observed in wet mounts. In iodine mounts, the trophozoite stains yellow-brown, and the macronucleus is easily visible. The cyst (40 to 60) μm in diameter) can also be recognised by the large macronucleus.
2. *Biopsy:* Diagnosis can also be made by the examination of biopsy specimens taken with the help of a sigmoidoscope or by examination of scraping of an ulcer.
3. *Culture:* Culture is rarely attempted for diagnosis as the parasite are more easily detected in faces by microscopy and in tissues on histological examination.

Treatment

Tetracycline 500 mg four times a day for 10 days or metronidazole 750 mg three times a day may be used for the treatment of *B. coli* infection.

Chapter

55

Blood and Tissue Pathogens

NAEGLERIA FOWLERI

This is an amoebo flagellate that is in the amoeboid form in tissue. It is a rare cause of meningitis.

Morphology

N. fowleri has two stages: motile trophozoites and nonmotile cysts. The trophozoites occur in two forms: amoeboid and flagellate. Amoeboid form is elongate broad anteriorly and tapered posteriorly. It is actively motile by means of eruptive, blunt pseudopodia called lobopodia. It measures 15-30 μm (average 22 μm) in length. It has distinctive phagocytic structures known as amoebostomes. The nucleus is small 3 mm in diameter and has a large central karyosome and no peripheral nuclear chromatin. Reproduction in *Naegleria* is by simple binary fission of the amoeboid form (Fig. 55.1).

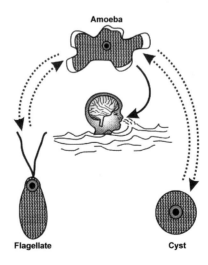

Amoeba

Flagellate

Cyst

Fig. 55.1: Life cycle and pathogenicity of *Naegleria fowleri*

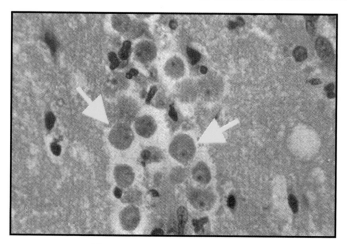

Fig. 55.2: *Naegleria fowleri* in brain

The cysts of *N. fowleri* are uninucleate, spherical, 7 to 15 μm in diameter and are surrounded by a relatively thin cyst wall. Amoebae encyst when conditions are appropriate and later excyst in a favourable environment. Cysts and flagellate forms of *N. fowleri* have never been found in tissues or CSF (Fig. 55.2).

Laboratory Diagnosis

The diagnosis can be made by microscopic identification of living or stained amoebae in CSF. Motile amoebae with characteristic morphology can be readily demonstrated in simple wet-mount preparation of fresh CSF specimen. CSF smear may be stained with wright or Giemsa's stains. With these stains amoebae have considerable amount of sky blue cytoplasm and relatively small, delicate pink nuclei.

In centrifuged deposit of CSF the amoebae tend to be rounded and flattened without pseudopodia. Amoebae can also be demonstrated by fluorescent antibody staining of the CSF and in the histologic sections of the brain biopsied tissue by immunofluorescence and immunoperoxidase methods.

N. fowleri may be cultivated by placing some of the CSF on non-nutrient agar (1.5%) spread with a lawn of washed *Escherichia coli* or *Enterobacter aerogenes* and incubated at 37°C. The amoebae will grow on the moist agar surface and will use the bacteria as food producing plaques as they clear the bacteria.

Treatment

Antibacterial antibiotics and antiamoebic drugs are ineffective. Amphotericin B a drug of considerable toxicity is the antinaeglerial agent for which there is evidence of clinical effectiveness.

Trypanosoma species: Two distinct forms of disease resulting from trypanosomes occur in man. African sleeping sickness is caused by *Trypanosoma brucei.* It is transmitted by the bit of tsetse fly.

TRYPANOSOMA BRUCEI

Morphology

In the blood of the vertebrate host, *T. brucei* exists in trypomastigote form. It occurs in three forms a long slender form having a flagellum, a short stumpy one without a flagellum, and an intermediate one. Because of these morphologic differences in the blood of the vertebrate host this trypanosome is said to be polymorphic.

In fresh blood the trypanosomes may be seen as motile colourless, spindle-shaped bodies with a blunted posterior end and a finely pointed anterior end (Fig. 55.3). The long slender and short stumpy forms measure 20 × 3 mm and 10 × 5 mm respectively. The nucleus is large oval and central in position. Kinetoplast (Parabasal body and Blepharoplast) is situated on the posterior end. From the blepharoplast arises the flagellum. It curves round the body in the form of an undulating membrane and then continues beyond the anterior end as a free flagellum. The intracellular portion of the flagellum is known as axoneme (Fig. 55.4).

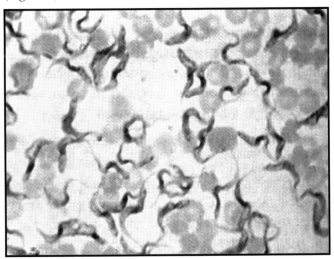

Fig. 55.3: Trypanosoma brucei-gambiense in peripheral blood film

Life cycle: T. brucei- gambiense passes its life cycle in two hosts. The vertebrate hosts are man, game and domestic animals and invertebrate host is the tsetse fly of the genus Glossina (*G. palpalis, G.fuscipes and G.tachinoides*). Both male and female flied bite man and may serve as vectors.

Fig. 55.4: *Trypanosoma brucei*

Pathogenicity: African trypanosomiasis is chronic in nature lasting up to 4 years. Man develops infection by the bite of infected tsetse fly. Both male and female tsetse flies suck blood and transmit infection. They bite during day usually in the early morning and evening metacyclic forms are injected into the subcutaneous tissue of man at the time of bite. A chancre develops at the site of bite. It is hard and painful nodule and fluid withdrawn from it contains actively dividing trypomastigotes.

Then the trypomastigotes spread throughout the entire body. They move through the blood and lymphatic vessels and multiply rapidly. Patients first experience intermittent recurring fever associated with lymphadenopathy. Hepatosplenomegaly may also be evident during the early stage of infection. If untreated central nervous system is involved. At this stage, patient develops severe headache and a wide array of behavioural changes ranging from aggressiveness to sleep-like states. Sleepiness becomes so pronounced that the patient falls asleep while eating, standing or sitting (sleeping sickness) and dies.

Laboratory Diagnosis

The diagnosis of African trypanosomiasis depends upon demonstration of the parasite (trypomastigote forms) in blood, lymph node aspirates, sternal bone marrow, cerebrospinal fluid or fluid aspirated from the trypanosomal chancre by direct microscopic examination of unstained and stained films, cultivation and animal inoculation. Trypomastigotes are present in largest numbers in the blood during febrile periods. Serologic techniques include immunofluorescence, complement fixation, enzyme-linked immunosorbent assay and card agglutination test. These tests detect antibody in the sera of infected individuals and utilise antigens from blood stage trypanosomes.

Treatment

Suramin is used to treat patients with primary stage infections that do not involves the CNS pentamidine isethionate, like suramin, does not cross the blood-brain barrier, therefore, it can also be used in initial stages. Melarsoprol can cross the blood-brain barrier and is used to treat patients in the last secondary CNS stage of trypanosomiasis.

TRYPANOSOMA CRUZI

It lives as trypomastigote in the blood and as an amastigote in reticulo-endothelial cells and other tissue cells of man and many mammals. In man the most frequent locations of the parasites are reticuloendothelial cells of spleen, liver, lymph nodes, bone marrow, and myocardium. Parasites may also occur in cells of striated muscles, nervous system, histiocytes of cutaneous tissue, cells of the epidermis, and in the intestinal mucous membrane (Figs 55.5 and 55.6).

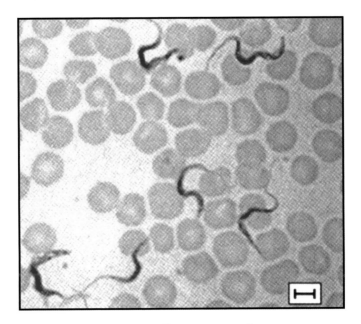

Fig. 55.5: *Trypanosoma cruzi*

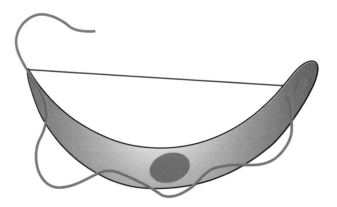

Fig. 55.6: *Schematic diagram of Trypanosoma cruzi*

Morphology

Two main morphological form are seen in human hosts:
- Trypomastigote form
- Amastigote forms.

Trypomastigote forms: These are present in the blood of the patient during the early acute stage and at intervals. It measures 20 μm in length, has a central nucleus and a large kinetoplast situated at the posterior end. Two forms occur in the blood, a long slender one and a short broad one.

Amastigote forms: These are round or oval in shape measure 1.5 to 4.0 μm in diameter, have a large nucleus and a kinetoplast. In myocardium and neuroglial cells, the amastigote forms are collected within a cyst like cavity in the invaded cells.

Development in man: Humans become infected when the trypanosomal infected faecal matter is discharged into the wound when the bug feeds. The bite is quite painful and the infected faeces are rubbed into the wound. The infection can also be transmitted by contamination of abraded skin, the conjunctiva and oral and nasal mucosa. The metacyclic trypomastigotes thus introduced, invade cells of reticuloendothelial system and other tissues particularly muscle and nervous tissue and are transformed into amastigote form. These multiply by binary fission and after passing through promastigote and epimastigote forms are again transformed into trypomastigote forms which are liberated in the blood. This leads to dissemination of infection and infects fresh triatomine bugs when they next feed. Infection may also be acquired by blood transfusion where blood is not screened for presence of *T. cruzi* antibodies and congenitally via placenta.

Pathogenicity

Acute infections are more common and more severe in children. At this state patient develops fever hepatosplenomegaly, generalized lymphadenopathy, facial or generalised oedema, rash, vomiting diarrhoea, anorexia and ECG changes like sinus tachycardia, increased P-R interval, t-wave changes and low QRS voltage, and patient may be die of acute mycarditis. It may also lead to meningoencephalitis mainly in infants and AIDS patients.

Patients surviving the acute infection, develop chronic disease in which cardiac changes are most common with arrhythmias, palpitations, chest-pain oedema, dissiness, syncope and dyspnoea. Patient may also develop dilatation of oesophagus and colon loss of regurgitation, dysphagia and severe constipation.

Congenital transmission can occur in both the acute and chronic stage of the disease.

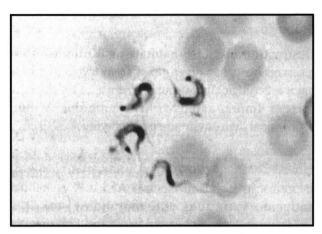

Fig. 55.7: *Trypanosoma cruzi* in peripheral blood film

Laboratory Diagnosis

Microscopic examination: In acute infection *T. cruzi* may be found transiently in the peripheral blood by direct microscopy of Giemsa's stained and unstained, wet blood film (Fig. 55.7). If the parasites are scanty and cannot be detected in thin stained and unstained film, thick film or concentration methods may be employed. The latter include: (i) haematocrit centrifugation (ii) allowing the blood to coagulate and searching for trypomastigotes in centrifuged serum.

Culture

Blood and other specimens are inoculated in NNN medium and incubated at 22-24°C and subcultured every 1-2 weeks.

Polymerase Chain Reaction

PCR has been used to detect positive patients.

Serodiagnosis

Antibodies against *T. cruzi* may be detected in patient serum by complement fixation test, indirect fluorescent antibody test, enzyme-linked immunosorbent assay and indirect haemagglutination test.

Treatment

Nifurtimox and benzonidazole may be used for the treatment of American trypanosomiasis.

LEISHMANIA SPP

Kala Azar

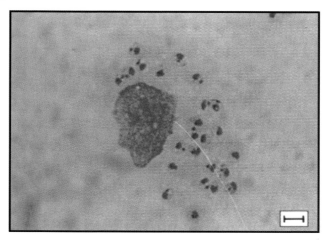

Fig. 55.8: Macrophage containing amastigotes

Morphology

It is an obligate intracellular parasite of reticuloendothelial cells, predominantly of liver, spleen, bone marrow and lymph nodes of man and other vertebrate hosts (dog and hamster) where it occurs in amastigote form (Fig. 55.8).

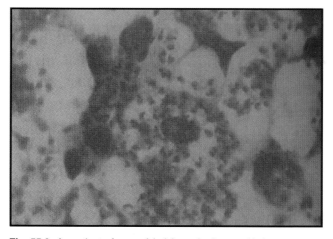

Fig. 55.9: Amastigote forms of *Leishmania donovani* in bone marrow

Amastigote: In the amastigote form the parasite resides in the cells of reticuloendothelial system. It is non-motile, round or oval body measuring 2-4 μm in length along the longitudinal axis (Fig. 55.9). Nucleus is round or oval less than 1 μm in diameter. It is situated in the middle of the cell or along

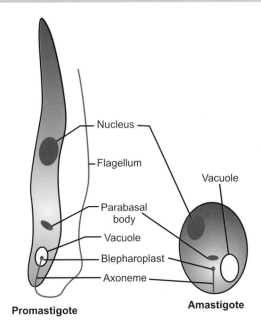

Fig. 55.10: Morphological forms of *Leishmania donovani*

the side of the cell wall kinetoplast consists of parabasal body and blepharoplast which are connected by one or more delicate fibrils. It lies tangentially or at right angle to the nucleus. The axoneme arises from the blepharoplast and extends to the margin of the body. In preparations stained with Giemsa or Wright stain, the cytoplasm appears pale blue the nucleus red, parabasal body deep red and kinetoplast bright red (Fig. 55.10).

Promastigote: Promastigotes are found in the digestive tract of insect vector and in the culture media. These are elongated motile, extracellular stage of the parasite. Fully developed promastigotes measure 15 to 25 μm in length and 1.5 to 3.5 μm in breadth. Nucleus is situated centrally. Kinetoplast lies transversely near the anterior end. From the blepharoplast arises the axoneme which projects from the anterior end of the parasite as free flagellum.

Pathogenicity

L. donovani causes visceral leishmaniasis or kala azar. The parasite spreads from the site of inoculation to multiply in reticuloendothelial cells, especially in the spleen, liver, lymph nodes, and bone marrow. This leads to progressive enlargement of these organs. The spleen and liver become markedly enlarged and hypersplenism contributes to the production of anaemia. Lymphadeno-pathy is also produced. The disease manifests clinically with fever, malaise, headache, progressive enlargement of spleen, liver and lymph nodes. Death in kala azar is due to secondary infections.

Laboratory Diagnosis

Various tests which can be carried out for the laboratory diagnosis of kala azar.

Non-specific Laboratory Tests

These include
1. *Blood count:* Total and differential leucocyte count reveals pancytopenia, mainly neutropenia and decreased erythrocyte count. The average total count of leucocytes is 3,000/µl of blood. During the course of the disease, the count may fall to 1,000/µl of blood or even below. Erythrocytes are also decreased in number.
2. *Haemoglobin estimation:* It reveals anaemia
3. *Estimation of serum proteins:* It reveals raised serum proteins with reversal of the albumin: globulin ration due to greatly raised IgG levels.

Parasitological Diagnosis

Diagnosis of leishmaniasis can be confirmed by:
1. *Peripheral blood film:* Amastigote form of the parasite may be demonstrated inside circulating monocytes and less often in neutrophils in the stained peripheral blood film by thick film method.
2. *Needle biopsy/aspiration:* Deeper tissues, e.g. lymph node, bone marrow, liver and spleen may be sampled by needle biopsy/aspiration. Amastigote forms of the parasite can be demonstrated in smears stained with Giemsa's stain. Spleen aspirate is the most reliable material for demonstrating parasites in kala azar.
3. *Culture:* Whatever material is collected (blood and biopsy/aspiration material from various organs) it should be inoculated on NNN medium or Hockmeyer's medium and incubated at 22 to 25°C and examined microscopically daily for the presence of promastigotes.

Immunological Tests

These include non-specific and more specific tests:
1. **Non-specific tests**
   - Aldehyde test
   - Antimony test
   - Complement fixation test with WKK antigen.

1. *Aldehyde test:* A drop of full strength (40%) formalin is added to 1 ml of serum. A positive test is indicated by the rapid and complete coagulation of the serum. This serum test merely indicates greatly increased serum gamma globulin and thus is non-specific.

2. *Antimony test:* This test also depends upon a rise of serum gamma globulin. When a 4 percent urea stibamine solution in distilled water is mixed with serum from a patient with kala azar it leads to the formation of a profuse flocculent precipitate.

3. *Complement fixation test with WKK antigen:* Complement fixation test may be carried out for detection of serum antibodies in visceral leishmaniasis. The antigen was prepared from human tubercle bacillus by Witebsky, Kleingenstein and Kuhn hence, known as *WKK* antigen since the antigen is not prepared from *L. donovani* therefore this test is non-specific.

More specific tests: More specific tests which become positive earlier in kala-azar include direct agglutination test (DAT) indirect haemagglutination (IHA) test and indirect fluorescent antibody test (IFT). Enzyme linked immunosorbent assay (ELISA) using species-specific monoclonal antibodies and DNA probes have been successful in the direct detection of *Leishmania* antigen.

MALARIAL PARASITES

These parasites belong to the genus plasmodium. This includes *Plasmodium falciparum, P. vivax, P. malariae and P. ovale.* These protozoa have a complex life cycle in the mosquito and in man. They grow in the anopheline mosquito's intestine, sporozoites being transferred from the female mosquito's to man. Malarial parasites cause life-threatening protozoan disease called malaria.

The sporozoites are the infective form of the parasite. They are present in the salivary glands of female anopheles mosquito. Man gets infection by the bite of infected mosquito. The cycle in man comprises of following stages.

Primary exoerythrocytic schizogony

Erythrocytic schizogony

Gametogony

Secondary exoerythrocytic schizogony.

Primary exoerythrocytic schizogony: The sporozoites leave the blood stream and enter into liver parenchyma cells. The sporozoites which are elongated, spindle-shaped bodies become rounded inside the liver cells. They undergo process of multiple nuclear division, followed by cytoplasmic division and develop into primary exoerythrocytic schizont. When primary exoerythrocytic schizogony is complete, the liver cell ruptures and releases merozoites into blood stream (Fig. 55.11).

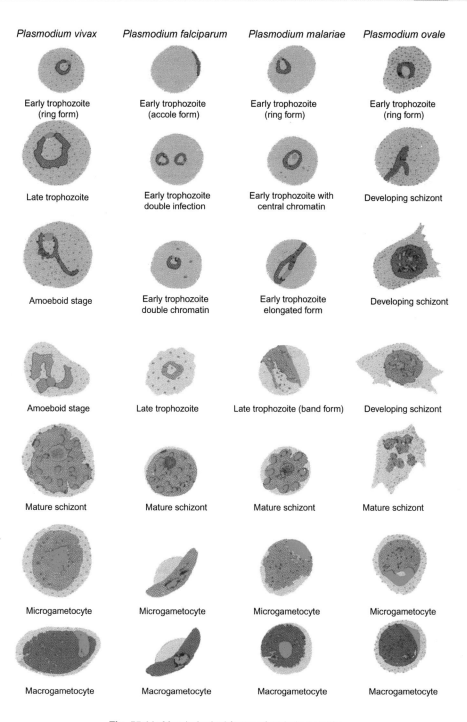

Fig. 55.11: Morphological forms of malaria parasite

Erythrocytic schizogony: The merozoites liberated from primary exoerythrocytic schizogony enter the blood stream and invade red blood cells. Here they pass through the stages of trophozoites, schizonts and merozoites. After malaria parasites have undergone erythrocytic schizogony for certain period, some merozoites develop within red cells into male and female gametocytes known as microgametocytes and macrogametocytes respectively. Only mature gametocytes are found in the peripheral blood.

Secondary exoerythrocytic schizogony: In case of *P. vivax* and *P. ovale*, some sporozoites on entering into hepatocytes enter into a resting (dormant) stage before undergoing asexual multiplication while others undergo multiplication without delay. The resting stage of the parasite is rounded, 4 to 6 μm in diameter, uninucleate and is known as **hypnozoite.** After a period of weeks, months or years hypnozoites are reactivated to become secondary exoerythrocytic schizonts and release merozoites which infect red blood cells producing relapse of malaria.

Clinical Features

Plasmodium falciparum causes malignant falciparum malaria: Fever occurs every 36 to 48 hr. An untreated primary attack lasts 2 to 3 weeks, but infection can persist for 6 to 11 months. The major problems are of cerebral malaria and anaemia.

Plasmodium vivax causes vivax or benign tertian malaria after an incubation period of 10 to 17 days. Fever occurs every 48 hr, and an untreated primary attack lasts 3 to 8 weeks or more, although infection can last for 5 to 7 years. Anaemia is the major complication, and mortality is low.

Plasmodium malariae causes quartan malaria after an incubation period of 18 to 40 days. Fever occurs every 72 hr. An untreated primary attack lasts 3 to 24 weeks, but infection persists for over 20 years with recrudescences. Proteinuria and even frank nephrotic syndrome can be a complication.

Plasmodium ovale causes ovale malaria after an incubation period of 10 to 17 days. Fever occurs every 48 hr. An untreated primary attack lasts 2 to 3 weeks, but infection persists for up to 12 months. It is usually a mild disease.

Laboratory Diagnosis

Microscopy

Diagnosis of malaria can be established by demonstration of malaria parasites in the blood. Thick and Thin smears of the blood are prepared on the same or different slides. Both thick and thin smears are stained with Leishman stain.

The smears are then examined under oil immersion lens. All asexual erythrocytic stages, as well as gametocytes can be seen in peripheral blood in infection with *P. vivax, P. malariae* and *P. ovale*, but in *P. falciparum* infection, only the ring forms and crescent-shaped gametocytes can be seen. Late trophozoite and schizont stages of *P falciparum* are usually confined to the internal organs and appear in peripheral blood only in severe or pernicious malaria.

The occurrence of multiple rings in an individual red blood cell with accole forms is diagnostic of *P. falciparum* infection.

RAPID DIAGNOSTIC TESTS (RDT)

RDTs are based on the detection of antigens derived from malaria patients in lysed blood, using immunochromatographic methods. Most frequently they employ a dipstick bearing monoclonal antibodies. The tests can be performed in about 15 minutes. Several commercial test kits are currently available.

General Test Procedure

Test strip consists of a sample pad, three detection lines containing antibodies specific for *P. falciparum*, all *Plasmodium* spp. and control antibody respectively and an absorbent pad. 2 to 50 µl of finger prick blood specimen is collected. The blood specimen is mixed with specific antibody that is labeled with a visually detectable with a visually detectable marker such as colloidal gold. If the malaria antigen is present, an antigen-antibody complex is formed. The labelled antigen-antibody complex migrates up to the test strip by capillary action towards the detection lines containing capture antibodies. A washing buffer is then added to remove the haemoglobin and permit visualisation of any coloured line on the strip. If the blood contains malaria antigen, the labelled antigen-antibody complex will be immobilised at the corresponding pre-deposited line antibody and will be visually detectable. The complete test run time varies from 5 to 15 minutes.

QUANTITATIVE BUFFY COAT (QBC) TEST

Method: The QBC tube is a high-precision glass haematocrit tube pre-coated internally with acridine orange stain and potassium oxalate. It is filled with 60 µl of blood. A clear plastic closure is then attached. A precisely made cylindrical float, designed to be suspended in the packed red blood cells, is inserted. The tube is centrifuged at 12,000 rpm for 5 minutes. The components of the buffy coat separate according to their densities, forming discrete bands. QBC tube is placed on the tube holder and examined using a standard white light microscope equipped with the UV microscope adapter, and epi-illuminated microscope objective. Fluorescing parasites are then observed at the red blood cell/white blood cell interface. The parasites contain DNA which takes up the acridine orange stain, they appear as bright specks of light among the non-fluorescing red cells.

OTHER TECHNIQUES

Other diagnostic methods include microscopy using fluorochromes such as acridine orange, polymerase chain reaction and antibody detection by serology.

Treatment

Chloroquine was the standard treatment for acute malaria. Quinine is the most reliable alternative to chloroquine for the treatment of malaria caused by chloroquine-resistant strains. Tetracycline and clindamycin exhibit some antimalarial activity and are used as an adjunct to quinine therapy. Mefloquine and halofantrine are also active against chloroquine-resistant strains.

Chloroquine and quinine do not eliminate exoerythrocytic parasites in the liver. For this primaquine should be used.

Toxoplasma Gondii

This coccidian parasite is found worldwide, its definitive host being the cat. The cat persistently excretes a large number of oocysts in the faeces; following maturation, these can infect other species, including man. There are two forms of trophozoites; tachyzoites(Figs 55.12 and 55.13), which are rapidly growing, and bradyzoites, which grow very slowly and form cysts. Man may also

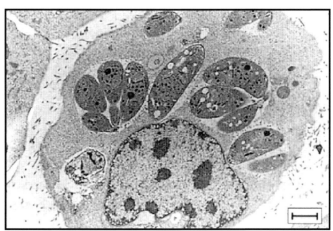

Fig. 55.12: *Toxoplasma gondii* tachyzoites

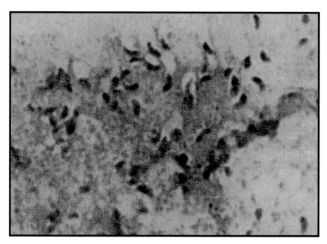

Fig. 55.13: Tachyzoites of *Toxoplasma gondii* in mouse peritoneum

become infected by eating undercooked meat containing bradyzoites. Infection is asymptomatic. When clinically apparent, it causes a glandular fever-like illness and more rarely encephalomyelitis. In immunocompromised patients, it is more likely to produce encephalomyelitis. *Toxoplasma gondii* can also cross the placenta to infect the fetus. The major problem in the child is chorioretinitis and subsequent blindness. Cerebral damage with intracerebral calcification and microcephaly can also occur.

Chapter

56

Multicellular Parasites

The multicellular parasites are subdivided into platyhelminthes flat worms which contain two classes—Cestodes and Trematodes and Aschelminthes, of which the class Nematoda contains human pathogens.

CESTODES

The cestodes are tapeworms that are acquired by the ingestion of improperly cooked fish (*Diphyllobothrium latum*), beef (*Taenia saginata*) or pork (*Taenia. solium*) containing larvae. The dog tapeworm (*Echinococcus granulosus*) is found particularly in sheep rearing areas. Ingestion of the eggs lead to Hydatid disease in man.

Taenia saginata and *Taenia solium*: Morphology

Adult Worm

The adult worms consist of scolex (head), neck and strobila which is made up

	T. saginata	T. solium
	Table 56.1: Differentiating features of T saginata and T solium	
Length	4-6 metres or more	2-4 metres or more
	Large, quadrate without rostelium and hooklets. Possesses four suckers which may be pigmented.	Small, globular, with rostelium armed with a double now of 25-30 alternating large and small hooklets. Possesses four suckers which are not pigmented.
Neck	Long	Short
Proglottids		
Number	1,000-2,000	800-1000
Explulsion	Expelled singly	Expelled in chains of 5 or 6
Number of lateral branches of uterus	15-30	5-10
Vaginal sphincter	Present	Absent
Ovaries	Two without any accessory lobe	Two with an accessory lobe
Testes	300-400 follicles	150-200 follicles
Measurement of gravid segment	20 mm in length and 5 mm in breadth	12 mm in length and 4 mm in breadh
Larva	Cysticercus bovis, present in cow and not in man	*Cysticercus cellulosae*, present in pig and may also develop in man.

of a large number of proglottids (segments) (Fig. 56.1). The differentiating features of *T. saginata* and *T. solium* given in Table 56.1.

Eggs

Eggs of both species are indistinguishable. They are spherical, brown in colour (bile stained) and measure 31 to 43 μm in diameter. They are surrounded by

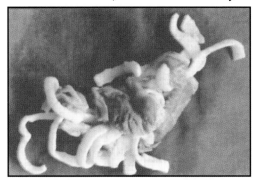

Fig. 56.1: *Taenia solium* in small intestine

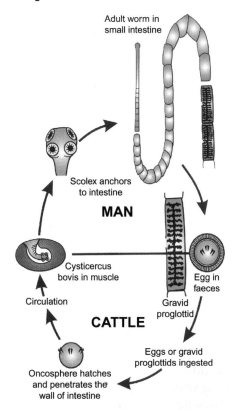

Fig. 56.2: Life cycle of *Taenia solium*

embryophore which is brown, thick walled and radially striated. Inside the embryophore is present hexa-canthembryo (oncosphere) with three pairs of hooklets. The eggs of *T. solium* are infective to pig and also to man while those of *T. saginata* are infective only to cattle (Figs 56.2 and 56.3).

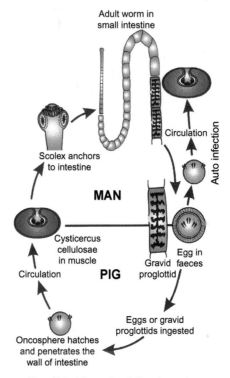

Fig. 56.3: Life cycle of *Taenia saginata*

Clinical Features of T. saginata and T. solium

Man acquires infection *Taenia saginata* by eating raw or under cooked beef, containing encysted larval stage (cysticercus bovis). Man acquires *Taenia solium* infection by eating raw or under cooked pork containing encysted larval stage (cysticercus cellulosae).

Cysticercus cellulosae can also develop in man as follow:

By ingesting the eggs with contaminated water and food. A man harbouring adult worms may auto–infect oneself either by unhygienic personal habits or by reverse peristaltic movements of the intestine whereby the gravid segments are thrown into the stomach, equivalent to the swallowing of thousands of eggs.

Clinical features abdominal discomfort, indigestion, persistent diarrhoea or diarrhoea alternating with constipation and loss of appetite.

Cysticercus cellulosae (larval form of *T. solium*) may develop in any organ and the effects produced depend on the location of cysticerci. They usually occur in large numbers, sometimes they may occur singly. They usually develop in the subcutaneous tissues and muscles forming visible nodules. It may also develop in brain leading to epileptic attacks and in anterior and vitreous chambers of the eye.

Laboratory Diagnosis

1. The diagnosis of *T. saginata* and *T. solium* worm infection can be carried out by:
   i. Demonstration of characteristic eggs in the stool by direct smear and concentration method by sedimentation technique.
   ii. Since eggs of both *T. saginata* and *T. solium* are similar, therefore, for the species diagnosis, the demonstration of gravid proglottids and scolices is essential.
2. The diagnosis of cysticercosis can be carried out by:
   i. Biopsy of subcutaneous nodule: It may reveal cysticerci.
   ii. X-ray of skull and soft tissue: It may reveal calcified cysticerci.
   iii. CT scan of the brain: It can accurately locate the lesion in the brain.
   iv. Differential leucocyte count: It reveals eosinophilia.
   v. Serological tests such as indirect haemagglutination (IHA), indirect fluorescent antibody (IFA) and enzyme-linked immunosorbent assay (ELISA) can be used for demonstration of specific antibodies in the serum.

Treatment

Praziquantel and niclosamide can be used for the treatment of human tapeworm infection. A single does of four tablets (each of 500 mg) of niclosamide is effective against adult *T. saginata* and *T. solium* in the intestine.

DIPHYLLOBOTHRIUM LATUM

Morphology

Adult Worm

D. latum is the longest tapeworm found in man. It consists of scolex, neck and strobila.

Scolex: The scolex is almond-shaped or spoon-shaped. It bears two slit-like grooves (bothria) demarcated by lateral lip-like folds, one dorsal and the other ventral.

Neck: It is situated immediately behind the scolex. It is thin, unsegmented and several times the length of the head.

Strobila: It has 3,000 or more proglottids (segments), consisting of immature, mature and gravid segments.

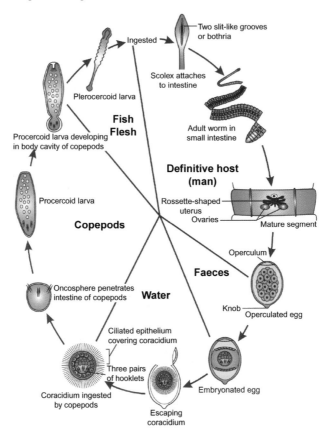

Fig. 56.4: Life cycle of *Diphyllobothrium latum*

Egg

Egg of *D. latum* is yellowish-brown in colour (bilestained), oval or elliptical in shape. It contains an immature embryo. It does not float in saturated solution of common salt. The egg is not infective to man.

Larval Stages

The egg develops into first-second, and third-stage larva. The first-stage larva is known as coracidium. The second-stage larva is known as procercoid. The third-stage larva is known as plerocercoid (Fig. 56.4).

Clinical Features

Humans become infected when they eat under-cooked, raw or lightly salted meat or roe from infected freshwater fishes. Patient may develop fatigue, weakness, diarrhoea and numbness of the extremities. Some patients develop mechanical obstruction of the extremities. Some patients develop mechanical obstruction of the bowel by a large number of the worms that become tangled together. In a few cases pernicious anaemia may develop due to manifest vitamin B_{12} deficiency.

Laboratory Diagnosis

The laboratory diagnosis of diphyllobothriasis in humans can be made by identification of the characteristic operculated eggs or proglottids in the faeces.

Treatment

Niclosamide and praziwquantel are the drugs of choice for the treatment of diphyllobothriasis.

Echinococcus Granulosus (Dog Tapeworm)
Adult Worm

It is a small tapeworm measuring 3 to 6 µm in length. It consists of a scolex, neck and strobila.

Eggs

Eggs are indistinguishable from those of other *Taenia species*. Larval form is found within the hydatid cyst which develops in the intermediate host (Fig. 56.5).

Clinical Features

E. granulosus causes cystic echinococcosis or hydatidosis or hydatid disease or hydatid cyst in man. The disease is generally acquired during childhood hydatid disease in humans is potentially dangerous:

Liver Cysts

The majority of hydatid cysts occur in the liver, causing symptoms that may include chronic abdominal discomfort, occasionally with a palpable or visible abdominal mass (Fig. 56.6).

Lung cysts: Cysts in the lungs are usually asymptomatic until they become large enough to cause cough, shortness of breath, or chest pain.

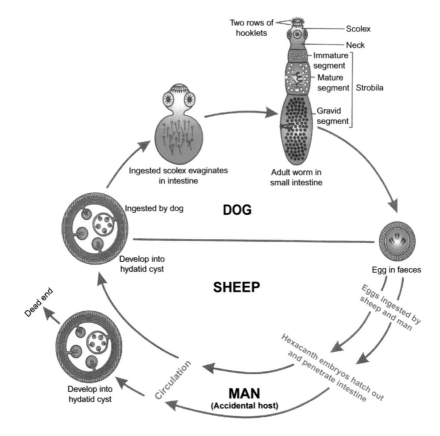

Fig. 56.5: Life cycle of *Echinococcus granulosus*

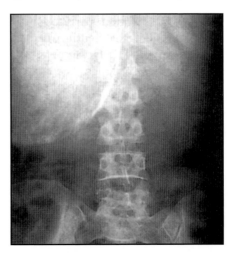

Fig. 56.6: Hydatid cysts on liver

Other sites: Other organs which may also be involved include spleen central nervous system and heart. Kidneys, bones, muscles, female genital tract, eyes, etc. leading to visible swelling and pressure effects.

Laboratory Diagnosis

It can be carried out by the following methods:

1. *Casoni test:* It is an immediate hypersensitivity skin test. Antigen for the Casoni test is sterile hydatid fluid drawn from unilocular hydatid cysts from sheep, pig, cattle or man. 0.2 ml of the antigen is injected intradermally in one arm for control an equal. Amount of sterile normal saline is injected intradermally on the other arm. In positive case there develops a large wheal measuring 5 cm or more in diameter with multiple pseudopodia within 30 minutes.

2. *Differential leucocyte count:* Differential leucocyte count may reveal eosinophilia.

3. *Serological tests:* Serodiagnosis of hydatid cyst may be carried out by enzyme-linked immunosorbent assay (ELISA), radioimmunoassay (RIA) complement fixation, indirect haemagglutination (IHA) bentonite flocculation and latex agglutination tests.

4. *Examination of cyst fluid: Examination of cyst fluid reveals* scolices, brood capsules and hooklets.

5. *Radiodiagnosis: X*-ray ultrasound and CT Scan are also helpful in the diagnosis of hydatid cyst.

Treatment

Surgical removal of the hydatid cyst. The cyst should be removed in toto and consequences of spilling its contents, could lead to anaphylactic shock. Therefore, postoperative chemotherapy may be given for at least two years after radical surgery. Praziquantel and albendazole are the chemotherapeutic agents for the treatment of hydatid cyst.

Chapter

57

Trematodes

Trematodes or flukes have a complex life cycle involving an intermediate host that is a mollusca usually a snail. The adult develops in man, who excretes ova in which a larva develops. The larva or miracidium infects the mollusca. In the mollusca, the trematode goes through a series of generations , finally liberating more larvae, known as 'cercariae'.

1. These cercariae infect man by penetrating the skin, e.g. Schistosoma spp.
2. By being eaten in a second intermediate host such as fish, e.g. *Clonorchis sinensis*.
3. By ingestion of vegetable matter such as water cress that is then eaten, e.g. *Fasciola hepatica.*

   Tremlatodes are classified as:
   1. INTESTINAL FLUKES
   2. BLOOD FLUKES

Intestinal Flukes: Two types of flukes have been grouped under the Intestinal flukes, They are (1) Fasciolopsis buski, (2) Fasciolopsis hepatica.

Fasciolopsis buski: (Fasciola buski): This is **commonly** known as large or giant intestinal fluke. The adult worm lives attached to the **mucosa** of the duodenum

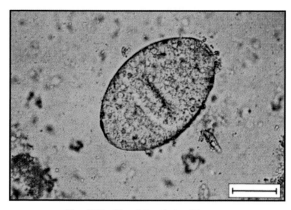

Fig. 57.1: *Fasciolopsis buski* ovum

and jejunum of man and pig where pig serves as the reservoir of infection for man (Fig. 57.1).

Morphology: Adult worm: It is fleshy and ovoid, the anterior end being narrower than the posterior. Oral suckers are present on the ventral surface of the worm.

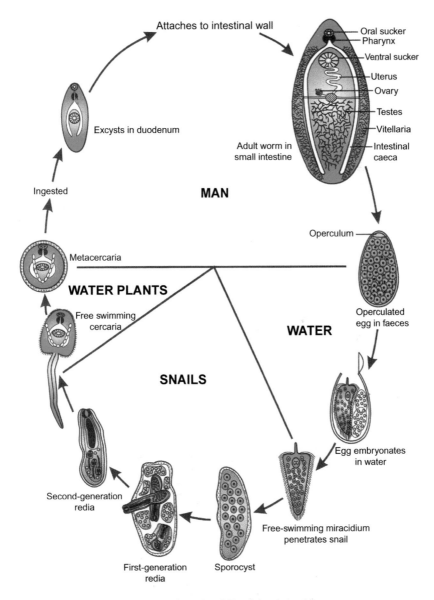

Fig. 57.2: Life cycle of *Fasciolopsis buski*

EGGS: These are large, oval, yellowish, brown and contains an immature larva, the miracidium. The miracidium undergoes maturation in the Intermediate host (Snail) and is known as cercariae-an encysted form (Fig. 57.2).

Metacercariae: The free swimming cercariae attach to the Aquatic vegetations which are the second intermediate host. Metacercariae are the excysted forms of the cercariae and these are the infective forms.

Clinical features: Man acquires infection from peeling off the skin of these infested plants. The metacercariae excyst in the duodenum, become attached to the intestinal wall and develop into adult worms in about 3 months. Infection caused by *F. buski* is known as fasciolopsiasis. Large number of worms provoke increased secretions of mucus and may lead to partial obstruction of the bowel. Patient complains of diarrhoea, initially alternating with constipation and persistent thereafter, abdominal pain, anorexia, nausea, vomiting, generalised toxic and allergic symptoms usually in the form of oedema particularly of the face, abdominal wall, and lower limbs. Ascites, anaemia, and asthenia are common and patient may die of profound toxaemia.

Laboratory diagnosis: Diagnosis of fasciolopsiasis can be made by detection of operculated eggs in the faeces. Adult worms may be recovered and identified after a purgative or an antihelminthic. A marked eosinophilia and leucocytosis are commonly seen.

Treatment: Praziquantel and niclosamide can be used for the treatment of casciolopsiasis.

Fasciolopsis Hepatica

Fasciolopsis hepatica are commonly known as the sheep liver *fluke* and this is prevalent in most sheep-raising areas (Fig. 57.3).

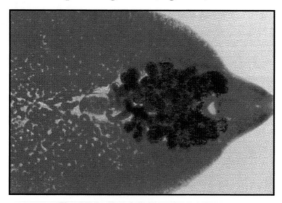

Fig. 57.3: *Fasciola hepatica* anterior

Morphology: The adult worm is large leaf-shaped and is brown to pale grey in colour. The oral suckers are situated at the anterior end.

EGGS: These are large elliptical to oval operculate, light yellowish-brown and contains an immature larva, the miracidium (Fig. 57.4).

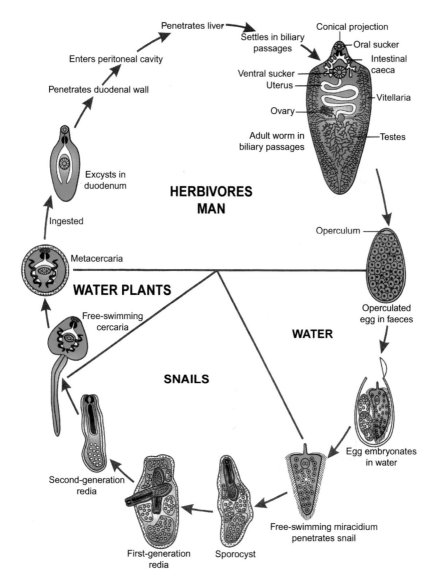

Fig. 57.4: Life cycle of *Fasciola hepatica*

Metacercariae: The miracidium undergoes changes in the snails and liberates free-swimming cercariae. These then become metacercariae and insist on water plants that are ingested by humans. These are the infective forms which result in fascioliasis.

Clinical features: The metacercarial larvae of *Fasciola hepatica* escape from cysts in the duodenum and migrates through the duodenal wall into the peritoneal cavity. The young migrating flakes enter the liver parenchyma and cause the disease.

Patient complains of fever, mild to severe abdominal pain, urticaria, on examination. **Hepatosplenomegaly,** ascites and jaundice are found however, some patients develop chronic cholecystitis, cholangitis and cholelithiasis, which may be accompanied by **biliary** colic, epigastric pain jaundice, nausea, prurits and upper right quadrant pain. In heavy infections liver abscesses occur.

Laboratory diagnosis: Laboratory diagnosis of fascioliasis can be made by:
- Detection of eggs in stool or in bile obtained by duodenal intubation
- Immunodiagnostic test based on antibody detection. A variety of immunological tests have been used. ELISA is sensitive.

Treatment: **F. hepatica** is not sensitive to praziquantel. Drugs currently used for this infection are bithionol, triclabendazole and dehydroemetine.

Blood flukes: Schistosoma mansoni, Schistosoma japonicum and *Schistosoma haematobium* are the three species among blood flukes. Schistosomes, unlike other trematodes, are not hermaphroditis. **Miracidia** hatch from eggs and are excreted in the stools-in the case of *S. mansoni* and *S. japonicum* and the eggs are excreted in urine in the case of *S. haematobium. S. mansoni* lives in the branches of the inferior mesenteric vein draining the lower colon, *S. japonicum* in the superior **mesenteric** vein (small intestine) and *S. haematobium* in the vesical, uterine and prostatic plexuses. Adult males and females mate in the respective veins, eggs being deposited there. They penetrate into the intestine or bladder and are thence excreted.

Clinical Feature

Acute schistosomiasis or **katayama fever** occurs in about a month after infection with *S. japonicum* and *S. mansoni* and rarely with *S. haematobium.* Characteristic symptoms include high fever, hepatosplenomegaly, lymphadenopathy, eosinophilia, and dysentery. *S. mansoni* and *S. japonicum* female worms lay eggs in the mesenteric branches of the portal vein. These eggs are carried into the liver and other organs (Figs 57.5 and 57.6). The

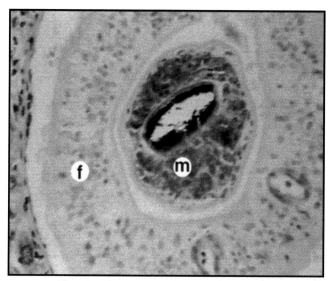

Fig. 57.5: *Schistosoma mansoni* in liver section

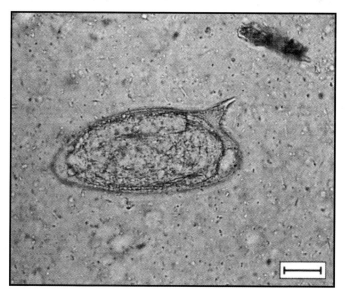

Fig. 57.6: *Schistosoma mansoni* ovum

remainder eggs stay in small venules. Acute inflammation results in the rupture of the vascular wall and escape of eggs. This inflammation leads to recurrent daily fever, pain abdomen, enlarged tender liver and spleen accompanied by dysentery or diarrhoea. The chronic phase of manifestations in *S. mansoni* and *S. japonicum* infections is characterised by hepatospleno-megaly and may therefore be called **hepatosplenic schistosomiasis** patient may develop ascites due to portal hypertension and hypoalbuminaemia bleeding from the varices may cause sudden death. In case of infection with *S. japonicum*, acute or chronic cerebral involvement may occur. Leading to headache, Jacksonian epileptic seizures, paraesthesia and poor vision. In *S. haematobium* infection. The main complaint is recurrent painless haematuria resulting from ulcers of the bladder. A burning sensation on micturition, frequency and suprapubic discomfort or pain may precede or be associated. *S. haematobium* may also cause infection of vulva, vagina, cervix, ovaries, fallopian tubes and uterus.

Laboratory Diagnosis

Diagnosis is made by demonstrating eggs in the faeces, urine or tissues, *S. mansoni* has ovoid egg with a lateral spine near one pole, *S. japonicum* is smaller with a small lateral spine and *S. haematobium* a terminal spine.

Immunodiagnostic techniques such as immunofluorescent antibody test, **ELISA, RIA** and complement fixation test may be used as indirect methods for the diagnosis.

Treatment: Praziquantel is the drug of choice for the treatment of schistosomiasis.

Chapter

58

Nematodes

Nematodes are non-segmented roundworms, most of which have a free living state.

INTESTINAL WORMS

Ascaris lumbricoides is a large (20 to 35 cm) intestinal worm (Fig. 58.1). Infection is acquired via the ingestion of mature embryonated eggs. The eggs are excreted in the faeces. The eggs hatch in the duodenum, the larvae penetrating the intestinal wall, entering the blood or lymphatics and passing to the liver. They then pass to the pulmonary circulation, break into the alveoli, ascend to the pharynx and descend the oesophagus to the intestine. There they mate and release eggs (around 2,00,000 per day). The infection is usually asymptomatic, except during the migratory phase, when asthmatic disease may occur. Heavy intestinal infestation may cause obstruction or failure to thrive. Diagnosis is by the detection of eggs in the faeces (Figs 58.1 and 58.2). Treatment is with mebendazole.

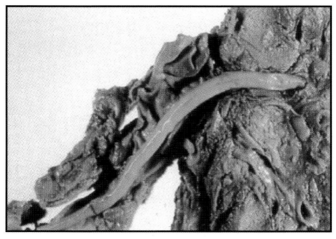

Fig. 58.1: *Ascaris lumbricoides* in a renal vein

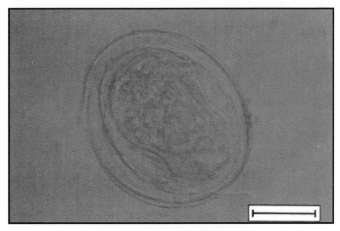

Fig. 58.2: *Ascaris lumbricoides* ovum

Fig. 58.3: Female *Enterobius vermicularis*

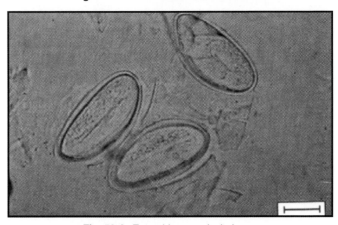

Fig. 58.4: *Enterobius vermicularis* ova

Enterobius vermicularis, the threadworm (Fig. 58.3), is a common infection of children. The adult worm resides in the caecum and adjoining areas. Gravid females migrate to the anus, where they deposit eggs (Fig. 58.4) on the anal verge. This is intensely irritant and causes the child to scratch, the eggs then being transferred to fingers and thence to the mouth to auto infect or infect others. The clinical features vary according to the amount of irritation. They may also migrate to the vagina, causing vulvovaginitis. Diagnosis is by the cellophane tape method. The tape is applied to the anal verge early in the morning and picks up eggs. The tape is then stuck to a microscope slide and examined using a 10 X (using a Ten X) objective, under which the eggs are clearly visible. Treatment is with mebendazole or pyrantel pamote.

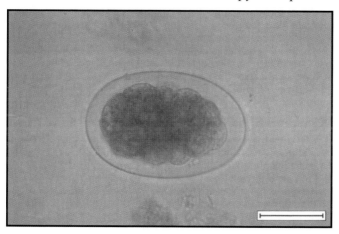

Fig. 58.5: *Ancylostoma duodenale* ovum

The hookworms *Ancylostoma duodenale* and *Necator americanus* both infect man by penetrating the intact skin, usually that of the feet. Man appears to be the only host of *A. duodenale.* Hookworm eggs are excreted in faeces and hatch preferable on moist sandy soil to produce larvae (rhabditiform) (Fig. 58.5). They then molt to become filariform or infective larvae, and enters human through skin. They are carried to the lungs by the bloods tream, where they burrow into alveoli. They then ascend to the pharynx and descend the esophagus to the small intestine. Penetration of the skin is accompanied by pruritus, and when they enter the lung they may cause pneumonitis. Once in the intestine, they may cause abdominal pain, but with persistent heavy infestation the major problem is one of severe iron deficiency anaemia. Diagnosis is by demonstrating eggs in the faeces. Treatment is with mebendazole.

Trichuris (Fig. 58.6) the whip worm and is acquired faeco-orally. It mostly produces asymptomatic infection but may rarely give rise to abdominal

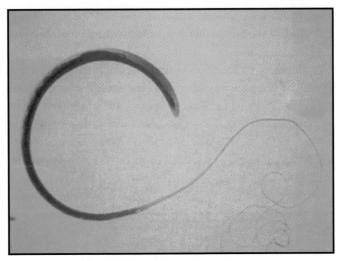

Fig. 58.6: Trichuris trichium

distension, bloody mucoid diarrhoea, weight loss and anaemia if the infection is heavy. Diagnosis is by demonstrating the characteristic barrel-shaped eggs in the stools. Treatment, is with mebendazole.

Blood and Tissue Nematodes

The filaria (*Wuchereria bancrofti, Loa loa, Onchocerca volvulus, Brugia malayi*) are widely distributed in the tropics and subtropics. They all have an insect (mosquito, mango fly, black-fly) as an intermediate host and are deposited on humans when they are bitten. The filarial produce lymphatic blockage leading to lymphoedema, cutaneous larva migrans or ocular damage depending on the particular worm.

Wuchereria bancrofti

Morphology

Adult Worms

Adult worms are transparent, creamy white, long, hair like structures (Fig. 58.7). The posterior end of the female worm is straight, while that of the male is curved ventrally. Both male and female worms remain coiled together. The female is viviparous and liberates sheathed embryos (microfilariae) into lymph from where they find their way into blood.

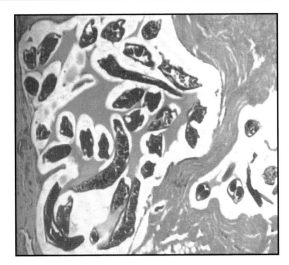

Fig. 58.7: *Wuchereria bancrofti* in lymph node

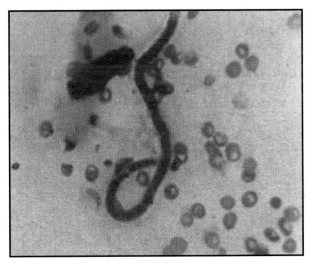

Fig. 58.8: *Microfilaria bancrofti* in peripheral blood

Microfilaria

It is transparent and colourless with blunt head and pointed tail (Fig. 58.8). It can move forwards and backwards within the sheath.

Clinical Features

Infection caused by *W. bancrofti* is known as *Wuchereriasis or Bancroftian filariasis*. It is mainly due to the presence of adult worms in the lymph nodes and vessels. The lymph nodes become enlarged, firm and fibrotic. In males, hydrocele, orchitis, funiculitis and epididymitis are common. The development of lymph scrotum results in chyluria, with lymph getting into the urine. In many patients, acute attacks of 'filarial fever' occur. Patient develops intermittent recurrent fever lasting 3 to 15 days, with headache, malaise, localised pain and tenderness with oedema and erythema above lymph vessels and glands, accompanied by acute lymphangitis and lymphadenitis of the groin or axilla. Examination of blood often shows high eosinophilia.

Laboratory Diagnosis

Bancroftian filariasis can be diagnosed by:

Detection of Microfilariae

Microfilariae of *W. bancrofti* (Fig. 58.9) circulate in the peripheral blood with a regular nocturnal periodicity. Therefore, to diagnose bancroftian filariasis, blood must be taken during night, optimally between 10 pm and 2 am. Thin and thick smears are prepared. The thick smear is dehaemoglobinised and both the smears are stained with haematoxylin or Giemsa's stain. The smears are then examined under microscope for the presence of characteristic microfilariae. Acridine orange-microhaematocrit tube technique can also be used for the detection of microfilariae. A microhaematocrit tube incorporating heparin, EDTA, and acridine orange serves the basis for this test. After centrifugation, parasites become concentrated in the buffy coat and can be visualised through the clear glass wall of the tube. The acridine orange stains

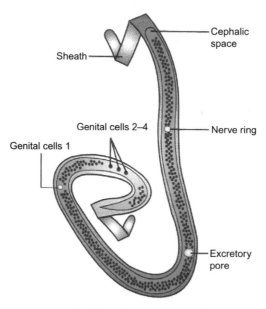

Fig. 58.9: *Microfilaria bancrofti*

the DNA of the parasites, and the morphologic characteristics, can be examined by fluorescence microscopy in making a species identification. Microfilariae may also be demonstrated in the chylous urine, exudates of lymph varix and in the hydrocele fluid.

Detection of Adult Worms

Adult worms can be seen in the biopsied lymph node and the calcified worm may be seen on X-ray examination.

Immunodiagnosis: Filarial antigen may be detected in the patient serum by enzyme immunoassays using monoclonal antibodies against microfilarial larval surface antigens.

DNA Probes

DNA probes have been developed for *W. bancrofti* and *B. malayi*. These may be used for the diagnosis of infection caused by these parasites.

Treatment

Diethylcarbamazine (DEC) is the drug of choice for the treatment of bancroftian filariasis. It is given orally in a dose of 6 mg per kg body weight daily for 12 days.

DRACUNCULUS MEDINENSIS—THE GUINEA WORM

This causes dracunculiasis

Morphology

Adult Worms

The mature female of *D. medinensis* is a slender, long worm. It resembles a piece of long twine thread. The body is cylindrical, smooth and milk-white in colour.

Embryos

These are coiled bodies with rounded heads and long slender tapering tails. The embryos are set free only at the time of parturition when the affected part if submerged in water. *D. medinensis* causes dracunculiasis. It is typically a disease of rural communities which obtain their drinking water from ponds or from large step wells where cyclops can breed. Migrating worm (Fig. 58.10) in the subcutaneous regions may cause some tenderness, the gravid female

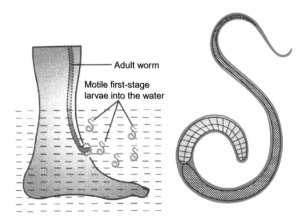

Fig. 58.10: First stage larvae of *Dracunculus medinensis*

worm reached the skin, and produced a blister at the outer end of the tunnel, and the blister has burst at the outer end of the tunnel, and the blister has burst in contact with water, there may also be pronounced systemic symptoms including erythema and urticarial rash, with intense pruritus, nausea, vomiting, diarrhoea, giddiness, and syncope. In a few hours the local lesion develops. It is a reddish papule with a vesicular centre and an indurated margin. It may be painful and may cause considerable inflammation. These lesions are most common between the metatarsal bones of the soles of the feet or on the ankles, but may be found on the hands or arms, trunk, buttocks, scrotum, knee joint, calf, thigh, shoulder, or even the angle of the jaw.

Laboratory Diagnosis

Diagnosis of dracunculiasis can be made by:

Detection of adult worm: This is possible when the gravid female worm appears at the surface of the skin.

Immunological tests: Antibodies to *D. medinensis* may be detected in patient serum by ELISA, Western blot and fluoresecent antibody test using deep frozen first-stage larvae.

Intradermal test: Injection of dracunculus antigen intradermally in patients suffering from dracunculiasis causes a wheal to appear in the course of 24 hours.

Treatment

Thiabendazole, niridazole, metronidazole, mebendazole and albendazole are effective.

Chapter

59 *Laboratory Procedures in Parasitology*

Laboratory diagnosis for the infection of intestinal parasites is based on gross examination and microscopic examination of the stool.

Stool is the single most important specimen submitted to the laboratory for parasitological examination. The fresh stool is used to see the presence of protozoa in the active (trophozoite) stage. The container to be used for the collection of stool specimens largely depends on the quality of stool. For example, dry stool can be collected in a cardboard box or (match box) but liquid stool must collected in a container. Collect the stool in clean, wide mouthed containers or card board box empty tin plastic container or glass jar. Collect sufficient quantity at least 4 ml by volume. All specimens must be clearly identified with patient's name, physician's name and the date and time of specimens collection.

GROSS EXAMINATION

Gross macroscopic examination of the stool with the naked eye can be helpful in the presumptive diagnosis of several intestinal disorders.

Consistency of the stool (formed, soft, liquid) may give an indication of protozoan stages. Trophozoites are seen mostly in soft or liquid stool and cyst staged in formed or semi formed stools. Adult helminthes may be seen on the surface of the stool specimen (Fig. 59.1).

1. Large roundworm (*Ascaris lumbricoides*) pinkish in colour, 0.3 to 0.5 cm thick and about 15 cm long (females are slightly longer).
2. Pin worm (*Enterobius vermicularis*) white in colour and 1 cm long (males are smaller and less common).
3. Other adult helminthes (whipworm or *Trichuris trichiura*, hookworks (*Ancylostoma, Necator*) and dwarf tapeworms (*Hymenolepis nana*) may be found in the stool.
4. Proglottids or mature segments of tapeworms (*Taenia*). The presence of proglottids in the stool specimen can be recognised by gross examination of the surface of the stool.

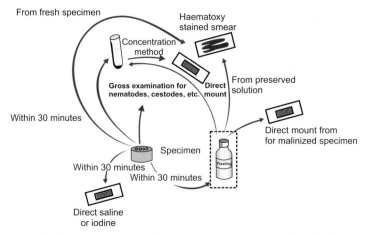

Fig. 59.1: Processing of stool specimen for intestinal parasite infection

MICROSCOPIC EXAMINATION

Microscopic examination of the stool is necessary to identify helminth eggs and larvae as well as protozoan cysts and trophozoites. This can be done by direct wet mount of fresh stool, preserved specimen, or faecal concentrates.

Preparation for Microscopic Examination

Wet Mount

It can be performed on fresh specimens, preserved specimens or concentrated specimens.

1. *Direct wet mount:* Specimen direct mounted the microscopic slide with out any mounting medium.
2. *Saline wet mount:* Specimen mounted on physiological saline.
3. *Saline wet mount:* Specimen mounted on dilute Lugol's iodine solution. Iodine stains some of the cysts.

    Direct wet mount of fresh specimens with or without any mounting medium, helps to look for the presence of protozoan trophozoites and cysts and also, helminth eggs, larvae and other parasitic elements.

Procedure for Direct Wet Mount

1. Put a drop of the specimen on a clean microscopic slide.
2. Put the coverslip and examine under low light intensity.
3. Use low power objective for screening and high dry objective for examining.

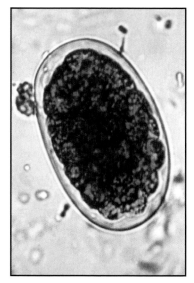 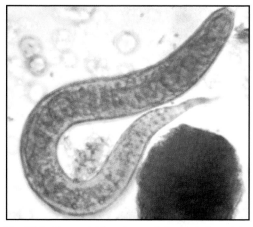

Fig. 59.2: Hookworm egg, iodine stain   **Fig. 59.3:** *Strongyloides stercoralis* larva, iodine stain

Saline and Iodine Preparations

Reagents

Mounting fluids
1. Sodium chloride solution (0.85%, w/v)
   Dissolve 8.5 gm of sodium chloride (NaCl) in a final volume of 1000 ml distilled water
2. Lugol's iodine solution.
   a. Stock solution

Iodine	1 gm
Potassium iodide	2 gm
Distilled water	100 ml

   Dissolve 2 gm of potassium iodide in 30 ml of distilled water in a 100 ml graduated cylinder. Add 1 gm of powdered iodine mix until dissolved, then add distilled water to a final volume of 100 ml.
   b. Working solution of iodine (Figs 59.2 and 59.3).
   Iodine solution should be diluted and stored in amber coloured bottle.

Procedure

Take a clean microscope slide and put a drop of sodium chloride in the middle of the left half and a drop of iodine on the middle of the right half. This provide a double mount of the specimen, one unstained and the other stained (Iodine).

Transfer a small amount of stool specimen on the slide and mix the sample with a drop of the sodium chloride solution. Use the wooden applicator stick for transfer. The volume of the stool will be about twice the size of pin head. Air bubbles should always be avoided.

If the specimen contains blood or mucus or is liquid, take the faecal material with blood stained mucus or from the surface of the liquid stool to look for amoebae. Place a coverslip over each drop. Avoid the formation of air bubbles. Examine first the saline preparation under the low power objective to screen. Switch to the high dry objective for a closer look. Identification of protozoa should be done with the high dry or 40X objective. The oil-immersion (100 X objective) is rarely used to view wet mounts. Examine the iodine mount in the same way as described for the saline mount.

CONCENTRATION PROCEDURE

The concentration technique increases the ability to detect protozoan cyst and helminth eggs and larvae. Direct examination, however, should be done first before proceeding to faecal concentration. Motile forms of protozoa (trophozoites) are not founding concentrated preparations. Two methods are generally employed sedimentation and flotation.

Sedimentation Method

The sedimentation method (using gravity or centrifugation) allows the recovery of all protozoan cysts and helminth eggs and larvae.

Simple Sedimentation Method by Gravity

Take about 2 ml of fresh faecal material in bottle containing 10 ml of tap water or 10 per cent of formaldehyde. Crush the specimen thoroughly to obtain a uniform suspension . Transfer the suspension to a 15 ml test tube or a graduated cylinder let it stand for 1 hr after 1hr carefully remove the top 2/3 of the supernatant; again add 10 ml (approximately) of water, re-suspend the faecal matter and let it stand for another hour. Repeat the process until the supernatant is relatively clear. Finally with the help of a Pasteur pipette or a long pipette, remove a bit of the sediment and prepare a wet mount.

The formalin ether centrifugal sedimentation method:

It is more efficient and sensitive than the simple sedimentation method.

Reagents
Formaldehyde solution 10%

Commercial formalin (with 37% formalin)	100 ml
Distilled water	300 ml

Ether: Use ether or ethyl acetate.
Saline: 0.85% (w/v) sodium chloride solution
Disinfectant: 5% phenol.

Procedure

Transfer a portion of stool about half teaspoon into 10 ml of 10 per cent formalin. Thoroughly crush and mix the stool specimen. Let the suspension stand for 30 mts for adequate fixation.

Filter the faecal suspension through two layers of wet gauze kept on a funnel, into a 15 ml glass centrifuge tube. Add physiological saline to within half inch of the top and centrifuge for 2 min at 1500 rpm. Decant the supernatant and discard it into a disinfectant solution. Finally 1 to 1.5 ml of sediment should be present. Add about 10 ml of 10 per cent formalin re-suspend the sediment. Add approximately 3 ml of ether to the above suspension. Stopper the tube tightly with a cork turn it on its side and shake vigorously for 30 seconds. Examine the sediment directly under the microscope after re-suspending it in the residual formalin solution.

Flotation Method

Reagents
Saturated solution of sodium chloride (Willis solution)

Sodium chloride	125 gm	
Distilled water(qs)	500 ml	

Dissolve the salt by heating the mixture to boiling point in a beaker (1,000 ml) filter and keep in a corked bottle.

Zinc sulfate solution (33% w/v): Dissolve 330 gm of zinc sulphate in 670 ml of distilled water and then make it to 1000 ml volume with distilled water.

Procedure

Place a portion of stool, approximately 2 ml by means of a wooden applicator stick in a bottle. Fill the bottle with saturated salt solution. Crush the stool and mix well, using the wooden applicator stick. Place a grease-free coverslip over the mouth of the bottle. Check that the coverslip is in contact with the liquid and leave for 10 minutes. Remove the coverslip with care after 10 minutes. Make sure that the drop of liquid adhering to the coverslip remains attached to the coverslip until it is placed on the slide. Place the coverslip on a clean slide and examine under the microscope. Examine the wet mount preparation under the low power objective (10X). Switch to the high-dry objective (40X) as soon as an object of interest is visible. Observe under the oil-immersion objective only when the object is too small.

REPORTING OF STOOL EXAMINATION

Report any worm segment seen such as proglottids or tapeworms. Consistency of stool-hard and dry, firm and formed, soft and formed, soft and unformed, semi liquid, or watery. Abnormal features-flakes of mucus, blood-stained, streaks of pus present, bright red coloured and pale coloured.

Microscopic examination: Identify the helminth eggs or larvae or protozoan cyst or trophozoites.

Cellotape Method

Specimen should be taken in the morning before taking bath.

Procedure

Take a strip of cellotape, sticky slide down, on a slide. Place the wooden tongue depressor flat against the underside of the slide. Now gently pull the tape away from the slide and loop it over the end of the wooden tongue depressor. Hold the completed tape swab in the right hand, pressing the slide firmly against the wooden tongue depressor. Press the sticky surface of the tape firmly on the skin around the anus. The eggs on the skin will stick to the cellotape. Take the slide and fold the tape back on to it, sticky side down. Examine under the low power. The eggs should be visible if present. They are transparent, colourless, oval-shaped with one side slightly flattened, there is a granular mass inside with a curled up embryo; the shell is smooth and thin; a double line is visible on the shell.

Staining of Faecal Smears and Blood Films

Trichrome stain: It is a simple procedure that produces well-stained smears.

Reagents
1. Trichrome stain

Chromotrope 2R	0.6 gm
Light green SF	0.3 gm
Phosphotungstic acid	0.7 gm
Glacial acetic acid	1 ml
Distilled water	100 ml

Add the stains and the acids in a beaker. Cover the beaker and allow the mixture "to ripe" or 15 to 30 minutes. Then add 100 ml of distilled water. Store the stain in a Coplin jar. It should be purple in colour.

2. Iodine solution (D' Antoni's)

Potassium iodide	1 gm
Iodine crystals (powdered)	1 gm
Distilled water	100 ml

Mix the ingredients in water in presence of potassium iodide. Store in brown glass-stoppered bottles in dark. Ethanol: Absolute (100%), 90 per cent and 70 per cent mounting fluid.

Procedure

Prepare fresh faecal smears on a labelled microscope slide.
Put the fresh smears in 70 percent ethanol for 5 minutes.
Place in 70 per cent ethanol with iodine solution for 2-5 minutes.
Place in 70 per cent ethanol for 5 minutes and repeat this for 3 minutes.
Place in trichrome stain for 10 minutes
Place in 90 per cent ethanol, acidified with 1 per cent glacial acetic acid (v/v) for upto 3 seconds.
Dip in 100 per cent ethanol followed by two changes in 100 per cent ethanol (3 minutes each).
Mount in any mounting fluid with a thin coverslip.

Iron Haematoxylin Stain

1. Haematoxylin solution

Solution A : 10 gm haematoxylin in 1,000 ml absolute ethanol. Keep in stoppered
Flask and allow to ripen in sunlight for at least a week.

Solution B :

Ferrous ammonium sulphate	10 gm
Ferric ammonium sulphate	10 gm
Hydrochloric acid (concentrated)	10 ml
Distilled water (qs)	1000 ml

Mix the ingredients in a stoppered bottle

Solution C : Mix solution A and B in equal parts.

Ethanol and iodine solutions: See trichrome stain
Xylene
Mounting fluid

Procedure

Prepare fresh faecal smears an a slide and add 70 per cent ethanol for 5 min and then place in 70 per cent ethanol with iodine solution. Place the smear in

70 per cent ethanol for 5 minutes. Wash under running tap water for 10 minutes. Place in the working solution of iron-haematoxylin for 4 to 5 minutes. Wash in running tap water for 10 minutes. Dehydrate the smear by placing in 70 per cent ethanol, 90 per cent ethanol, 100 per cent ethanol and xylene each for 5 minutes. Put the mounting fluid and place a thin coverslip.

Staining of Blood Films

Giemsa's, stain is ideal for the examination of blood films in search of blood parasites. The stain is commercially available. Blood films used for differential count is unsuitable for the study of blood parasites.

PROCESSING OF SPECIMENS OTHER THAN STOOL

There are three other specimens other than stool-sputum, urine, and urogenital exudates (Urethral and vaginal).

Sputum: Sputum specimens are used to observe the nematode larvae which pass through the lungs for their development (e.g.) *Ascaris*. The specimen is examined under the low power objective or by naked eye. The larvae are large up to 35 mm for *Ascaris lubricoides* and 12 mm for hookworm.

Urine: A urine specimen is submitted for the diagnosis of blood fluke infections. *Schistosoma haematobium:* Collect the urine specimen during the noon hours (11 am to 2 pm) when the eggs are found in greater concentration that at any other time of the day.

Concentrate the eggs in the urine specimen. If centrifuge is available, centrifuge (1500 rpm for 5 minutes) 10 ml of well-mixed urine specimen within 1 hour after collection. With centrifuged specimen discard the supernatant and suspend the sediment. This is done either by tapping the tube or by means of a Pasteur pipette. Draw the liquid up and blow it down gently several times. Make a wet mount of the homogeneous urine sediment and examine under a high power objective (40X). Eggs of *Schistosoma haematobium* are about 130 μm, oval with one rounded pole the shell is smooth and very thin. They have a terminal spine at the conical end which is characteristic of this parasite. The eggs are pale yellow coloured and the embryo can be seen which is well-formed with tiny cilia around the edges.

Urogenital Swab: Swabs of urogenital discharge, particularly the vaginal discharge in women (and occasionally the urethral discharge in men) are submitted for the diagnosis of *Trichomonas vaginalis*. This is one of the most common protozoan. Infections and is demonstrated by the actively motile flagellates in a saline suspension of the specimen. The preparation must be examined promptly (Fig. 59.4).

Collection of Specimen: Collect the vaginal exudates or urethral exudates on a wet swab placed in saline (0.85% NaCl).

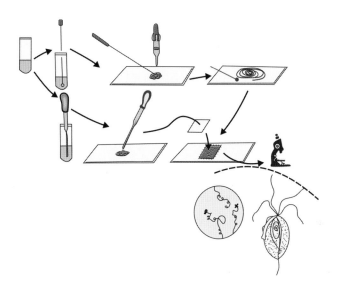

Fig. 59.4: Processing and examination of urogenital swab

Procedure: Place a drop of the fresh specimen (vaginal discharge) on a microscope. Add a drop of lukewarm saline solution. This increases the motility of the trophozoites. Look at the wet mount. First examine under low power with reduced light. The organisms will appear as tiny transparent bodies, the size of white cells, moving rapidly in jerks and loops. Switch to the high-dry objective for identification. Trophozoites are of 10 to 20 μm size, round and globular, motile with four fast moving flagella.

Report the presence of *Trichomonas vaginalis.*

Laboratory Diagnosis of Protozoan Infections

Protozoan infections in humans are commonly associated with gastrointestinal tract (GI) and genitourinary tract (CU). Laboratory diagnosis for these disorders is based on the examination of stool and vaginal or urethral exudates.

Entamoeba Histolytica

Amoebic dysentery, caused by *Entamoeba histolytica*, results in severe bloody diarrhoea. Microscopic observation shows the presence of abundant erythrocytes and leucocytes and charcot-leyden crystals. Trophozoites and cysts are diagnostic.

Trophozoites: 12 to 35 μm shapeless, has a directional movement progressive motility by means of pseudopods, transparent cytoplasm and red blood cells are present. Iodine staining stops the motility of amoeba and stains the nucleus without the central karyosome.

Cysts: 12 to 15 μm, round, 1 to 4 nuclei which are regular, circular and each with a compact central karyosome. Iodine staining makes the nucleus more prominent, yellow grey, granular cytoplasm. Chromatid bodies may be present which have rounded ends.

Giardia lamblia: Cysts of *Giardia lamblia* are found in soft and formed stool and can be seen in concentrated stool. Cysts do not have flagella, they are oval shaped 8 to 12 μm in size, thick shells with double lines, to 4 nuclei karyosome small central and faintly coloured. Clear refractile cytoplasam when unstained and yellowish-green or bluish cytoplasm in iodine solution.

Trichomonas vaginalis: A wet mount of the freshly collected specimen without iodine stain, is the best way to diagnose *Trichomonas vaginalis.* Any moving trophozoites of *Trichomonas vaginalis. Trichomonas vaginalis* does not have a known cyst phase.

Leishmania: Infection is diagnostic by the formaldehyde gel method.

Procedure

1. Take 1 ml serum in a small-size test tube (5 ml)
2. Add 1 drop of 40% formaldehyde.
3. Till the tube and watch for coagulation.
4. Complete coagulation occurs in few seconds in acute cases.
5. The reaction may be slow in the early stage of the disease.

The Helminths: Helminths are multicellular organisms. The adult stage is often visible to the naked eye. Larvae and eggs, are microscopic. The helminthes are primarily identified by the morphology of eggs, larvae, or parasite elements like proglottids. Laboratory diagnosis of most intestinal helminth infections is based on the detection of characteristic eggs and larvae in the stool. Occasionaly adult worms or portions of worms may also be found. No permanent stains are required, and most diagnostic features can be easily seen on direct wet mounts of fresh or concentrated faecal materials.

Identification of Helminth Eggs

Most Commonly Found as

Nonoperculated	
A. Without spine or knobs, eggs with 3 pairs of hooklets	
1. Eggs with 3 pairs of hooklets, small size (< 50 μm)	*Taenia*
a. Round, single thick dark brown shell with transverse striation, dark yellowish-grey content	
b. Pale grey, oval, double shell, no transverse striations, colourless inner shell	*Hymenolepis*
2. Eggs with no hooklets	
A. Double shell with dark brown external cell	
i. Barrel-shaped with Polar plugs	*Trichuris trichiura*
ii. No polar plugs, external shell corticated or uncorticated	
a. Medium size egg (50-70 μm)	Fertile egg of *Ascaris*
b. Large size egg (>70 μm):	Unfertile egg of *Ascaris*
B. Single shell which is thin, clear and transparent	
i. No eggs in stool	
a. Eggs medium size (50-70 μm) collected from Anus surface	*Enterobius*
b. Larvae in stool (viviparous	*Strongyloides*
ii. Eggs found in stool: Medium size (50-70 μm), shell	*Ancylostoma*
Very thin, contents visible (cellular, granular or embryonated.	*Necator*
B. With conspicuous spine or knob, Large size egg (>70 μm)	
Lateral spine	*Schistosoma mansoni*
Terminal spine	*Schistosoma haematobium*
Minute knob *Schistosoma japonicum*	
Operculated	
A. Small size egg (< 35 μm), distinct shape with clear operculum	*Clonorchis*
B. Very large egg (>130 μm), oval with rounded poles, shell smooth and fine with a double line, inner content with large indistinctive cells	*Fasciola*

Identification Characters of Helminths

Ascaris lumbricoides (Roundworm)

The adult *Ascaris lumbricoides* may be seen in or on the surface of the stool. It is long, smooth, unsegmented with pointed ends. The eggs have a double shell and the outer one is corticated (thick tuberculated capsule). Less frequently the external shell with the rough outer surface is missing. The fertilized eggs have a single round, granular central mass. The pale colourless eggs have a brown external shell. Though the shells are indistinct, the unfertilized eggs are slightly bigger than the fertilized eggs and more elongated (elliptical or irregular). The egg is full of large round, very refractile (shiny) granules.

Ancylostoma Duodenale and Necator Americans (Hookworms)

The ova of these two forms look the same although the adults have different morphology. Eggs are medium-sized (50 to 60 µm) oval, with rounded, slightly flattened poles; the shell is thin and appears as a black line; cells inside are pale grey.

Enterobius Vermicularis (Pinworm)

Special method of specimen collection cellotape technique is done. The egg is transparent and colourless oval but clearly asymmetrical (flattened on one side and rounded on the other) 50 to 60 µm size; the shell is thin and a double line is visible contents may be either a granular mass or embryonated, as indicated by a small curled up larva.

Trichuris trichiura (whipworm): Whipworm infection is worldwide. Trichuriasis leads to abdominal discomfort, anaemia, bloody stool, prolapse of the rectum (in case of massive trichuriasis) and allergic conditions. The eggs of *Trichuris trichiura* are easily recognised by their barrel shape with two polar plugs. Diagnosis can be made either with a direct smear in case of heavy infections of the stool specimen is concentrated for mild infections.

Wuchereria bancrofti (microfilaria): The parasites do not produce eggs or ova and reproduce through microfilariae. The diagnosis is primarily based on clinical symptoms and demonstration of microfilariae in the blood film. Blood is usually drawn at night because the organism exhibits nocturnal periodicity.

Procedure

Collect blood between 10 pm and 4 am by the finger prick method or by venipuncture. Put a drop of blood directly on a clean slide with a capillary

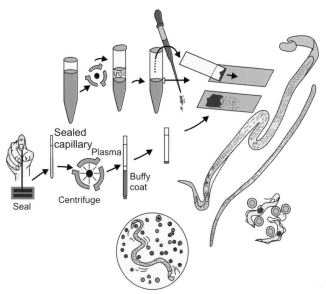

Fig. 59.5: Examination of *Microfilaria*

tube or Pasteur pipette and add a drop of solution. Mix the blood specimen with the saline and then make a wet mount. Examine the wet mount under reduced illumination the first sign of the presence of microfilariae is the rapid movement of the red cells. Prepare a thick smear of blood on another microscope slide. To prepare the thick smear, place three drops of blood in the centre of a microscope slide spread the blood out in a circular manner with the help of an inoculating loop. Stain the blood smear with field rapid stain or Giemsa's stain. Examine the stained blood film first under low power and then under the high-dry objective. Confirm the findings under the oil-immersion objective. Look for the size (length and diameter), presence of a sheath, and the tail (tapered, rounded or hooked Fig. 59.5).

Observation

Presence of microfilariae is easily recognised from the wet mount but it can only be identified on stained smears. Pathogenic microfilariae are thicker (almost the diameter of red cell, 7 µm) and longer (250 µm) than the non-pathogenic ones.

Blood Staining Procedure for Examining Microfilaria

Prepare a thick smear of blood taken from finger stick. Air dry. Place the slide vertically in the staining trough (or beaker) filled with clean water. Wait for 10 minutes for complete haemolysis. Take slides out, drain them and air dry. Stain for 30 minutes using a 1:10 dilution of Giemsa. Cover the slide with

Giemsa's stain and wait for 30 minutes. Wash the stain off and air dry. Screen under the low power objective. Identify *Wuchereria bancrofti* under oil-immersion objective.

Blood Haemolysis Method

Take 1 ml of citrated blood in bottle and mix with 2 per cent formaldehyde solution in a conical centrifuge tube. After waiting for 5 minutes to haemolysis the red cells, centrifuge for 5 minutes (2,500 rpm). Save the sediment and pour off the supernatant. Suspend the deposit in the residual liquid. Take a drop of the suspension on a microscope slide spread the drop to form a thin smear, air dry, fix with ether: alcohol (1:1) and leave to dry for 2 minutes. Stain immediately with Giemsa's stain and examine the slide for microfilariae which stain well.

Microhaematocrit method: Fill a capillary tube with the blood specimen. Seal one end. After centrifugation, lay the capillary tube on a slide and secure the two ends with an adhesive tape. Examine the capillary tube at the buffy coat line under the low power objective with reduced illumination. Infected blood may show the motility of microfilariae at the top layer of the buffy coat. If movement is seen make a smear with the first drop from each piece of the broken tube. A thick film can be prepared. Stain with Giemsa's stain and examine under the oil-immersion objective identify the organism.

SECTION

6

MYCOLOGY

60 *Introduction, Classification and General Properties of Fungus*

The study of fungi is called Mycology, which is derived from the Greek word *'Mykos'* meaning mushroom. The simplest type of fungus is the unicellular budding yeast. Elongation of the cell produces a tubular, thread like structure called hypha. A tangled mass of hyphae constitutes the mycelium. Fungi which form mycelia are called moulds or filamentous fungi.

Depending on cell morphology, fungi can be divided into four classes; yeasts, yeast like fungi, moulds and dimorphic fungi.

INTRODUCTION TO PARASITIC FUNGI

1. *Moulds:* These are filamentous, mycelial fungi. The septate mycelium is the vegetative part of the organism that spreads out over the substrate. Other hyphae develop into the aerial mycelium. The ringworms or the dermatophytes belong to this group.
2. *Yeasts:* The most important pathogenic yeast is *Cryptococcus neoformans*. It is the causal agent of mycotic meningitis is and other systemic disorders. It occurs mainly as a single spherical or ellipsoidal cell (Fig. 60.1), which reproduces by budding or fission. Laboratory culture on Sabouraud dextrose medium yields creamy-coloured colonies, which are mucoid in consistency.

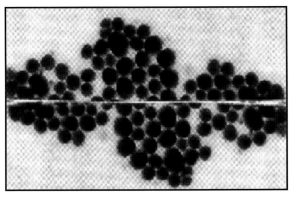

Fig. 60.1: Microscopic appearance of yeasts

3. *Yeast-like fungi:* Some of the fungi like *Candida* are monomorphic, produce pseudomycelia, (Figs 60.2A and B) and also produce yeast like blastospores. The pseudomycelia are long elongated, filamentous cells which are joined end-to-end giving a false appearance of septate, hyphae-like structures. *Candida albicans* causes thrush in the mouth, and vulvovaginitis. *Torulopsis glabrata* is now renamed as *Candida glabrata.* Blastospores of *Candida albicans* produce germ tubes when incubated with human serum. In addition, when subcultured on high carbohydrate medium, *Candida* produce chlamydospores borne on the pseudomycelia.

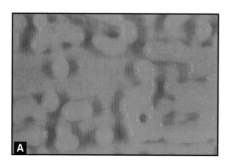 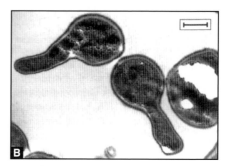

Figs 60.2A and B: Appearance of yeast-like fungi

4. *Dimorphic mycotic agents:* These organism change their morphological form with changes in temperature. At room temperature, they take the filamentous form, while at body temperature (37°C) they change to the yeast form. The most important pathogenic dimorphic fungi are *Blastomyces dermatidis, Histoplasma capsulatum, Coccidioides immitis* and *Sporothrix schenckii.* Except for *Sporothrix schenckii,* all are causes of systemic infections. They infect the lungs and other parts of the body, like spinal fluid, bone marrow, and the renal system. *Sporothrix schenckii* causes subcutaneous infections (sporotrichosis) which commonly presents clinically as an ulcerative lesion, usually on the hand, with nodules developing along the lymphatic vessels draining the area. Tissue forms (Yeast-phase) are seen in a wet mount or biopsy specimen, or when cultured in the laboratory on brain heart agar medium.

Opportunistic Fungi and Filamentous Bacteria

The opportunistic fungi are those which normally lead a saprophytic life. They often appear as contaminants in laboratory culture, but may infect patients with impaired defense systems the compromised, those under steroid therapy, and those with long-term antibiotic treatment, some of these opportunistic fungi belong to Deuteromycetes (*Candida, Geotrichum*), **while others belong** to *Ascomycetes* (*Aspergillus* and *Penicillium*) or *Phycomycetes* (Rhizopus, mucor, Absidia) **Actinomyces** and **Nocardia** are the most important pathogenic filamentous bacteria that have characters of both fungi and of bacteria. Hence, they are known as filamentous bacteria. Both the organisms are gram-positive bacilli.

SPECIMEN COLLECTION

The location of the infection site considerably helps in the diagnosis of a mycotic infection. Always examine the specimen promptly and if laboratory culture is needed, inoculate as quickly as possible. Refrigerate the specimen if delay is anticipated.

Skin Scrapings, Nails, and Hair

Specimens for dermatophyte infection are skin scrapings, nail scrapings, and hair. Attempts should be made to minimise contamination, by first cleansing the area with 70 per cent alcohol and then with sterile water. From skin lesions, scales preferably 2 to 3 mm in diameter, should be scraped outwards with a blunt scalpel from the active periphery of the lesions (ringworm) and the domes of any vesicles snipped off. In external otitis, scrapings from the external ear will be needed. For mycotic keratitis, corneal scrapings are submitted to the laboratory for diagnosis. Scrape the nail with a scalpel, and discard the first superficial scrapings. Some of the friable material from under the nail may be used. Infected hairs are carefully selected which have dusty stumps of ectothrix infections. These hairs may fluoresce under ultraviolet light (Wood's lamp), which suggests the aetiologic agent may be *Microsporum*. Pluck the hair with a sterile fine forceps, and collect in a sterile petri dish. Both the hair shaft and the hair root should be obtained. Non-fluorescent hair should also by examined under a wet mount, and cultured, if a fungus is suspected clinically.

Most likely mycotic agents in various specimens.

Common specimens	Likely organism
Skin scrapings	Microsporum, *Trichophyton, Epidermophyton*
Corneal scrapings	*Aspergillus* and other saprobic molds
External ear scrapings	*Aspergillus*
Sputum (respiratory secretions)	*Histoplasma, Coccidioides, Blastomyces, Candida, Aspergillus, Nocardia, Sporothrix*
Sinal fluid	*Cryptococcus*
Pus, exudates	*Madurella, Actinomyces, Nocardia*
Mucocutaneous lesions	*Candida, Histoplasma, Coccidioides, Blastomyces*
Vaginal exudates	*Candida*
Urine	*Candida*, systemic mycotic agents
Bone marrow	*Histoplasma*
Throat swab	*Geotrichum, Candida*

Sputum

Sputum specimen is submitted in case of systemic mycotic infections. An early morning specimen is desirable. The patient must rinse his mouth thoroughly and raise the sputum by coughing. Collect the sputum in a clean. Sterile container with a tight-fitting lid.

Pus and Exudates

Pus specimens are taken from mucocutaneous ulcers, draining subcutaneous sinuses, and from abscesses. Cotton swabs are seldom satisfactory. A draining pus specimen helps considerably in examining and diagnosing the causal agent by direct microscopic examination. The abscess is first cleansed with 70 per cent alcohol, then cleaned with tincture of iodine, and then cleaned with sterile water. Pus is first aspirated with a sterile needle and syringe before the incision is made. Remove the aspirate with a sterile Pasteur pipette, and place it in a sterile labelled test tube.

Cerebrospinal Fluid

The spinal fluid is spun at low speed, and the sediment is used for microscopic examination and culture.

Other Specimens

Blood specimens, used for bacterial culture may show the growth of organisms which stay in the yeast phase within the body. Urine sediment is used to examine some of the yeast forms of systemic mycoses.

61 *Laboratory Diagnosis of Fungal Infections*

Laboratory diagnosis of mycotic infections is based on the following:
1. Histopathological studies
2. Macroscopic examination
3. Microscopic examination by wet mount, i.e. direct examination of the specimen
4. Laboratory culture
5. Biochemical and other tests.

HISTOPATHOLOGY OF FUNGAL INFECTIONS

The histopathologist plays an important role in the diagnosis of mycotic diseases. Particularly in pulmonary infections and mycotic infections of the haematopoietic system.

MACROSCOPIC EXAMINATION

Hair and skin infected with the dermatophyte *Microsporum* fluoresce when placed under an ultraviolet light (wood's lamp) in a dark room. The organism produces a yellow green fluorescence of infected hair. Skin infected with *Malassezia furfur* (tinea versicolor) also fluoresces under the ultraviolet light The fluorescent area should be scraped and examined under the microscope and then subjected to laboratory culture.

DIRECT MICROSCOPIC EXAMINATION

Direct microscopic examination by wet mount is probably the most important method of examining and diagnosing mycotic infections in the laboratory.

WET MOUNT IN WATER

A wet mount in water, with or without staining, is recommended for certain specimens, such as spinal fluid, urine sediments and other fluid material.

Parker's super chrome blue black ink or methylene blue is used for staining. These impart a blue colouration to the hyphae. Place a drop of water (with stain in equal proportion, 1:1) on a clean slide. Suspend a small portion of the colony in the drop of water with the help of a Nichrome inoculating wire. Make a light suspension. Place a thin cover slip over the drop of water and examine under the microscope. Screen under the low power, and examine closely under high power. The thin aqueous preparation permits examination of the shape, size and morphology of the organism.

WET MOUNT IN ALKALI SOLUTION

Use 10 per cent KOH (or 10% NaOH) solution as the mounting medium for opaque and hard specimens. The alkali helps in digestion, and acts as a clearing agent and disinfectant. With the clearing of the debris, the fungal elements become more prominent.

Reagents

10 per cent KOH (or NaOH): Dissolve 10 gm of potassium hydroxide (or sodium hydroxide) in water and make up to 100 ml.

Alkali-glycerine Solution

Glycerin	10 ml
Sodium hydroxide	20 gm
(or Potassium hydroxide)	
Distilled water	90 ml

Procedure

Place a drop of mounting fluid (alkali or alkali-glycerin solution) on the centre of a clean microscope slide.

Wet a stiff teasing needle with KOH solution, and pick up the material to be examined for, e.g. hair, scrapings of skin and nails; or a portion of a colony cultured in the laboratory.

Tease the material into a thin preparation.

Add a coverslip

Apply gentle heat by passing through a flame a few times.

Wait for 5 minutes.

Results

Fungal elements appear as clear hyaline structures. Examine the preparation under low power for screening, and confirm under the high power objective.

Unstained wet preparations are good enough to identify the hyphae and arthrospores.

Wet Mount in Lactophenol Cotton Blue

This is the most commonly used fungal stain which can also serve as a mounting medium. The method is used for yeasts as well as moulds regeant

Lactic acid	20 gm
Phenol crystals	20 gm
Glycerin	40 gm
Distilled water	20 ml

Dissolve the ingredients by heating gently over a steam bath. Add 0.5 gm of cotton blue dye.

Procedure

1. Place a drop of lactophenol cotton blue on a clean slide.
2. Place a small amount of specimen in this drop, and suspend it. If a laboratory culture is to be examined, remove a piece of the medium with the embedded growth.
3. Tease the specimen if opaque.
4. Place a coverslip on the mounting fluid and press down gently.
5. Examine under the low power to screen, and switch to the high power objective for closer examination.
6. If a permanent mount is required, seal the edge with nail polish.
7. Fungal material appears pale to dark blue.

INDIA INK PREPARATION OF CSF

This technique is used in the identification of *Cryptococcus neoformans* that causes mycotic meningitis. CSF is examined using the technique.

Procedure

1. Centrifuge the spinal fluid and use the sediment for the India ink preparation.
2. Transfer a drop of India ink to a clean slide. Check under the microscope (low power) and if the preparation is too dark, dilute with a drop (or more) of sterile distilled water.
3. Flame sterilise a Nichrome wire loop, cool completely. Transfer a loopful of CSF sediment to the India ink, and mix thoroughly.

4. Place a clean thin coverslip over the preparation. Gently press to make a thin film.
5. Examine under the high-dry and the oil-immersion objective, using reduced light for differentiation.
6. *Cryptococcus neoformans* appears as a clear disc against a black background. The faintly visible unstained cell is surrounded by a wide, clear, capsular space. The yeast-like organism inside the capsule is oval to spherical.

Chapter

62

Laboratory Culture of Fungi

The presence of spores of filaments in clinical material is not enough to identify the organisms with certainty. Hence, all specimens should be cultured in the laboratory.

Media Requirements and Incubation Conditions

The commonly used media for the culture of pathogenic microbes is given in Table 62.1. Sabouraud dextrose agar medium is the medium of choice for most fungi. The antibiotic cycloheximide is added to the medium for the culture of dermatophytes and others, because contaminating bacteria are likely to be present. The antibiotic inhibits the growth of the bacteria, without affecting the fungus. For the study of sporulation, special carbohydrate rich media are needed cornmeal agar, chlamydospore agar.

Table 62.1: Common culture media used in the laboratory for various mycotic infections	
Media used	**Infection**
Sabouraud dextrose without antibiotics	Most mycotic infection and opportunistic fungi.
SAB broth	*Candida* and other yeast-like organisms
Sabouraud dextrose with antibiotics*	Dermatophyte infection and for specimens where bacterial contamination is suspected (skin scrapings, sputum, pus and others)
Littman oxgall medium	Same as SAB
Cornmeal agar, rice grain, chlamydospore agar potato-dextrose agar	Chlamydospore formation for *Candida*
Brain heart infusin agar (with * and without antibiotics)	Fastidious organisms, yeast-phase growth of dimorphic fungi
Blood agar (with * and without antibiotics)	*Candida*, dimorphic organisms (Yeast-phase)

Note: Antibiotics are not suitable for the growth of fastidious organisms (*Cryptococcus*), yeast forms of dimorphic fungi, and for opportunistic fungi.

Preparation of Culture Media for Mycotic Agents

Sabouraud Dextrose Broth

This is suitable for the growth and differentiation of *Candida* and other species of the yeast-like fungi

Dextrose	40 gm
Peptone	10 gm
Distilled water (qs)	1000 ml

1. Dissolve the ingredients with gentle heating and stirring
2. Dispense in 10 ml amounts in culture tubes (18 mm × 150 mm)
3. Autoclave (121°C for 10 minutes).

Sabouraud (Dextrose Agar) (SAB)

This is the most useful selective medium for the culture of mycotic agents, particularly the filamentous moulds. This helps in the culture of skin scrapings.

Ingredients

Dextrose	40 gm
Peptone	10 gm
Agar	15 gm
Distilled water (qs)	1000 ml
pH 6.0	

1. Dissolve agar in 1000 ml of distilled water by heating.
2. While hot, add peptone and dextrose
3. Boil gently until dissolved
4. Adjust the pH to 6.0.
5. Dispense into culture tubes (20 ml) with cotton plugs or caps.
6. Sterilise by autoclaving (121°C for 15 minutes).
7. Cool the culture medium in slants, or when the temperature of the medium reaches 50°C pour in sterilised plates.
8. Store the plates, tubes and slants in the refrigerator after leaving overnight at room temperature.

Sabouraud Dextrose Agar with Antibiotics

This medium is recommended for growing mycotic agents. When the specimen is contaminated with bacteria. The antibiotics are added by the following procedure.

1. Remove solidified SAB medium in tubes from storage in the refrigerator, melt by keeping in a boiling water bath and then keep at the room temperature to cool down to 50°C. Add the antibiotics into the molten agar (50°C) in the tube and mix.

2. The antibiotics may be added after dissolving the medium by heating and before autoclaving. The following amounts are recommended for 1000 ml of the medium

Cycloheximide	400 gm
Chloramphenicol	50 gm

Add chloramphenicol dissolved in 5 ml of 95 per cent alcohol and the cycloheximide dissolved in 5 ml of acetone. Mix well and distribute into tubes or bottles as indicated. Autoclave at 121°C for no longer than 10 minutes.

Cornmeal Agar

This medium is used for suppressing vegetative growth, and stimulating the production of chlamydospore formation in *Candida albicans*.

Ingredients

Yellow cornmeal	40 gm
Agar	15 gm
Distilled water	1000 ml

1. Heat cornmeal in 500 ml of distilled water by boiling, or place in a 60°C water bath for 1 hour.
2. Filter and add water up to a 1000 ml volume in a beaker (2 litres) with graduation.
3. Add 15 gm of agar, and dissolve with gentle heating and agitation.
4. Dispense in culture tubes
5. Autoclave at 121°C for 15 minutes.

Potato Dextrose Agar

Potato dextrose agar contains dextrose and potato infusion. It is used for slide culture of yeasts and fungi. It stimulates spore production, and thus aids in the study of sporulation.

Ingredients

Potatoes	200 gm
Infusion from glucose (dextrose)	20 gm
Agar	15 gm
Distilled water (qs)	1000 ml.

1. Cook 200 gm of peeled and diced potatoes in 500 ml of distilled water.
2. Filter the infusion through cotton, and add distilled water up to a 1000 ml volume.
3. Add to the infusion (1000 ml), glucose (20 gm), and agar (15 gm)
4. Dissolve the ingredients by heating.
5. Sterilise by autoclaving (121°C for 15 minutes).

DIAGNOSTIC TESTS FOR MYCOTIC AGENTS

Germ Tube Test

The blastospores of *Candida albicans* produce germ tubes within 3 hours when incubates (37°C) in human serum. The germ tube is a non-septate tubular process that eventually give rise to pseudohyphae. This is typical of *Candida*. Transfer a small inoculum from a young colony to a small tube containing human serum. Incubate for 3 hours at 37°C. Examine the suspension of blastospores by making a wet mount after 3 hours. Look for the germ tube production with a high-dry objective. Germ tube production by the blastospores confirms the identification of *Candida albicans*.

Ureas Test

Cryptococcus neoformans and a few other mycotic agents are urease positive. Thus the urease test can differentiate *Cryptococcus* from other yeasts.

Inoculate Christensen urea agar with the organism to be tested. Use a fresh colony, and take the inoculums from the top of an isolated colony suspected to be cryptococcus. Incubate for 5 days at 30°C (examine daily). If the medium turns pink, the organism is urease positive.

Carbohydrate Utilisation (Assimilation) Test

Paper discs impregnated with various sugars are placed on a uniformly seeded plate of the yeast form mycotic agent. Grow the organism for 48 hours in Sabouraud dextrose broth and inoculate. Growth around a specific disc indicates that the fungus utilises that specific carbohydrate. The test helps in differentiation of various species of *Candida*.

Chapter

63

Diagnostic Mycology

Laboratory diagnosis of mycotic infections (diagnostic mycology) is based on both direct microscopic examination, laboratory culture and performing the various diagnostic tests.

Laboratory Diagnosis of Dermatomycosis

Clinically important species of dermatophytes are *Epidermophyton, Microsporum* and *Trichophyton*. These infect hair and skin causing ringworm or tinea. Laboratory diagnosis of a dermatomycosis is based on the direct examination of skin scrapings, nail scrapings and hair, microscopic examination of these specimens by the wet mount technique on 10 per cent KOH (use 20% KOH for tougher specimens like hair), and their laboratory culture. The hair preparation may show and ectothrix or endothrix pattern of spore growth, and the skin scrapings may show branching, septate, hyaline hyphae. When filamentous structures are found in the specimens, laboratory culture of the organism should be done. Sabouraud dextrose with antibiotics (cycloheximide and chloramphenicol) is the appropriate medium. Culture are maintained at 30°C under aerobic conditions. Colonies are visible within 5 days, but may take as long as 2 weeks. For the sporulation study, subculture on carbohydrate rich agar medium.

Trichophyton

Most species of *Trichophyton* invade hair, sometimes skin and nails may also be affected. When grown in the laboratory on sabouraud dextrose agar with antibiotics, it is slow growing and takes about 2 weeks to grow into cottony, velvety or powdery colonies that have different colours. Macroconidia are scanty and when present, are smooth-walled, elongated and septate (2 to 6 segments). Microconidia are abundant, teardrop to club shaped , and grow in clusters. Hyphae may show spirals in some species. The growth pattern in and on hair, divides *Trichophyton* into endothrix and ectothrix groups (Figs 63.1 to 63.4).

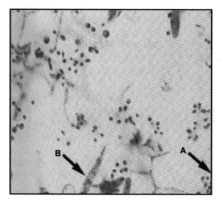

Fig. 63.1: *T. mentagrophytes*

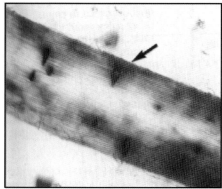

Fig. 63.2: *T. mentagrophytes* hair perforation

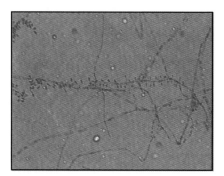

Fig. 63.3: *T. rubrum*

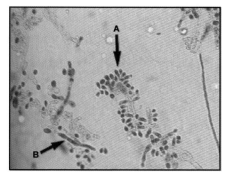

Fig. 63.4: *T. tonsurans*

Microsporum

Microsporum is more commonly isolated from hair and skin but not from nails. It fluoresces with a yellow green colour under ultraviolet light. Thus *Microsporum* is diagnosed directly from the infected areas of skin or hair by

Fig. 63.5: *M. canis*

Fig. 63.6: *M. gypseum*

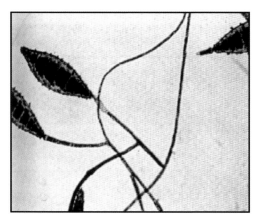

Fig. 63.7: Macroconidia

exposure to wood's lamp. Colonies are whitish, pinkish or other colours (Figs 63.5 to 63.6). The macroconidia are numerous, large, multiseptate (5 to 15 segments) spindle-shaped (Fig. 63.7). Microconidia are few or absent.

Epidermophyton

Epidermophyton infects skin and nails, but not the hair. Skin scrapings will indicate the presence of arthospores. Colonies are fast growing (1 week) characterised by a velvety to powdery appearance. Macroconidia are numerous in the laboratory culture, large, smooth multiseptate (2 to 4 segments) clavate, club-shaped (Fig. 63.8). Macroconidia are not seen.

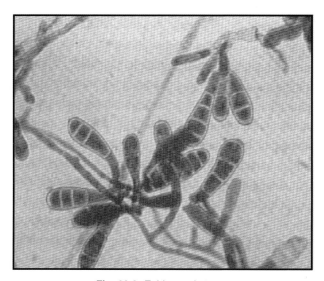

Fig. 63.8: Epidermophyton

64

Laboratory Diagnosis of Subcutaneous Mycosis

This includes such clinical conditions as sporotrichosis, maduromycosis and chromoblastomycosis.

Sporothrix Schenckii

This organism is the causal agent of sporotrichosis, which is a subacute and chronic subcutaneous lymphatic mycosis, characterised by abscesses and granulomatous nodules. It is a dimorphic fungus.

Diagnosis

Specimen: Pus or exudates from an ulcerated lesion is examined for diagnosis. Aspirated material from a fluctuating nodule is a preferred specimen.

Direct microscopic examination: This is of little value.

Laboratory culture: Primary plating is done on sabouraud dextrose agar with antibiotics, and on brain heart infusion agar with and without antibiotics. Incubation at 30°C for 3 to 7 days yields moist, white smooth, leathery colonies with wrinkled and folded areas. A wet mount will show septate hyphae, and small pyriform conidia arranged in clusters on the ends of lateral conidiophores. The yeast form show fusiform yeast cells that resemble cigar bodies in shape and reproduce by budding.

MADUROMYCOSIS

Maduromycosis is a chronic granulomatous infection, called mycetoma, that usually involves the soft tissues and bones of the feet, and rarely other parts of the body (Madura foot). A number of mycotic agents (**Madurella, Petriellidium, Exophialas**) are involved in this clinical condition. Most of them are monomorphic filamentous fungi. *Nocardia* can also be an important causal agent for maduromycosis.

Diagnosis

Specimen: Pus and curetting.

Direct microscopic examination: Microscopic examination of the specimen reveals the presence of grains or granules. The granules are the dense collection of hyphae with or without spores.

Laboratory culture: When cultured on Sabouraud dextrose agar with anti-biotics, they yield a characteristic filamentous growth.

CHROMOBLASTOMYCOSIS

Chromoblastomycosis is a chronic mycosis of the skin characterised by ulcerated verruciform lesions.

Specimen: Scrapings and biopsy tissue.

Direct microscopic examination: The tissue form (found in the body) of all chromoblastomycotic agents is characterised by yellowish-brown, spherical septate bodies, found in abscesses within giant cells, or are free.

Laboratory culture: When cultured in the laboratory on sabouraud dextrose medium with antibiotics, the organisms produce dark colonies with black reverse sides and short grey aerial mycelia. Species are identified by the morphology of condiophores.

MYCETOMAS

A mycetomas is a subcutaneous infection caused by a number of opportunistic fungi including *Madurella mycetomi* and filamentous bacteria *Actinomyces* and *Nocardia*. It common affects the foot, following trauma to the skin, and leads to swelling and deformity. The chronic ulcers have several sinuses discharging pus which may contain coloured granules.

Diagnosis

Specimen: Pus is the most common specimen. During specimen collection, allow the sinuses to discharge into a piece of surgical gauze; examine the discharged granules under the microscope.

Direct microscopic examination: Direct examination of the specimen is most helpful. If sulphur granules are present, actinomycosis is suspected. Other granules may suggest mycotic infection (*Madurella mycetomic*). If granules other than sulphur granules are present, place one granule on a clean slide, and view under the low power objective. Crush one granule between two slides, take them apart and stain with the Gram stain, and Kinyoun's acid-fast stain weak. A weakly acid-fast reaction suggests *Nocardia*.

Laboratory culture: Wash the granules in several changes of saline, crush, and use as the inoculums. When granules are absent, culture the pus. Inoculate the following media-sabouraud dextrose agar without antibiotics, brain heart infusion agar and thioglycollate broth. *Actinomyces* will grow only in the thioglycollate broth while the others *Nocardia* and *Madurella*, will grow under aerobic conditions. Filamentous bacteria do not have typical fungal hyphae while the mycotic infections will show their presence. The size of the hyphae, morphology of granules and their staining reaction provide a tentative diagnosis of the causal agent of mycetoma. This is confirmed by laboratory culture.

65 *Laboratory Diagnosis of Systemic Mycosis*

Systemic mycoses involve infection by pathogenic fungi. The organism may be localised in some organs, e.g. cryptococcal meningitis or disseminated all over the body. The causal agents for systemic mycoses can be grouped under the following four classes.

1. *Yeasts: Cryptococcus*
2. *Yeast-like organisms: Candida, Geotrichum*
3. *Dimorphic organisms: Histoplasma, Coccidioides, Blastomyces.*
4. *Opportunistic fungi: Aspergillus, Mucor, Rhizopus* causing pulmonary infection and infection of eye and outer ear. These are monomorphic and always found in the mycelial form.

YEASTS

Cryptococcus Neoformans

Cryptococcal infections cause mycotic meningitis which is often fatal. The organism is widely spread in nature and present abundantly in pigeon-faeces. The organism is only found in the monomorphic yeast form.

Diagnosis

Specimen: Spinal fluid is the principal specimen. Other specimens include respiratory secretions, urine, blood and serous exudates for the diagnosis of pulmonary and disseminated cryptococcosis.

Direct microscopic examination: This is done by a wet mount of the specimen using an India ink preparation. *Cryptococcus* is encapsulated, and so India ink cannot penetrate through the capsule. This results in negative staining of the yeast-form of the organism.

Laboratory culture: Inoculate 15 to 20 ml of uncentrifuged spinal fluid into a series of culture tubes containing blood enriched agar or brain heart infusion agar. The organism is present in very small numbers, and the centrifugation may destroy the more fragile cells. *C. neoformans* will also grow on Sabouraud dextrose medium without cycloheximide. *C. neoformans* is sensitive to cycloheximide, hence this should not be used in any of the media. The colonies are wrinkled and whitish in the early stages, but they become slimy mucoid and cream-coloured or brownish as they grow older. The creamy colony tends to flow down to the bottom of the slant. Microscopic examination by India ink wet mount will show the budding cells which do not have capsules in young cultures but are encapsulated in older cultures. An additional confirmatory test for the identification of *Cryptococcus neoformans* is the urease test. Inoculate a urea agar slant. (Christensen urease agar) with the growing organism; a positive reaction-red colouration will be seen after 1 to 2 days incubation at room temperature.

YEAST-LIKE ORGANISMS

These organism produce both budding cells (blastospores) as well as mycelia which are different from the filamentous forms, called pseudomycelia.

Candida Albicans

Candidiasis (or candidosis) is caused by *Candida albicans*, which results in acute, subacute or chronic superficial infections like thrush or moniliasis in the mucosa (mouth or vagina). *Candida* is normally present in the mouth, intestine and vagina as a commensal and behaves as an opportunistic invader.

Diagnosis

Specimens: Swabs from the mucosal surface or infected areas-vaginal and oropharynx-are the usual specimens submitted for thrush and moniliasis. Scrapings of skin are submitted for cutaneous candidiasis. Sputum, blood, urine, and spinal fluid are submitted for disseminated candidiasis.

Direct microscopic examination: The presence of oval budding yeast-like cells in the wet mount of a fresh specimen mounted with 10 per cent KOH, suggests the possibility of candidiasis. Gram staining of the smear shows the presence of gram-positive, yeast-like cells, and hyphae.

Laboratory culture: Candida albicans grows well on sabouraud dextrose medium with antibiotics. The organism is fast growing and the colonies are soft, white to tan or cream-coloured smooth and pasty. Microscopic examination of the wet mount of the material taken from the colony, will reveal thin-walled budding cells along with elongated pseudohyphae. The pseudohyphae are not true hyphae; they are elongated daughter cells originated from budding blastospores. These are referred as pseudomycelia.

Germ tube test: This is specially applicable for the identification of *Candida albicans*. In addition, the urease test helps to differentiate *Candida* from *Crytococcus*. *Candida albicans* is urease negative while *Cryptococcus neoformans* is urease positive.

Geotrichum Candidum

Geotrichum: It is a saprophyte, that may cause an infection of the mouth, respiratory tract, or the gastrointestinal area, leading to the clinical condition known as geotrichosis.

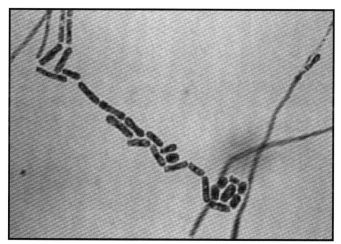

Fig. 65.1: *G. candidum*

Diagnosis

Specimen: The usual specimens submitted to the laboratory for the diagnosis of geotrichosis are sputum, scrapings from oral mucosa and faeces.

Direct microscopic examination: Wet mount of the specimen and laboratory culture indicate the presence of oval, barrel-shaped to round cells, with septate hyphae, and arthrospores (Fig. 65.1).

Laboratory culture: Sabouraud dextrose agar with antibiotics is the preferred medium for the laboratory culture of *Geotrichum candidum*. The colony is white to gray, flat with spreading undulated edges. Aerial hyphae will be seen at the centre with a wrinkled appearance. Microscopic examination reveals that the mycelium consists of hyphae containing rectangular arthrospores.

DIMORPHIC ORGANISMS

Three organisms are primarily discussed here—*Histoplasma capsulatum, Coccidioides immitis* and *Blastomyces dermatitidis.* Infection by all these organisms occurs when the spores are inhaled. The primary infection site is therefore, the respiratory tract. Inside the body these dimorphic organisms take the yeast form, leading to disseminated systemic disorders

Histoplasma Capsulatum

Histoplasmosis, caused by *Histoplasma capsulatum* affect the reticuloendothelial system and central nervous system. Infection is acquired by inhalation. The infections resemble tuberculosis leaving behind an area of Miliary calcification. The reticuloendothelial system is involved with resultant lymphadenopathy, hepatosplenomegaly, fever, anaemia. Granulomatous and ulcerative lesions may develop on the skin and mucosa.

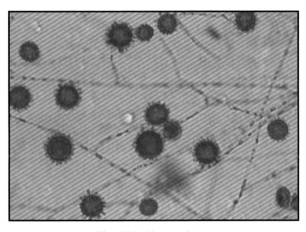

Fig. 65.2: *H. capsulatum*

Diagnosis

Specimen: A sputum specimen is submitted for the diagnosis of suspected cases of pulmonary histoplasmosis.

Direct microscopic examination: A wet mount of the specimen reveals the tissue phase (Fig. 65.2) (yeast) of the organism. The individual yeast cell is small, oval to pear-shaped, is single or in clusters.

Laboratory culture: Laboratory culture can only provide a confirmed report. Inoculate promptly with a centrifuged specimen (sputum). The recommended media are sabouraud dextrose agar medium with an antibiotic and on brain heart infusion agar without an antibiotic. *H. capsulatum* is recognised as a raised, white fluffy mould, becoming tan to brown with age. A wet mount of the isolates from the fluffy colonies will show the presence of septate mycelia, smooth microconidia (2 to 4 mm) and the characteristic tuberculate macroconidia. These spores are large (7 to 25 mm) round, thick walled, and covered with knob-like or spike-like projections (tuberculate).

Coccidioides Immitis

Coccidioidomycosis, like histoplasmosis, is a respiratory tract infection, which is acute, benign and self-limiting. *C. immitis* occurs as a saprophyte, in the mycelial form. It produces arthrospores or arthroconidia from the mycelia and these are inhaled from contaminated soil.

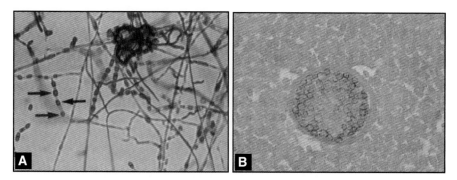

Figs 65.3 A and B: *C. immitis* **(A)** Mycelial form **(B)** Tissue form

Diagnosis

Specimen: Sputum and gastric washings are the common specimens. Other specimens, like a swab from mucocutaneous ulcers, spinal fluid and urine may also be submitted for the disseminated caeses.

Direct microscopic examination: Microscopic examination of the infected material shows the presence of spherules which contain small numerous endospores. These spherules or sporangia releases the endospores (Tissue phase) by rupture of the cell wall.

Laboratory culture: Sabouraud dextrose medium with antibiotics is used for the laboratory culture. Incubate at 30°C or at room temperature. Filamentous growth on sabouraud dextrose medium will be seen in 3 to 5 days as a white, cottony mycelium turning from buff to brown with age. Microscopic examination of young cultures will show the presence of a branching, septate mycelium forming chains of thick-walled, rectangular or barrel-shaped arthrospores (Figs 65.3A and B). In lactophenol cotton blue mounts, the mycelia show only alternate deeply stained arthrospores, with dried out transparent cell tags on either side. The presence of arthrospores confirms the diagnosis of *Coccidioides immitis*.

Blastomyces Dermatitidis

Blastomycosis is a chronic granulomatous and suppurative disease, caused by the dimorphic fungus, *B. dermatitidis*. It primarily affects the lungs, but may also involve the skin, and other organs in the disseminated form.

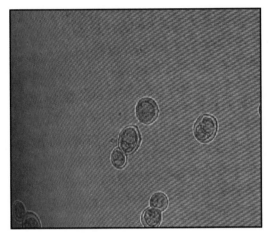

Fig. 65.4: *B. dermatitidis*

Diagnosis

Specimen: Sputum and scrapings or pus from the periphery of cutaneous lesions are the specimens.

Direct microscopic examination: This is based on a wet mount using 10 per cent KOH (apply mild heat on a flame, with opaque material). View under the low power objective, and switch to the high power objective, for a closer view of the yeast-like organisms (Fig. 65.4).

Laboratory culture: Culture of the organism is done by inoculating the specimen on sabouraud dextrose agar with antibiotics. Incubate at room temperature or 30°C colonies will be visible in about 2 weeks and are white with cottony appearance.

OPPORTUNISTIC FUNGI

Opportunistic fungi are those which normally live as saprophytes. The opportunistic fungi include two genera of *Phycomycetes* (*Rhizopus* and *mucor*), and one genus of *Ascomycetes* (*Aspergillus*). Aspergillosis is a bronchopulmonary infection external ear (otomycosis) and eye (mycotic keratitis) infections may involve *Aspergillus, Penicillium, Rhizopus, Mucor* and *Absidia* (Figs 65.5 and 65.6).

Aspergillus Fumigatus

Aspergillus fumigatus is commonly associated with pulmonary aspergillosis, that results in granulomatous lesions in the lungs (Fig. 65.7).

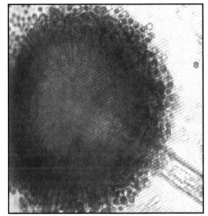

Fig. 65.5: *A. niger*

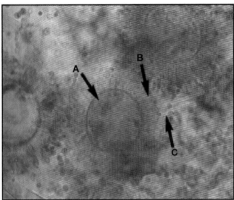

Fig. 65.6: *A. flavus*

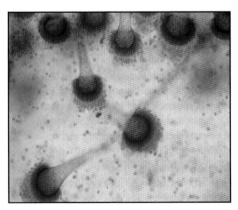

Fig. 65.7: *A. fumigatus*

Diagnosis

Specimen: Usually submitted specimens for the laboratory diagnosis include sputum from bronchopulmonary infection and skin scrapings from the external ear canal for ear infections.

Direct microscopic examination: Wet mount of the specimen in 10 per cent KOH shows white septate hyphae and condiophores.

Laboratory culture: Aspergillus is cultured in the laboratory on sabouraud dextrose agar without an antibiotic. A flat white colony appears in the initial stage of mycelial growth. This rapidly becomes bluish-green and powdery in older cultures. Microscopic examination of the material will show septate hyphae that give rise to conidiophores that expand into a large, inverted flask-shaped vesicle covered with small sterigmata.

Penicillium notatum is rarely isolated from sputum. When inoculated on sabouraud dextrose agar without antibiotics, grey colonies are seen. The colonies are heaped and wrinkled with a white periphery. The white aerial hyphae give a soft, cottony appearance. Microscopic examination reveals branching septate hyphae.

Chapter

66

Rhizopus, Absidia and Mucor

Zygomycosis is caused by Phycomycetes (*Rhizopus, Absidia* and *Mucor*). These are saprophytes like other opportunistic fungi and can cause infections of the eye, brain and lungs. The mycelia have aseptate hyphae.

Specimen: A variety of specimens are submitted to the laboratory, which include sputum, nasal exudates, and corneal scrapings.

Direct microscopic examination: This may show the presence of non-septate hyphae in the wet mount (Fig. 66.1).

Laboratory culture: Inoculate on SAB without antibiotics and incubate at room temperature. The growth is coarse, woolly and fills the test tube rapidly with a loose, grayish mycelium dotted with brown or black sporangia.

Rhizopus: This has a rich wooly mycelium, that is white at first, and later becomes grey with black and brown dots which are the sporangia. Microscopic study shows the characteristic features of *Rhizopus*. They have large, broad,

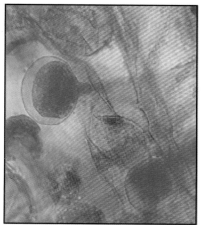

Fig. 66.1: *Zygomycetes*

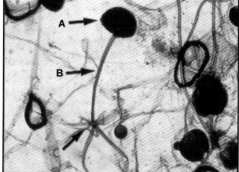

Fig. 66.2: *Rhizopus*

Fig. 66.3: *Mucor*

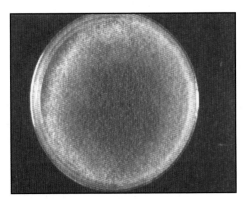

Fig. 66.4: *Rhizopus* colony

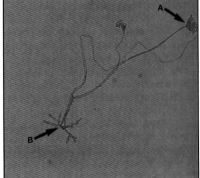

Fig. 66.5: *Absidia*

non-septate, hyaline mycelia that produce horizontal runners (called stolons) which attach at contact points with the substrate by root-like structures called rhizoids (hence the name *Rhizopus*) (Fig. 66.2).

Mucor: This grows rapidly and produces aerial mycelia that are soft and white at first and later become grey to brown at the sporulation stage microscopic examination reveals non-septate hyphae that lack rhizoids and give rise to tall single, erect sporangiophores (Fig. 66.3).

Absidia: This bears broad, aseptate, irregularly branching, ribbon like hyphae with non-parallel opposing walls (Fig. 66.5). The fungus resembles *Rhizopus* and *Mucor* in morphological characteristics (Figs 66.4 and 66.5).

Chapter 67

Antifungal Chemotherapy

Polyenes: The polyenes nystatin and amphotericin B are lipophilic and have affinity to the sterols present in fungal membranes. Amphotericin B is insoluble in water and must be administered intravenously as a colloidal suspension. Almost all fungi are susceptible to amphotericin B, and the development of resistance is too rare.

Azoles: The important antifungal azoles are imidazole, ketoconazole, and the triazoles, fluconazole, and itraconazole.

Ketoconazole was the first azole to be useful in systemic infections but it is substantiated by either fluconazole or itraconazole for most systemic mycoses, including aspergillosis and candidiasis, for which ketoconazole is not effective ketoconazole and itraconazole are given orally and fluconazole either orally or intravenously azoles are also effective for superficial and subcutaneous mycoses. Two other azoles are clotrimazole and miconazole. These are used as topical preparations.

Voriconazole: Provide a broad spectrum of antifungal activity against some yeasts and molds that are resistant to other azoles. It can be given intravenously or orally.

Allylamines: These are a group of synthetic compounds and includes an oral agent, terbinafine, and a topical agent, naftifine. Both are used in the treatment of dermatophyte (ring worm) infections.

Flucytosine: Flucytosine is well absorbed after oral administration. It is active against most clinically important yeasts including C. albicans and C. neoformans, but has little activity against molds or dimorphic fungi. Treatment in combination with amphotericin B is used for systemic infections.

Caspofungin: This has good activity against Candida and Aspergillus.

Griseofulvin: It is active only against the agents of superficial mycoses. Griseofulvin is administered orally as it is well absorbed from the gastrointestinal treat.

Potassium iodide: It is the oldest known oral chemotherapeutic agent for fungal infection. It is effective only for cutaneous sporotichosis.

Tolnaftate: It has activity against dermatophytes but not against yeasts. It has been effective in topical treatment of dermatophytoses.

Mycotoxicosis

Fungi can generate substances with direct toxicity for humans and animals. Such toxins are secondary metabolites that are synthesised and secreted directly into the environment and include a variety of mycotoxins elaborated by mushrooms. Exposure to these toxins results in a disease termed mycetismus, whose severity depends on the amount and type of mycotoxin ingested.

Amatoxins and Phallotoxins

The amatoxins and phallotoxins represent two important families of mycotoxins. The amatoxins are among the most potent. The phallotoxins, which are not absorbed by the gastrointestive tract and are not a cause of

Summary of Human Mycetismus

Aetiology	Mycotoxin	Site of involvement	Symptoms
Boletus satans Lactorius torminosus Lepiota morgani	Unidentified	Gastrointestinal tract	Nausea and diarrhoea; mild to severe
Amanita phalloides Amanita verna	Amatoxins Phallotoxins	Gastrointestinal treat and parasympathetic nervous system	Diphasic 1. Violent vomiting diarrhoea dehydration, muscle cramps (2) Renal and hepatic failure, confusion, perspiration lacrimation salivation, jaundice, coma
Clitocybe species Inocybe species	Muscarine	Gastrointestinal tract and parasympathetic nervous type	Violent gastrointestinal upset perspiration, salivation, delirium, hallucinations or coma, cardiac and respiratory failure
Helvella esculata	Gyromitrin	Gastrointestinal track	Nausea, vomiting, diarrhoea, jaundice
Psilocybe cubensis	Psilocybin	Central nervous system	Hallucination

mycetismus. Both toxins are produced by poisonous mushrooms, such as *Amanita*.

Aflatoxins and other Tumourogenic Mycotoxins

Fungi may elaborate a variety of mutagens and carcinogens: Although these toxins can be lethal or tumourogenic for animals. The most potent example is aflatoxin, of which eight varieties are produced by certain swains of *Aspergillus flavus* and other molds. Other mycotoxins with demonstrated carcinogenesis include ochratoxin, sporidesmin, zearolenone and sterigmatocystin.

Chapter

69

Summary of Procedures in Mycology

The laboratory procedures in fungal diagnosis include:
1. Demonstration of fungal elements in the specimen and culture.
2. Detection of fungal antigens in body fluids.

Laboratory investigations include the following:
a. Macroscopic examination of the material.
b. Microscopic examination after preliminary procedures (Figs 69.1 to 69.3).
c. Growth on appropriate media to note the macroscopic and microscopic morphology of isolates.
d. Biochemical and special tests for confirmation.
e. Pathogenicity tests for some agents.

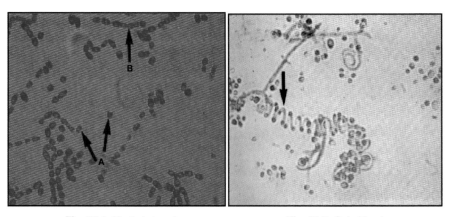

Fig. 69.1: Hyphal strand   **Fig. 69.2:** Spiral hyphae

MATERIALS

a. **An inoculation hood:** It is preferable to carry out all techniques inside an inoculation hood to avoid contamination of the atmosphere with fungal spores.

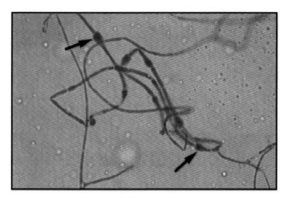

Fig. 69.3: Racquet hyphae

b. **A stiff straight nichrome wire** (22-gauge) and another one bent at an angle to remove adherent cultures for further procedures.

c. **A pair of short stiff teasing needles,** used in pulling apart dense masses of mycelium on the slide prior to microscopic examination.

d. **A scalpel knife** and a pair of scissors and forceps to help in cutting tissues and teasing out clinical materials.

e. **Large test-tubes with cotton plugs** are preferred over petri dishes for specimens from sterile areas, e.g. CSF.

    Petri dishes are used where isolation from mixed flora is necessary, e.g. sputum, urine, etc.

f. **Incubators** adjusted both for 37°C and a BOD incubator for 28°C.

TYPE OF CLINICAL SPECIMENS EXAMINED AND MEDIA REQUIRED

Details on these are summarised for ready reference as given below:

Specimens	Direct microscopy	Fungi most likely	Media and temp. of Incubator	
1. CSF	WP Mod India Ink with 2% mercurochrome G/S	*Cryptococcus neoformans* Rarely *Trichosporon,* *Candida*	25°C 1 SFSA slope 1 SDA slope	37°C 1 SDA slope and Thio BHIA slope
2. All fluids Body fluids	G/S WP	*Candida* *Aspergillus* sp. Aerobic actinomycetes	1 SDA slope 1 BHIA slope	1 SDA slope 1 BHIA slope and Thio
3. Urine	WP/GS Before and after centrifugation	Candida Rarely *Aspergillus* sp. If *Cryptococcus* suspected IF *Histoplasma* suspected	Colony count SDA SFSA BIBA	1 SDA 1 SAB

Contd...

Contd...

Specimens	Direct microscopy	Fungi most likely	Media and temp. of Incubator	
4. Cornical scraping	LPCB G/S	*Aspergillus* sp., *Fusarium* Dematiaceous and other filamentous fungi BHIA,	SAB SDA, Thio	Bacteria include BA, CA, *Nocardia.*
5. Acqueous tap Vitreous tap	W/P G/S	*Aspergillus* sp. Filamentous fungi	SAB SDA	If bacterial include BA, Thio CA, RMA
6. Corneal button	KOH	*Aspergillus* sp. other Filamentous fungi	SDA	SDA
7. Blood	–	*Candida* sp. Rarely *Cryptococcus* filamentous fungi	2 SBPM	
8. Bone marrow	Giemsa	*Histoplasma capsulatum* Rarely *Candida, Cryptococcus*	I SBPM BHIA, BIBA SFSA	
9. Bronchial wash sputum	WP/KOH G/S, KAF	*Candida, Aspergillus* sp. Aerobic actinomycetes Rarely *Cryptococcus*	SDA, SAB	SDA, SAB BHIA, BIBA, BIAB (Thio)
10. Pus	WP, KOH G/S, KAF	*Candida, Cryptococcus* (rare) Aerobic actinomycetes	SDA, SAB, SFSA	BHIA BIBA, BIAB, Thio
11. Lymph node	Giemsa	Rarely *Cryptococcus* *Histoplasma capsulatum*	SFSA, SDA SAB	BHIA BIBA, BIAB, Thio'

Microscopic Procedures

Examination of Clinical Materials

Specimens: Specimens must be processed as soon as possible (Fig. 69.4).

Direct microscopy and culture: After direct microscopy, the bulk of or all the remaining material should be used for culture as numerous hyphae may be seen on direct microscopy. If 2 swabs are received one can be used for Gram's stain for yeast like organism (YLO) and the other for culture. If only one swab is received, use it for culture only.

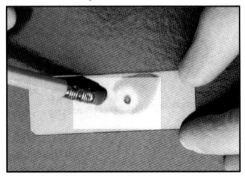

Fig. 69.4: Wet mount preparation

Cultural Procedures

Using sterile techniques, remove a small portion of the colony with a stiff nichrome wire straight or bent. Place a small bit of the growth in a drop of lactophenol cotton blue mounting fluid. They can be teased using sterile needles before placing the cover slip. Examine under low power of the microscope and then with the higher power, to ascertain the morphology to identify it.

Observation of Colony Morphology

a. Rate of growth
b. General topography: Flat, heaped, regularly or irregularly folded
c. Texture: Yeast-like, glabrous, powdery granular, velvety, or cottony
d. Surface pigmentation and type of pigmentation
e. Pigmentation on reverse.
    Slide cultures must be done for microscopic examination.

Slide Culture Technique

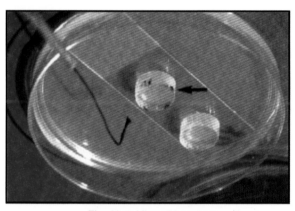

Fig. 69.5: Microslide culture

On a bent glass rod place slide and cover slips in a Petri dish. Cover and sterilise (Fig. 69.5).

Prepare sabouraud dextrose agar medium (about 15 ml) in a Petri dish. (Cornmeal agar or potato Dextrose agar may also be used. Diluted SDA is used for sporulation of dematiaceous fungi). Allow to solidify.

Cut agar blocks about 1 cm square and 2-3 mm deep from the agar plate using a template. Place the block of agar, using sterile techniques on the sterile slide in the sterilie Petri dish. Inoculate the centre of four sides of agar block with the test fungus.

Cover the inoculated block with the sterile cover slip. With a sterile pipette, add 8 ml sterile water, to which 2 drops of 10 per cent glycerin has been added, to the bottom of the Petri dish to set it up as a moist chamber. Incubate at 28°C until sporulation occurs, usually within 48 hours to few days for rapid growers. Inspect with the cover slip intact and when spores appear, carefully lift off cover slip and lay aside with fungus growth upward. Lift agar block from slide and discard. Place drop of lactophenol cotton blue on the slide and cover with a clean cover slip. Use lactophenol without cotton blue for dematiaceous fungi. Examine under low power or high dry objective.

LABORATORY STUDY OF SUPERFICIAL AND CUTANEOUS MYCOSES

Materials Received

Skin scales, skin scrapings, hair, nail scrapings and swabs or pus from mucosal lesions or from vaginal lesions as in candidiasis.

Preliminary Examination

Wood's Lamp Examination

Keep the material in a dish and examine it in a dark room under Wood's lamp. If fluorescent hairs are present, pick these up for further examination, e.g. *Microsporum* species and *T. schoenleinii* fluoresce.

KOH, 10% preparation: Keep a bit of the skin scraping, hair, etc. on a clean slide add a drop of 10 per cent KOH. Put on a clean cover slip and leave for 10 minutes. If scales or scrapings are thick, carry out a preliminary teasing (Fig. 69.6).

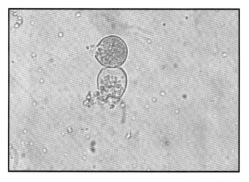
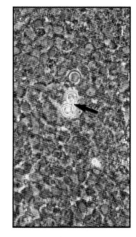

Fig. 69.6: KOH

Chloral lactophenol mounting: Alternatively hair can be examined with chloral lactophenol mounting fluid with spores intact.

Spores inside the hair are called "endothrix" and spores outside the hair shaft are called "ectothrix".

KOH 40% preparation: In the case of hard scrapings as in nail, digest with 40 per cent KOH and leave for 20 to 30 minutes before examination.

Gram's stain: When candidiasis is suspected, make smears on a clean slide with the moist material/scrapings swabs or pus, dry and fix the smear and examine after Gram's stain.

Microscopic Examination

a. Examine first under the low power with reduced light. Search for fungal elements. Note the size, branching, presence and distribution of spores, etc. Examine under high power for confirmation.
b. Gram stained smears are examined under oil immersion for the presence of gram-positive budding cells and pseudomycelium.

CULTURAL PROCEDURES

Materials are inoculated on SDA with and without antibiotics.

Inoculate materials suspected of dermatophytes such as hair, skin or nail scrapings with a thick needle or scalpel deep into the media on a slant; push material partially into the medium. When materials are suspected of *Candida*, streak on SDA in a Petri dish for isolation.

Incubate cultures of dermatophytes at RT (25 to 30°C) Cultures for *Candida* are incubated at 37°C.

Examination of Cultures

1. Observe colony morphology
2. Make a preliminary preparation for microscopic examination
3. Make subcultures on
    Littman agar
    Sabouraud Dextrose agar
    Sporulating media, e.g. potato Dextrose agar, Czapex-Dox agar, Corn meal agar.
4. Examine slide cultures for colony morphology and microscopic morphology.

SUBCULTURE

Microsporum Group (*M. audouini, M. canis, M. gypseum*)
Inoculate on Rice Grain Medium

After 7 to 10 days growth on rice grain media, M. *audouini* produces no aerial mycelium, but shows brownish discolouration of medium.

M. *canis* produces heavy white cottony aerial mycelium with pinkish-buff discoluration of medium.

M. *gypseum* produces powdery aerial mycelium of cinnamon colour.

M. *canis* and M. *gypseum* can be distinguished fairly well by macroscopic cultural characteristic and morphology of macroconida.

Trichophyton Group

Subculture on:
1. Sabouraud's Dextrose agar.
2. Casein medium and casein-thiamin medium to compare growth on each
3. Cornmeal agar with 10 per cent glucose (CMA).
4. Other sporulation promoting media, e.g. potato Dextrose agar (PDA)
5. Hair penetration test: To differentiate non-pigmented strains of T. *rubrum* from T. *mentagrophytes*.

   Place short strands of hair in a Petri dish, and sterilise by autoclaving at 120°C for 10 minutes. Add 25 ml sterile distilled water to which 3 drops of 10 per cent sterilised yeast extract has been added. Small fragments of the fungus to be tested should be put in the fluid in the Petri dish. Incubate at 26°C.

   Remove piece of hair with sterile forceps place in a drop of lactophenol cotton blue on a slide with a coverslip. Heat gently. Examine for the presence of "perforators".

Final Confirmation of Macroscopic and Microscopic Morphology

a. Pigmentation of T. *rubrum* and T. *violaceum* is better on Cornmeal agar with dextrose.
b. Growth 4+ on Casein-thiamin medium as against 1+ on Casein medium, e.g. T. *tonsurans*, T. *violaceum*, T. *mentagrophytes* and T. *rubrum* give 4+ growth on both media.
c. Hair penetration

   Remove piece of hair with sterile forceps, place in a drop of lactophenol cotton blue on slide with a coverslip. Heat gently. Examine for the presence of "perforators".

   For example, hair penetration.

Positive: *T. mentagrophytes* and *M. gypseum*

Negative: *T. rubrum*

d. Microscopic appearance

1. Note the presence or absence of septation, presence of racquet hyphae, spiral hyphae, favic chandeliers, pectinate bodies, etc.

   Note also the presence and nature of chlamydospores, microconidia and macroconidia.

2. Confirmation of identity is based on nature and presence of macroconidia.

Characteristics of macroconidia of dermatophytes					
Fungus	**Number**	**Shape**	**Septa tions**	**Walls**	**Arrangement**
Microsporum	Numerous	Spindle	3-15	Thick rough	Single
Trichophyton	Usually rare	Blunt smooth	2-8	Thin,	Single
E. floccosum	Numerous	Club-shaped in groups	2-4	Inter-mediate smooth	Groups of 2 or 3

Laboratory Study of Subcutaneous Mycoses

The important subcutaneous mycoses include:

1. Mycetoma
2. Subcutaneous phycomycosis
   a. Basidiobolomycosis
   b. Conidiobolomycosis
3. Sporotrichosis:
4. Chromoblastomycosis
5. Rhinosporidiosis.

MYCETOMA

Aetiologic Agents of Mycetoma

As mycetoma is caused by both actinomycotic bacterial and eumycotic agents, it is very important to ascertain the aetiologic agent.

Whatever the type of material received an intense search for the presence of granules must be made. Actinomycotic granules are generally small and rarely coloured. Eumycotic granules are usually large and coloured.

Proceed to examine pus or exudates received for the presence and details of granules. If pus or granules are not received and if sinuses are present the clinician should be requested to obtain scrapings from the swalls of the sinus canal and to send the material.

Squeeze granule between two slides and take them apart. Dry both slides. Fix and stain one by Gram's stain. The presence of slender, gram-positive, branching filaments and in some species fragmentation into bacillary and coccoid forms, indicate "actinomycotic" mycetoma. If gram-positive branching filaments are seen, fix and stain the second slide by Kinyoun's acid-fast technique.

Example: Nocardia are acid-fast streptomyces are non-acid-fast.

If no gram-positive filaments are seen, take another part of a granule and press between slides and take apart. Add a small drop of saline and place a coverslip and press gently. Examine microscopically. Search for eumycotic mycelial elements, with 3 to 5 μ width. Sometimes rounded bigger sized structures of diameter 10 μ are found which are the chlamydospores.

Culture Procedures

Break-up granules, which are well washed in many changes of sterile saline and inoculate the pieces in different areas on two sets of SDA and BHIA, with and without antibiotics. Incubate one set each at 25 to 30°C and another set at 37°C.

If actinomycotic agents are suspected, proceed to look for colonial morphology of nocardia, *Streptomyces.*

Laboratory Study of Systemic Mycoses

The systemic mycoses involve any or all of the internal organs of the body including the bones and subcutaneous tissues. In some diseases the skin may also be involved at some stage. Infections may be asymptomatic or they may manifest with varying degrees of severity. The agents of systemic mycoses include the yeast-like fungi, dimorphic fungi, and filamentous fungi. The laboratory studies of important clinical diseases will be dealt with under the following sections
1. Nocardiosis
2. Mycoses by yeast-like fungi
3. Mycoses by dimorphic fungi.

Clinical Materials Examined

Early morning specimen of sputum, bronchial washings, CSF or pus from abscesses, scrapings from sinus tracts and biopsy materials from mycetoma.

Preliminary Examination

Granules are not usually present in sputum. In other materials, granules are very small and not easily seen macroscopically.

Microscopic Examination

Make smears with sputum, pus or other materials and stain by Gram and modified Kinyoun's. *Nocardia* are Gram-positive, thin, branching filaments. Bacillary or coccoid forms are conspicuous as different from *Streptomyces*.

Cultural Procedures

1. As granules are not present culture sputum. Bronchial washings, gastric juice or other materials with and without concentration. *Nocardia* withstand concentration procedures and even grow on LJ medium thus facilitating its isolation from the above materials, even when specimens are sent for *M. tuberculosis* and nocardiosis is not suspected.
2. Inoculate materials on SDA without antibiotics and incubate at 37°C and at 25 to 30°C for a period of 21 days. If after concentration inoculate LJ medium as well.

    Examine the cultures for development of characteristic colonies on SDA which are dry, folded, chalky white or orange, depending on the species.

    Moist glabrous colonies on LJ medium may simulate atypical mycobacteria.

    Check and confirm morphology of colonies after staining with Grams and Kinyoun's acid-fast stains.

CANDIDIASIS

A variety of clinical infections are caused by various species of *Candida* spp. aetiological agents are:

    C. albicans
    C. tropicalis
    C. glabrata
    C. krusei
    C. parapsilosis
    C. guillermondii

Clinical materials examined are the variety of materials from infectious sites for laboratory investigations.

Microscopic examination: Scrapings from skin or nails are examined after KOH mount; other materials are examined in lactophenol blue preparations of after Gram stain. If CSF centrifuge and do a Gram's stain with the sediment. *Candida* are seen in clinical specimens as gram-positive, budding cells; fragments of pseudo and true mycelium which are characteristically pinched off at septations are also seen.

Cultural Procedure

Candida are not inhibited by antibiotics but a number of *Candida* species are sensitive to cycloheximide. Materials may be inoculated on SDA and SAB. Incubate at 37°C and at 25 to 30°C for 24 to 48 hours. *Candida* grow as white, opaque. Soft colonies with submerged Mycelium. They are easily emulsified like bacteria. A yeast-like smell is characteristic. Confirm microscopic morphology after Gram's stain of smears from growth.

Special Test

Chlamydospore Formation

Inoculate suspected *Candida* on plates or slides with chlamydospore agar or cornmeal agar with 1 percent tween 80. Inoculate by cutting at angles into the medium. Include a known culture of *C. albicans*, as control. Incubate at room temperature for 48 to 72 hours. Examine slides, plates or slide mounts from plates, under the microscope at 10X and 40X for the presence of chlamydospores and characteristic growth of mycelia.

Germ Tube Test: Reynold-Braude Phenomenon

Inoculate lightly the test strain of *Candida* and a known *C. albicans* to act as control into two separate tubes containing 0.5 ml each of the same batch of human serum incubate at 37°C. Examine first the control *C. albicans* for satisfactory formation of germ tubes. This is seen as an outgrowth from the cell with no constriction at the base resembling a lady's hand mirror.

CRYPTOCOCCOSIS

Cryptococcosis produced by *C. neoformans* may manifest in different clinical forms such as meningitis, pulmonary cutaneous or osseous forms. It sometimes mimics tuberculosis and some chronic bacterial infections.

Clinical Material Examined

CSF, pus, tissue, sputum, urine.

Microscopic Examination

For CSF or fluids centrifuge at 1500 to 2500 rpm for 10 minutes. Examine by India Ink or make smears from sediment. Examine under reduced light. Look for capsulate, typically spherical, yeast-like cells with budding.

Smear the sediment, dry and fix lightly by heat, stain by Gram's and examine. They appear as gram-positive, round, yeast-like organisms with single budding with an unstained halo around suggesting presence of capsule.

Make smears from other materials received, fix lightly and stain by Gram's stain and examine.

Sputum and pus if thick or tissue may be examined after 10 per cent KOH treatment.

Soft gelatinous tissue got at biopsy or brain after postmortem (or animal tissue) may be examined by pressing a bit of the tissue between two slides, taking them apart, staining one by India ink quickly and their other by Gram's stain.

Cultural Procedures

a. Inoculate representative materials on SDA, sunflower seed agar (SFSA) and BHIA.
b. Sediments from CSF after centrifugation is cultured similarly and into thioglycollate broth.
c. Incubate cultures at 25 to 30°C for few days to two weeks. When grown colonies are large, opaque, white or tan, soft and mucilaginous indicating capsulation of the organism.

Confirm morphology after India Ink preparation and Gram's stain from the growth.

a. Test for urease production.

All cryptococci are urease positive. The large majority of other yeast like organisms are urease negative.

Animal Pathogenicity Tests

The animal of choice is the mouse

a. Inoculate 0.02 to 0.03 ml of a 1 in 100 dilution of a saline suspension from a 4 to 5 day old culture on SDA.
b. Routes may be intraperitoneal, intracerebral or subcutaneous.
c. In the case of *C. neoformans*, mice die within few days to 2 weeks. Subcutaneous route takes longer time.
d. At autopsy gelatinous material can be demonstrated in the peritoneum, inside the skull or subcutaneously according to the route of inoculation.

Take the gelatinous material or bits of tissue, press between two slides, take them apart, stain one by India Ink and the other by Gram's and examine. The gelatinous material may be directly examined by mixing with India Ink

Typical encapsulated budding yeast-like organisms demonstrated from the above sites as well as from liver, spleen, lungs or heart.

MYCOSES BY DIMORPHIC FUNGI

The fungi which cause most of the systemic mycosis are in general unrelated species that are morphologically and physiologically dissimilar. It is remarkable, however, that most fungi causing systemic mycoses develop a pathogenic phase that is entirely different from the saprophytic phase; hence, referred as dimorphic fungi. This "diphasic" phenomenon has been noted in the fungi which cause some subcutaneous mycoses as well as many systemic mycoses. All are considered in this section. Subcutaneous mycoses by dimorphic fungi.

- Sporotrichosis
- Aetiologic agent
- *Sporothrix schenckii* (*Sporotrichum schenckii*)

Microscopic Examination

1. Tissue phase or yeast phase
   Small gram-positive, oval or elongate, budding cells called "cigar bodies" sparse in exudates. When grown at RT the colonies show fine, branching septate mycelia with microconidia borne as flower-like or "sleeve-like" arrangement.

Cultural Characteristics

At room temperature, 25 to 30°C they grow on SDA with chloramphenicol and better still on BHIA or BIBA with Cycloheximide at RT in 3 to 7 days, as moist, white colonies, irregularly wrinkled at the centre, turning to black in 2 weeks. When incubated at 37°C they grow on protein rich medium, e.g. BHIA, BIBA as creamy white, yeast-like colonies.

Histoplasmosis
a. Aetiologic agent

Histoplasmosis capsulatum
a. Clinical material examined
   Adrenal pus, oral tissue, sputum.

Microscopic Examination

Tissue Phase or Yeast Phase

A small, round or oval, yeast-like cells with refractile walls found within mononuclear cells or polymorps. Stains well by Giemsa, PAS, Gridley or Gomori stains. Gram-positive, small budding, yeast-like cells seen in culture.

The mold-like growth show branching filamentous fungi with characteristic "tuberculate" macroconidia or Chlamydospores. Microconidia appear earlier.

At room temperature (25 to 30°C) on SDA but beeter on BHIA or BHIBA with or without antibiotics they grow as white, fluffy, silky growth in 10 to 14 days changes to tan colour on aging and sporulation.

At 37°C they grow on enriched, moist medium, e.g. BHIBA as moist, raised, convex cream colonies with 7 days.

LABORATORY STUDY OF MYCOSES BY OPPORTUNISTIC FUNGI

Aspergillosis

a. Clinical Lesions

Pulmonary infections	Sinusitis
Myocarditis	Cellulitis faece and orbit
Otomycosis	
Allergic disorders	Keratomycosis

Microscopic Examination

KOH preparation and lactophenol cotton blue mounts from clinical materials show characteristic septate, dichotomously branching, filaments, culture shows typical heads with chains of conidia arranged on vesicles borne on conidiophores.

Cultural Characteristic

Grows rapidly on SDA or SAB

A. fumigatus the most common pathogen, grow initially as white fuzzy colonies which later develop as dark bluish-green, powdery growth.

A flavus-lime yellow-green pigmentation

A-niger-quick growing colonies with brown or black heads.

Zygomycosis or Phycomycosis

a. Aetiologic Agents

Rhizopus, Mucor, Absidia, Basidiobolus and *Conidiobolus.*

b. Clinical Lesions.
c. Agent Disease

Absidia	Pulmonary zygomycosis
Rhizopus	Gastrointestinal ulceration
And Mucor	Orbital and CNS zygomycosis, etc.

Microscopic Examination

In KOH preparations or lactophenol mounts of clinical material they apper as branches hyphase of large diameter. On culture sporangiospores are enclosed within sporangium borne on sporangiophore.

Cultural Characteristic

Rhizopus: Most commonest agent of zygomycosis.

Growth is volumin and rapid. Long coarse and wooly mycelium first white later become grey sprinkled with many black or brown dots (sporangia)

Microscopically (slide culture better) the mycelia are broad non-septate and colourless. The fruiting body consists of long stalks (Sporangiophores) surmounted by spherical sporangia. Sporangiophores are unbranched, and clustered at nodes opposite rhizoids along a horizontal runner (stolon).

Mucor

Colony is rapid growing filling the test tube or petri dish in 2 to 3 days with a fluffy aerial mycelium that is at first white, but later becoming grey to brown. Microscopically the mycelium is broad, non-septate, colourless, and without rhizoids. Sporangiophores arise singly from the mycelium forming a thick, erect turf. They are either unbranched with terminal sporangia or branched with spherical many spored sporangia on all the branch ends. Columellae are always present.

Absidia

They are similar to thos of the genus *Rhizopus,* but its sporangiophore arise at the internodes of the stolon instead of at the nodes. The sporangia are slightly pear shaped.

SECTION

7

MEDICAL
ENTOMOLOGY

Introduction to Medical Entomology

The medically important insects and Ectoparasites are numerous and the study of these comes under Medical Entomology.

AGENTS DIRECTLY CAUSING DISEASE

Leeches

There are annelid worms with specialised chitinous mouth parts and secrete an anticoagulant, hirudin. Leeches can cause severe loss of blood. If the leech is pulled out from the skin, It may lead to secondary infection (Fig. 70.1).

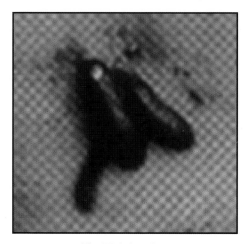

Fig. 70.1: Leeches

Lice

Three species of lice infect man the public louse, body louse and the head louse. Head lice infect the hair covered areas of the head. They feed by grasping the skin with the sucking mouth. After fertilisation, nits are deposited. Head lice are spread by close contact. This can be removed with a fine comb and treatment with appropriate insecticide (e.g. Malathion) (Fig. 70.2).

Fig. 70.2: Lice

Mosquitoes

Female mosquitoes are important vectors of disease. *Anopheles* are important biological vectors of malaria, filariasis, alphaviruses, flaviviruses and bunya viruses (Figs 70.3 A and B).

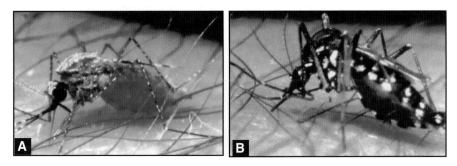

Figs 70.3 A and B: Mosquitoes

Aedes species also transmit filariasis and are major vectors of alphaviruses, flavi viruses and some bunya viruses.

Bugs

Bed bugs are the major parasite of man. Females lay up to 100 eggs in a life time. These are deposited in cracks and crevices and produce irritating bites and may be a vector of hepatitis B virus (Fig. 70.4).

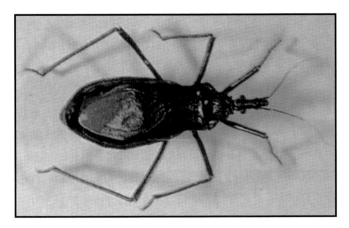

Fig. 70.4: Bug

Sarcoptes Scabiei

Sarcoptes scabiei causes scabies. It is spread person-to-person by close contact in families and can be spread by sexual contact (Fig. 70.5). The adult scabies mite is a small flattened disc with eight, short, squat legs. The fertilized female burrow into the skin for several millimeters. The sites chosen are the wrists, exterior surfaces of the elbows, axillae, penis, scrotum and under the breasts. Treatment is with benzyl benzoate or gamma benzene hexachloride.

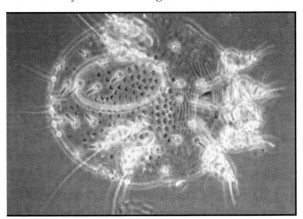

Fig. 70.5: Sarcoptes

Phlebotomus Sandflies

Only the females feed on blood. Bites may result in an urticarial reaction and the flies are vectors of cutaneous and visceral Leishmaniasis, oraya fever and phlebo virus (Fig. 70.6).

Fig. 70.6: Sand fly

Black Fly

It is the female simulium species that feeds on blood. These transmit the filarial parasite onchocerca volulus causing onchocerciasis (Fig. 70.7).

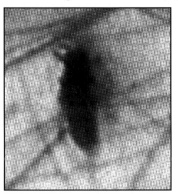

Fig. 70.7: Black fly

Tsetse Fly

The tsetsefly or *Glossina* spp are attracted by bright colours and powerful odours. They produce painful bites and are vector of sleeping sickness (*Trypanosoma brucei*) (Fig. 70.8).

Fig. 70.8: Tsetse fly

Ticks

All ticks are obligate, blood sucking parasites (Fig. 70.9). There are two forms. (i) Soft or argasid ticks and (ii) hard or ixodid ticks. The hard ticks are slow feeders.

Fig. 70.9: Ticks

Whereas, the soft ticks feed quite often overnight. Soft ticks transmit tick borne relepsing fever (Borrelia) to man. The hard tick transmit Lyme disease (*B. burgdorferi*) rickettsioses, rocky mounted spotted fever, babesiosis, flaviviruses and bunya viruses.

SECTION

8

LABORATORY INVESTIGATION OF MICROBIAL INFECTIONS

Laboratory Investigation of Microbial Infections

The laboratory investigation of microbial diseases involves:
Examining specimens to detect, isolate, and identify pathogens by:
- Microscopy
- Culture techniques
- Biochemical methods
- Immunological (antigen) tests.
  Testing serum for antibodies produced in response to infection.

Microorganisms can be examined microscopically for their motility, morphology, and staining reactions.

Examples

Motile *Vibrio cholerae* in a rice water faecal specimen from a person with cholera (Hanging drop method).

Treponema pallidum in chancre fluid (using dark-field microscopy), establishing a diagnosis of primary syphillis.

Fungal hyphae and arthrospores in a sodium hydroxide preparation of skin indicates ringworm.

Gram-negative reaction and characteristic morphology of *Neisseriae gonorrhoeae* (intracellular diplococci) in a urethral discharge reveals gonorrhoea.

Gram-positive reaction and morphology of pneumococci in cerebrospinal fluid represents pneumococcal meningitis.

Gram-positive reaction and morphology of yeast cells in a vaginal discharge from a woman with vaginal candidiasis.

Acid-fast reaction of *Mycobacterium tuberculosis* in Ziehl-Neelsen stained sputum from a person with pulmonary tuberculosis.

CULTURE TECHNIQUES

The culture of pathogens enables colonies of pure growth to be isolated for identification and, when required, antimicrobial sensitivity testing.

BIOCHEMICAL METHODS

Following culture biochemical tests are often required to identify pathogens including the use of substrates and sugars to identify pathogens by their enzymatic and fermentation reactions.

Example

Catalase test to differentiate staphylococci which produce the enzyme catalase from streptococci which are non-catalase producing.

Oxidase test to help identify *Vibrio, Neisseria, Pasteurella* and *Pseudomonas species*, all of which produce oxidase enzymes.

Coagulase test to help identify *Staphylococcus aureus* which produces the enzyme coagulase.

Fermentation tests to differentiate enterobacteria, e.g. use of glucose and lactose in Kligler iron agar medium to assist in the identification of *Shigella* and *Salmonella organisms.*

Indole test to detect those organisms that are able to breakdown tryptophan with the release of indole. It is mainly used to differentiate *Escherichia coli* from other enterobacteria.

Urease test to assist in the identification of organisms such as *Proteus species* which produce the enzyme urease.

IMMUNOLOGICAL (ANTIGEN) TEST

Antigen tests often enable an early diagnosis or presumptive diagnosis of an infectious disease to be made. They involve the use of specific antibody (antisera or labelled antibody).

To identify a pathogen that has been isolated by culture, e.g. identification of Salmonella species, *Shigella species*, and *Vibrio cholerae* by direct slide agglutination.

To identify pathogens in specimens using direct immunofluorescence, e.g. identification of respiratory viruses, rabies virus, cytomegalovirus, *Pneumocystis carinii*, and *Chlamydia.*

To identify antigens of microbial origin that can be found in serum or plasma, cerebrospinal fluid, urine specimen extracts and washings, or fluid cultures. Highly specific monoclonal antibody reagents are often used.

PRINCIPLES OF ANTIGEN TESTS

Direct Slide Agglutination

The is used to identify bacteria following culture on a carbohydrate-free medium. A bacterial colony of pure growth is emulsified in physiological saline on a slide and antiserum containing specific antibody is added. The antibody binds to the bacterial antigen, resulting in the agglutination of the bacterial cells.

Antiserum + Bacteria → Bacteria AGGLUTINATED

Latex Agglutination

Latex particles are coated with specific antibody. The specimen containing microbial antigen is mixed with the latex reagent, resulting in agglutination of the latex particles.

$$\left(\begin{array}{c}\text{Antibody latex}\\ \text{regent}\end{array}\right) + \left(\begin{array}{c}\text{Antigen in}\\ \text{specimen}\end{array}\right) = \begin{array}{c}\text{Latex particles}\\ \text{AGGLUTINATED}\end{array}$$

Coagglutination (COAG)

Specific antibody is bound to the protein A of staphylococci (Cowan type 1 strain of *Staphylococcus aureus*). Soluble microbial antigen in the specimen is mixed with the COAG reagent, resulting in the agglutination of the staphylococcal cells.

$$\left(\begin{array}{c}\text{Antibody COAG}\\ \text{regent}\end{array}\right) + \left(\begin{array}{c}\text{Antigen in}\\ \text{specimen}\end{array}\right) = \begin{array}{c}\text{Staphylococcal cells}\\ \text{AGGLUTINATED}\end{array}$$

Direct Immunofluorescence

Specific antibody is conjugated (joined) to a fluorochrome such as fluorescein isothiocyanate and applied to the specimen containing the pathogen on a slide. The fluorochrome antibody conjugate binds to the pathogen (antigen). When examined by fluorescence microscopy the pathogen is seen to fluoresce (e.g. yellow-green or orange) against a dark background.

$$\left(\begin{array}{c}\text{Fluorochrome}\\ \text{antibody reagent}\end{array}\right) + \left(\begin{array}{c}\text{Antigen}\\ \text{pathogen}\end{array}\right) = \begin{array}{c}\text{Pathogen}\\ \text{FLUORESCES}\end{array}$$

Enzyme Immunoassays (EIA) to Detect Antigen

Enzyme assays are also referred to an enzyme linked immunosorbent assays (ELISAS).

Antibody against the antigen to be detected is fixed to the well of a microtitration plate or membrane of an individual test device such as a plastic block. Soluble microbial antigen in the specimen binds to the antibody. After washing antibody conjugated to an enzyme (e.g. horseradish peroxidase) is added. This binds to the captured antigen. After another wash, a chromogenic (colour-producing) substrate such as hydrogen peroxide joined to an indicator is added. The enzyme hydrolyzes the substrate, producing a colour reaction. The colour can be read visually (membrane EIA) or spectrophotometrically (microtitration plate EIA).

1. Fixed antibody + $\left(\begin{array}{c}\text{Antigen in}\\ \text{specimen}\end{array}\right)$ = Antigen binds to antibody

2. Enzyme conjugated → Binds to antigen-antibody
   antibody added   complex

3. Chromogentic substrate added → COLOUR produced.

Dipstick Comb Immunoassays to Detect Antigen

These assays involve dipping a plastic comb in the specimen and reagent solutions. Each comb is designed for testing up to 6 specimens and controls although the comb can be cut when there are fewer specimens. When used for antigen detection, specific antibody is fixed to the ends of the comb teeth. The comb is dipped in the specimen and antigen in the specimen is captured by the antibody. After washing, the comb is dipped in a colloidal gold antibody conjugate. This binds to the antibody-antigen complex. After washing, a pink dot is produced, indicating a positive test. Although easy to perform, dipstick assays are not as rapid as most IC strip/card immunoassays.

1. Antibody on teeth of comb + Antigen in specimen → Antigen binds to antibody.
2. Colloidal gold conjugate applied → Binds to antigen-antibody complex. PINK DOT produced.

Testing Serum for Antibodies (Serological Tests)

In district laboratories, serological testing in which antigen is used to detect and measure antibody in a person's serum is used mainly:

To help diagnose a microbial disease when the pathogen or microbial antigen is not present in routine specimens or if present is not easy isolated and identified by other available techniques, e.g. dengue, brucellosis, rickettsial infections, syphilis, leptospirosis.

To screen donor blood for antibody to HIV-1 and HIV-2.

To measure antibody levels to determine the prevalence of infectious disease in a community and immune status of individuals.

To screen for rises in anti-streptolysin O, e.g. in the investigation of rheumatic fever, acute glomerulonephritis, and other complications of group A streptococcal infection.

To screen pregnant women for infections such as syphilis and HIV infection.

EXAMINATION OF URINE

Possible Pathogens
BACTERIA

Gram-positive	**Gram-negative**
Staphylococcus	*Escherichia coli*
saprophyticus	*Proteus species*
Haemolytic streptococci	*Pseudomonas aeruginosa*
	Klebsiella strains
	Salmonella typhi
	Salmonella paratyphi
	Neisseria gonorrhoeae

Notes on Pathogens

The presence of bacteria in urine is called bacteriuria. It is usually regarded as significant when the urine contains 10^5 organisms or more per ml ($10^8/1$) in pure culture.

E. coli is the most common urinary pathogen causing 60 to 90 per cent of infection.

UTIs caused by *Pseudomonas, Proteus, Klebsiella species* and *S. aureus*, are associated with hospital-acquired infections, often following catheterization or gynaecological surgery. *Proteus* infections are also associated with renal stones.

S. saprophyticus infections are usually found in sexually active young women.

Infection of the anterior urinary tract (urethritis) is mainly caused by *N. gonorrhoeae* (especially in men), staphylococci, streptococci, and *Chlamydia*.

Candida urinary infection is usually found in diabetic patients and those with immunosuppression.

M. tuberculosis is usually carried in the blood to the kidney from another site of infection. It is often suspected in a patient with chronic fever when there is pyuria but the routine culture is sterile.

Pyuria with a negative urine culture may also be found when there is infection with *Chlamydia trachomatis, Ureaplasma,* or *N. gonorrhoeae,* or when a patient has taken antimicrobials.

S. typhi and *S. paratyphi* can be found in the urine of about 25 per cent of patients with enteric fever from the third week of infection.

Typhoid carriers may excrete *S. typhi* in their urine for many years.

Commensals

The bladder and urinary tract are normally sterile. The urethra however may contain a few commensals and also the perineum.

With female patients, the urine may become contaminated with organisms from the vagina. Vaginal contamination is often indicated by the presence of epithelial cells (moderate to many) and a mixed bacterial flora.

COLLECTION AND TRANSPORT OF URINE

Whenever possible, the first urine passed by the patient at the beginning of the day should be sent for examination. This specimen is the most concentrated and therefore the most suitable for culture, microscopy, and biochemical analysis.

Midstream urine (MSU) for microbiological examination is collected as follows.

1. Give the patient a sterile, dry, wide-necked, leak-proof container and request a 10 to 20 ml specimen.

   Explain to the patient the need to collect the urine with as little contamination as possible, i.e. a 'clean-catch' specimen.

   Female patients: Cleanse the area around the urethral opening with clean water, dry the area, and collect the urine with the labia held apart.

   Male patients: Wash the hands before collecting a specimen (middle of the urine flow).

   When a patient is in renal failure or a young child, it may not be possible to obtain more than a few millilitres of urine.

2. Label the container with the date, the name and number of the patient, and the time of collection. As soon as possible, deliver the specimen with a request form to the laboratory.

   When immediate delivery to the laboratory is not possible, refrigerate the urine at 4 to 6°C. When a delay in delivery of more than 2 hours is anticipated, add boric acid preservative to the urine specimens.

   Normal freshly passed urine is clear and pale yellow-to-yellow depending on concentration.

Examine the Specimens Microscopically

Urine is examined microscopically as a wet preparation to detect:

- Significant pyuria, i.e. WBCs in excess of 10 cells/ml(10^9/1) of urine

- Red cells
- Casts
- Yeast cells
- *T. vaginalis* motile trophozoites
- *S. haematobium eggs*
- Bacteria

Preparation and Examination of a Wet Preparation

Aseptically transfer about 10 ml of well mixed urine to a labelled conical tube.

Centrifuge at 500 to 1000 gm for 5 minutes. Pour the supernatant fluid.

Remix the sediment by tapping the bottom of the tube. Transfer one drop of the well-mixed sediment to a slide and cover with cover glass.

Examine the preparation microscopically using the 10X and 40X objective with the condenser iris closed sufficiently to give good contrast.

A few pus cells are normally excreted in urine, Pyuria is usually regarded as significant when moderate or many pus cell are present, i.e. more than 10 WBC/ml Bacteriuria without pyuria may occur in diabetes, enteric fever, bacterial endocardits, or when the urine contains many contaminating organisms.

Pyuria with a sterile routine culture may be found with renal tuberculosis, gonococcal urethritis, *C. trachomatis* infection, and leptospirosis, or when a patient with urinary infection has been treated with antimicrobials.

Yeast cells: These can be differentiated from red cells by their oval shape and some of the yeasts usually show single budding.

Trichomonas vaginalis: Found in the urine of women with acute vaginitis. The *Trichomonas* are a little larger than white cells and are usually easily detected in fresh urine because they are motile.

Examination of a Gram Stained Smear

Prepare and examine a Gram stained smear of the urine when bacteria and/ or white cells are seen in the wet preparation.

Transfer a drop of the urine sediment to a slide and spread it to make a thin smear. Allow to airdry, protected from insects and dust. Heat fix or methanol fix the smear and stain it by the Gram's technique.

Examine the smear first with the 40X objective to see the distribution of material, and then with the oil immersion objective. Look especially for bacteria associated with urinary infections, especially gram-negative rods. Occasionally gram-positive cocci and streptococci may be seen.

Usually only a single type of organism is present in uncomplicated acute urinary infections. More than one type of organism is often seen in chronic and recurring infections. Vaginal contamination of the specimen is indicated by a mixed bacterial flora (including gram-positive rods) and often the presence of epithelial cells.

Neisseria Gonorrhoeae in Urine

In male patients with acute urethritis, it is often possible to make a presumptive diagnosis of gonorrhoea by finding gram-negative intracellular diplococci in pus cells passed in urine.

Test the Specimen Biochemically

Biochemical tests which are helpful in investigating UTI include:
- Protein
- Nitrite
- Leucocyte esterase.

Protein

Proteinuria is found in most bacterial urinary tract infections. Other causes include glomerulonephritis, nephrotic syndrome, eclampsia, urinary schistosomiasis, hypertension, and severe febrile illnesses.

Nitrite

Urinary pathogens, e.g. *E. coli* (the most common cause of UTI), *Proteus* species, and *Klebsiella* species, are able to reduce the nitrate normally present in urine to nitrite. This can be detected by the Greiss test or a nitrite reagent strip test.

The test is negative when the infection is caused by pathogens that do not reduce nitrate such as *Enterococcus faecalis, Pseudomonas* species, *Staphylococcus* species and *Candida* organisms.

Greiss Test to Screen for UTI in Pregnancy

A Greiss test to detect nitrite reducing pathogens such as *E. coli,* together with a protein test and the visual examination of urine for cloudiness, are useful ways of screening for UTI in pregnancy in antenatal clinics.

Greiss Test Using a Dry Reagent

Transfer 0.5 to 1.0 gm (pea-size) of well mixed dry Greiss reagent* to the well of a white porcelain tile or to a small test tube.

Moisten the reagent with a drop of urine (first morning urine). The urine should be tested within about 1 hour of being collected.

Look for the immediate development of a pink-red colour.

Pink-red colour ... Positive nitrite test.

Leucocyte Esterase (LE)

This enzyme is specific for polymorphonuclear neutrophils (pus cells). It detects the enzyme from active and lyzed WBCs. LE testing is an alternative method of detecting pyuria when it is not possible to examine fresh urine microscopically for white cell of when the urine is not fresh and likely to contain mostly lyzed WBCs.

LE can be detected using a reagent strip test such as the BM-Test-LN (Boehringer strip) which detects both nitrite and leucocytes (LE) or a multi test reagent strip with an area for leucocyte detection.

Culture the Specimen

It is necessary to estimate the approximate number of bacteria in urine because normal specimens may contain small number of contaminating organisms, usually less than 10,000 (10^4) per ml of urine. Urine from a person with an untreated acute urinary infection usually contains 1,00,000 (10^5) or more bacteria per ml.

The approximate number of bacteria per ml of urine, can be estimated by using a calibrated loop or a measured piece of filter paper.

Cystine Lactose Electrolyte-deficient (CLED)

— Mix the urine (freshly collected clean-catch specimen) by rotating the container.

— Using a sterile calibrated wire loop, inoculate a loopful of urine on a quarter plate of CLED agar. If microscopy shows many bacteria, use a half plate of medium.

— Incubate the plate aerobically at 35 to 37°C overnight.

DAY 2 AND ONWARDS

Examine and Report the Cultures

CLED Agar Culture

Look especially for colonies that could be:

• *Eschericha coli* (perform indole and beta-glucaronidase tests for rapid identification.

• *Proteus* species

- *Klebsiella strains*
- *Staphylococcus aureus*
- *Staphylococcus saprophyticus*
- *Enterococcus faecalis*

Appearance of Some Urinary Pathogens on CLED Agar

- *Escherichia coli*: Yellow (lactose-fermenting) opaque colonies often with slightly deeper coloured centre.
- *Klebsiella species:* Large mucoid yellow or yellow white colonies.
- *Proteus species:* Transluscent blue-grey colonies.
- *Pseudomonas aeruginosa:* Green colonies with rough periphery (characteristic colour).
- *Enterococci faecalis:* Small yellow colonies.
- *Staphylococcus aureus:* Deep yellow colonies of uniform colour.
- *Staphylococcus saprophyticus* and other coagulase negative staphylococci: Yellow to white colonies.

Reporting Bacterial Numbers

Count the approximate number of colonies. Estimate the number of bacteria, i.e. colony-forming units (CFU) per ml of urine. Report the bacterial count as:
- Less than 10,000 organisms/ml (10^4/ml), not significant.
- 10,000 to 100,000/ml (10^4-10^5/ml), doubtful significance (suggest repeat specimen)
- More than 100,000/ml (10^5/ml), significant bacteriuria.

Antimicrobial Sensitivity Testing

Perform sensitivity testing on urines with significant bacteriuria, particularly from patients with a history of recurring UTI. Cultures from patients with a primary uncomplicated UTI may not require a sensitivity test.

Summary of the Microbiological Examination of Urine		
Day 1		
1.　**Describe**	**Describe**	ADDITIONAL INVESTIGATIONS
Appearance	-　Colour	
	-　Whether clear or cloudy	
2.　**Examine**	**Wet preparation**	**Gram smear:** When bacteria
Microscopically	Report:	or WBCs (pus cells) are seen
	- WBCs (pus cells)	in wet preparation.
	- Red cells	

Contd...

Contd...

Day 1		
	- Casts - Yeast cells - *T. vaginalis* flagellates - *S. haematobium* eggs - Bacteria (fresh urine only) - Crystals of importance	
3. Test Biochemically	**Tests to help diagnose UTI** - Protein - Nitrite (Greiss test) - Leucocyte esterase (When microscopy for WBCs not possible)	**Glucose, ketones, bilirubin, urobilinogen:** As indicated
4. Culture Specimen	CLED agar When bacteria and, or pus cells are present: - Inoculate CLED medium - Incubate aerobically	
Day 2 and onwards		
5. Examine and Report Cultures	**CLED culture** Look particularly for: *E. coli* (common cause UTI) *Proteus species* *P. aeruginosa* *Klebsiella* *E. faecalis* *S. aureus* *S. saprophyticus* Report bacterial numbers: • Less than 10^4/ml, not significant • 10^4-10^5/ml, doubtful significance • More than 10^5/ml, significant bacteriuria.	**Antimicrobial sensitivity testing:** As indicated

LABORATORY EXAMINATION OF SPUTUM

Day 1

Describe the Appearance of the Specimen

Describe whether the sputum is:

Purulent: Green-looking, mostly pus
Mucopurulent: Green-looking with pus and mucus
Mucoid: Mostly mucus
Mucosalivary: Mucus with a small amount of saliva

When the sputum contains blood, this must also be reported.

Examine the Specimen Microscopically

Gram smear

Examine the smear for pus cells and predominant bacteria. Look especially among the pus cells for:

- Gram-positive diplococci (capsulated) that could be *Streptococcus pneumoniae*
- Gram-positive cocci in groups that could be *Staphylococcus aureus*
- Gram-negative rods and cocci-bacilli that could be *Haemophitus influenza*
- Gram-negative diplococci in and between pus cells that could be *Myloplasma catarrhalis.*

Gram-stained smears of sputum must be reported with caution. Cocci, diplococci, streptococci and rods may be seen in normal sputum because these organisms form part of the normal microbial flora of the upper respiratory tract.

Ziehl-Neelsen Smear to Detect AFB

AFB is sputum smears are significantly increased if the organisms are first concentrated by centrifugation. Sodium hypochlorite is recommended for liquefying the sputum because it kills *M. tuberculosis*, making the handling of specimens safer for laboratory staff.

Culture the Specimen

Blood agar and chocolate agar
- Wash a purulent part of the sputum in about 5 ml of sterile physiological saline.
- Inoculate the washed sputum on plates of:
  Blood agar
  Chocolate (heated blood agar)

Summary of Microbiological Examination of Sputum		
Day 1		
1. Describe Specimen	Report whether specimen: - Purulent, mucopurulent, mucoid, salivary - contains blood is satisfactory for further analysis.	
2. Examine microscopically	Gram smear: For pus cells and bacteria Zn smear: For AFB	Giemsa smear: When pneumonic plague or histoplasmois suspected KOH preparation: When *Aspergillus* infection suspected Toluidine blue-O and Giemsa smears: When pneumocystis pneumonia suspected Eosin preparation: When an allergic condition requires investigation Saline preparation: when paragonimiasis suspected

Contd...

Contd...

Day 1		
3. Culture specimen	Blood agar • Add an optochin disc • Incubate aerobically Chocolate agar - Incubate in CO_2	Culture for *M. tuberculosis*
Day 2 and onwards		
4. Examine and report cultures	Blood and chocolate agar cultures Report significant growth of *S. pneumoniae* *H. influenzae, S. aureus* *Less commonly found pathogens* *K. pneumoniae, P. aeruginosa, M.* *catarrhalis, S. pyogenes, Proteus,* *C. albicans*	Test H. influenzae for beta- lactamase production antimicrobial sensitivity tests as required

Key: ZN—Ziehl-Neelson, KOH—Potassium hydroxide, CO_2—Carbon dioxide

Day 2 and Onwards

Examine and report the cultures

Blood agar and chocolate agar cultures

Look especially for a significant growth of :

* *Streptococcus pneumoniae* sensitive to optochin
* *Haemophilus influenzae*
* *Staphylococcus aureus*

Antimicrobial Sensitivity Testing

Sensitivity tests should be performed only when the amount of cultural growth of a pathogen is significant. Strains of *S. pneumoniae* should be tested on blood agar for sensitivity to penicillin, tetracycline and erythromycin.

Examination of Throat and Mouth Specimens

Possible pathogens

* BACTERIA

  Gram-positive   **Gram-negative**

  Streptococcus pyogens   Vincent's organisms

  Corynebacterium diphtheriae

  Corynebacterium ulcerans
* Viruses
* Fungi

  Candida albicans and other yeasts.

COLLECTION AND TRANSPORT OF THROAT AND MOUTH SWABS

In a good light and using the handle of a spoon to depress the tongue, examine the inside of the mouth. Look for inflammation, and the presence of any membrane, exudate or pus.

Swab the affected area using a sterile cotton-wool swab. Taking care not to contaminate the swab with saliva, return it to its sterile container.

LABORATORY EXAMINATION OF THROAT AND MOUTH SPECIMENS

Culture the Specimen

Blood Agar

- Inoculate the swab on a plate of blood agar . Use the loop to make also a few stabs in the agar.
- Incubate the plate preferably anaerobically.
- Beta-haemolytic streptococci produce larger zones of haemolysis when incubated anaerobically. A minority of Group A *Streptococcus* strains will only grow anaerobically.

Examine the Specimen Microscopically

Gram's smear

Make an evenly spread smear of the specimen on a slide. Allow the smear to air-dry in a safe place. Stain by the Gram's technique.

Albert Stained Smear when Diphtheria is Suspected

Examine the smear for bacteria that could be *Corynebacterium diphtheriae*. Look for pleomorphic rods containing dark-staining granules. The pleomorphic rods tend to join together at angles giving the appearance of Chinese letters.

DAY 2 AND ONWARDS

Examine and Report the Cultures

Blood agar culture

Look for beta-haemolytic colonies that could be *Streptococcus pyogenes*. (Group A *Streptococcus*). The organism should be tested serologically to confirm that it belongs to Lancefield Group A or tested biochemically using the PYR test.

Isolation and Identification of C Diphtheriae

- Examine a Gram stained smear for variable staining pleomorphic rods.

- Subinoculate two slopes of Dorset egg medium or Loeffler serum agar. Incubate at 35 to 37°C for 6 hours or until sufficient growth is obtained.

  Examine an Albert stained smear of the subculture for pleomorphic rods containing volutin granules. Examine a Gram's stained smear also.
- Identify the isolate biochemically.
- Using the growth from the other subculture, test the strain for toxin production using the Elek precipitation technique.

Antimicrobial Sensitivity Testing

The major pathogens involved in bacterial pharyngitis are *S. pyogenes* and *C. diphtheriae*. Benzylpenicillin and erythromycin are considered as the antibiotics of choice to treat both types of infection. In cases of diphtheria, treatment with antitoxin is also indicated.

Summary of Microbiological Examination of Throat and Mouth Swabs		
Day 1		
1. Culture specimen	Blood agar - Add a bacitracin disc - Incubate, preferably anaerobically (or in CO_2)	ADDITIONAL INVESTIGATION MTM or TBA: When diphtheria suspected
2. Examine	Microscopically Gram smear Look for: • Pus cells and Gram-negative Vincent's organisms • Gram-positive pleomorphic rods when diphtheria suspected • Gram-positive yeast cells when thrush suspected	Giemsa or Wayson's smear: when diphtheria suspected
Day 2 and onwards		
3. Examine and Report cultures	Blood agar culture Look for beta-haemolytic streptococci, sensitive to bacitracin. Identify as *S. pyogenes* Lancefield group PYR test	MTM or TBA cultures Examine for growth of *C. diphtheriae*

Key: MTM—Modified tinsdale medium, TBA—Tellurite blood agar

EXAMINATION OF PUS, ULCER MATERIAL AND SKIN SPECIMENS

PUS

Possible pathogens

- *BACTERIA*

Gram-positive
Staphylococcus aureus
Streptococcus pyogenes
Enterococcus species
Anaerobic streptococci
Other *streptococci*
Clostridium perfringens
and other *clostridia*
Actinomycetes
Actinomyces israeli
Also *Mycobacterium tuberculosis*

Gram-negative
Pseudonomous aeruginosa
Proteus species
Escherichia coli
Bacterioides species
Klebsiella species
Pasteurella species

- *FUNGI*

Histoplasma C. duboisii, Fungi that cause, mycetoma.

- *PARASITES*

Entamoba histolytica
(in pus aspirated from an amoebic liver abscess).

ULCER MATERIAL AND SKIN SPECIMENS

BACTERIA

Gram-positive
Staphylococcus aureus
Streptococcus pyogenes
Enterococcus species
Anaerobic streptococci
Erysipelothrix rhusiopathiae
Bacillus anthracis

Gram-negative
Escherichia coli
Proteus
Pseudomonas aeruginosa
Yersinia pestis
Vincent's organisms

Also *Mycobacterium leprae, Mycobacterium ulcerans, Treponema carateum* and *Treponema pertenue.*

- *VIRUSES*

Poxviruses and herpes viruses

- *FUNGI*

Dermatophytes (ringworm fungi) *Malassezia furfur,* Fungi that cause chromoblastmycosis, *Candida albicans.*

- *PARASITES*
  Leishmania species
  Onchocerca volvulus
  Dracunculus medinensis.

Notes on Pathogens

- *Staphylococcus aureus* is the most common pathogen isolated from subcutaneous abscesses and skin wounds. It also causes impetigo.
- *Pseudomonas aeruginosa* is associated with infected burns and hospital-acquired infections.
- *Escherichia coli. Proteus species, P. aeruginosa,* and *Bacteroides* species are the pathogens most frequently isolated from abdominal abscesses and wounds.
- *Clostridium perfringens* is found mainly in deep wounds where anaerobic conditions exist.
- *Mycobacterium tuberculosis* is associated with 'cold' abscesses.
- *Actinomycetes* (filamentous bacteria) and several species of fungi cause mycetoma) specimens of pus from the draining sinuses contain granules examination of which helps differentiate whether the mycetoma is bacterial (treatable) or fungal (less easily treated).
- *Actinomyces israeli* and other species of *Actinomyces* cause actinomycosis. Small yellow granules can be found in pus from a draining sinus (often in the neck).
- Vincent's organisms (Borrelia vincenti with Gram-negative anaerobic fusiform bacilli) are associated with tropical ulcer.
- *Bacillus anthracis* causes anthrax, with the cutaneous form of the disease.
- *Mycobacterum leprae* can be found in skin smears in lepromatous leprosy.
- Skin diseases are common in those with HIV diseases. Bacterial infections include recurrent boils and folliculitis caused by *Staphylococcus aureus* and *Streptococcus pyogenes*.

COLLECTION AND TRANSPORT OF PUS, ULCER MATERIAL, SKIN SPECIMEN

Pus from an abscess is best collected at the time the abscess is incised and drained, or after it has ruptured naturally. When collecting pus from abscesses, wounds, or other sites, special care should be taken to avoid contaminating the specimen with commensal organisms from the skin.

Using a sterile technique, aspirate or collect from a drainage tube up to 5 ml of pus.

When mycetoma is suspected: Obtain a specimen from a draining sinus tract using a sterile hypodermic needle to lift up the crusty surface over the sinus opening.

When tuberculosis is suspected: Aspirate a sample of the pus and transfer it to a sterile container.

When the tissue is deeply ulcerated and necrotic: Aspirate a sample of infected material from the side wall of the ulcer using a sterile needle and syringe.

Fluid from pustules, buboes, and blisters: Aspirate a specimen using a sterile needle and syringe.

Serous fluid from skin ulcers, papillomas, or papules, that may contain treponemes: Collect a drop of the exudate directly on a clean cover glass and invert it on a clean slide. Immediately deliver the specimen to the laboratory for examination by dark-field microscopy.

LABORATORY EXAMINATION OF PUS, ULCER AND SKIN SPECIMENS

Day 1

Describe the appearance of the specimen: When from a patient with suspected mycetoma or actinomycosis report the appearance of the specimen and whether it contains granules.

Examine the Specimen Microscopically

Gram smear

Make an evenly spread smear of the specimen of a slide and stain by the Gram technique.

Examine the smear for bacteria among the pus cells Using the 40X and 100X objectives. Look especially for:

- Gram-positive cocci that could be *Staphylococcus aureus* or streptococci that could be *Streptococcus pyogenes* or other beta-haemolytic streptococci, anaerobic streptococci, or enterococci.
- Gram-negative rods that could be *Proteus* species, *Echerichia coli* or other coliforms, *Pseudomonas aeruginosa* or *Bacteroides* species.
- Gram-positive large rods with square ends that could be *Clostridum perfringens* or *Bacillus anthracis*.
- Large numbers of pleomorphic bacteria .
- Gram-positive yeast cells with pseudohyphae, suggestive of *Candida albicans*.

Ziehl-Neelsen Smear when Tuberculosis or Buruli Ulcer is Suspected

Polychrome Loeffler methylene blue smear when cutaneous anthrax is suspected.

Examination by dark-field microscopy to detect treponemes.

Potassium hydroxide preparation when ringworm or other (superficial fungal infection is suspected).

Culture the Specimen

Blood agar MacConkey's agar, cooked meat medium (or thioglycollate broth).

Inoculate the specimen:
- On blood agar to isolate *S. aureus* and streptococci. Add a bacitracin disc if streptococci are seen in the Gram smear.
- On MacConkey agar to isolate Gram-negative rods.
- Into cooked meat medium or thioglycollate broth.

Anaerobic culture
When an anaerobic infection is suspected or the Gram smear shows an 'anaerobic mixed flora', inoculate a second blood agar plate and incubate it anaerobically for up to 48 hours. The anaerobic blood agar plate-may be made selective by adding neomycin to it.

Day 2 and Onwards

Examine and report the cultures
Blood agar and MacConkey agar cultures
Look especially for colonies that could be:
- *Staphylococcus aureus*
- *Streptococcus pyogenes*
- *Pseudomonas aeruginosa*
- *Proteus species*
- *Escherichia coil*
- *Enterococcus species*
- *Klebsiella species.*

Summary of Microbiological Examination of Pus, Ulcer Material and Skin Specimens		
Day 1		
1. Describe specimen		ADDITIONAL INVESTIGATIONS Look for granules: When mycetoma or actinomycosis is suspected
2. Culture specimen	Blood agar Incubate aerobically MacConkey aerobically Cooked meat medium subculture at 24 h, 48 h, and 72 hr as indicated	Culture for *M. tuberculosis* or *M ulcerans* Requires facilities of a tuberclosis reference laboratory
3. Examine microscopically	Gram smear for pus cells and bacteria	Ziehl-Neelsen smear: when tuberclosis or Buruli ulcer is suspected KOH preparation: When a fungal or actinomycete infection is suspected Giemsa or Wayson's smear: When bubonic plague is suspected Polychrome methylene blue: When cutaneous anthrax is suspected Dark-field microscopy: To detect treponemes when yaws or pinta is suspected
Day 2 and onwards		
4. Examine and report cultures	Blood agar and MacConkey agar cultures Look particularly for: *S. aureus* *S. pyogenes* *P. aeruginosa* *Proteus species* *E. coli* *Enterococcus species* *Klebsiella species* Anaerobes: *C. perfringens* *Bacteroides fragilis group* *Peptostreptococcus species*	Antimicrobial sensitivity tests as indicated

Anaerobic Blood Agar Culture and Cooked Meat Culture

- *Clostridium pefringens:* Grows rapidly in cooked meat medium with hydrogen sulphide gas production (gas bubbles in turbid medium). On anaerobic blood agar, colonies are usually seen after 48 h incubation. Most strains produce a double zone of haemolysis.
- *Bacillus fragilis:* Grows in cooked meat medium producing decomposition with blackening of the meat. On anaerobic blood agar, non-haemolytic grey colonies (gram-negative pleomorphic rods) are seen.

Subculture on Cooked Meat Medium

Subculture the cooked meat broth after overnight incubation and when indicated also at 48 hr and 72 hr.

Antimicrobial Sensitivity Testing

Sensitivity testing may be required for *Staphylococcus aureus*, enterobacteria and non-fermentative Gram-negative rods. Only routinely used antibiotics should be tested.

Sensitivity tests should not routinely be performed on an aerobic bacteria by the disc diffusion technique.

Examination of Effusions

An effusion is fluid which collects in a body cavity or joint.

Effusions sent to the laboratory for investigation include

Fluid	Origin
Synovial	From a joint
Pleural	From the pleural cavity (Space between the lungs and the inner chest wall)
Pericardial	From the pericardial sac (Membranous sac surrounding the heart).
Ascitic (peritoneal)	From the peritoneal (abdominal) cavity
Hydrocoele	Usually from the sac surrounding the testes

Possible Pathogens

SYNOVIAL FLUID

Gram-positive	*Gram-negative*
Staphylococcus aureus	*Neisseria gonorrhoeae*
Streptococcus pyogenes	*Neisseria meningitidis*
Streptococcus pneumoniae	*Haemophilus influenzae*
Anaerobic streptococci	*Brucella* species
Actinomycetes	*Salmonella* species
	Escherichia coli
	Pseudomonas aeruginosa
	Proteus

PLEURAL AND PERICARDIAL FLUIDS

Gram-positive	*Gram-negative*
Staphylococcus aureus	*Haemophilus influenzae*
Streptococcus pneumoniae	*Bacteroides*
Streptococcus pyogenes	*Pseudomonas aeruginosa*
Actinomycetes	*Klebsiella* strains
	Other enterobacteria

Also *Mycobacterium tuberculosis*, fungi, and viruses especially coxsackie B virus.

ASCITIC FLUID

Gram-positive	**Gram-negative**
Enterococcus species	*Escherichia coli*
Streptococcus pneumoniae	*Klebsiella strains*
Staphylococcus aureus	Other enterobacteria
Strptococcus pyogenes	*Pseudomonas auruginosa*
Streptococcus agalactiae	*Bacteroides*
Viridans spreptococci	
Clostridium perfringens	

Also *Mycobacterium tuberculosis* and *Candida* species.

HYDROCOELE FLUID

Occasionally *Wuchereria bancrofti* microfilariae and rarely *Brugia* species can be found in hydrocoele fluid.

After aspiration, aseptically dispense the fluid as follows:

- 2 to 3 ml into a dry, sterile, screw-cap tube or bottle to observe for clotting.
- 9 ml into a screw-cap tube or bottle which contains 1 ml of sterile trisodium citrate (Mix the fluid with the anticoagulant. Trisodium citrate prevents clotting).

Label, and as soon as possible deliver the samples with a completed request form to the laboratory.

LABORATORY EXAMINATION OF EFFUSIONS

Day 1

Describe the appearance of the specimen
Report:
- Color of the effusion
- Whether it contains blood
- Whether it is clotted.

Purulent effusion: When the specimen is pus or markedly cloudy, examine and report a Gram's stained smear as soon as possible.

Blood-stained effusion: Culture the specimen and examine a Gram's stained smear.

Examine the Fluid for Cells

Estimate the number of white cells in the fluid.

Culture the Specimen

Centrifuge the citrated sample in a sterile tube at high speed for about 20 minutes to sediment the bacteria. Culture the sediment as follows:

Chocolate Agar, Blood Agar and MacConkey Agar

- Inoculate the sediment on chocolate agar, blood agar and MacConkey's agar.
- Incubate the chocolate agar plate in a carbon dioxide enriched atmosphere at 35 to 37°C for up to 48 hours, checking for growth after overnight incubation.

    Incubate the blood agar plate and MacConkey's agar plate aerobically at 35 to 37°C for up to 72 hours, examining for growth after overnight incubation.

Examine the Specimen Microscopically

Gram Smear

Make a thin evenly spread smear of a purulent effusion or sediment from a centrifuged non-purulent sample.

Look especially for
- Gram-positive that could be *S. aureus*
- Gram-positive streptococci that could be *S. pyogenes*, or possibly enterococci
- Gram-positive diplococci or short chains that could be pneumococci
- Gram-negative rods that could be enterobacteria, *Pseudomonas*, or *H. influenzae*
- Gram-negative intracellular diplococci that could be gonococci when the fluid is from a joint
- Gram-positive branching threads that could be actinomycetes.

Ziehl-Neelsen Smear

Make a smear on a slide using several drops of sediment from the centrifuged fluid. Fix the dried smear and stain by the Ziehl-Neelsen technique.

Summary of Microbiological Examination of Effusions		
Day 1		
1. Describe specimen	Describe colours and whether: • Clear, cloudy, or purulent • blood stained • Contains clots (non-citrated sample)	ADDITIONAL INVESTIGATIONS
2. Examine for cells	Estimate cell numbers Report % of cells that are: • Neutrophilis • Lymphocytes	
3. Protein	Report total protein in g/L	
TRANSUDATE	Clear unclotted fluid with few cells and protein below 30 g/l: No need to test further	
EXUDATE	When cloudy fluid contains many pus cells and protein over 30 g/l: Proceed to steps 4 and 5 Note: When the fluid contains many pus cells, examine as described for pus.	
4. Culture specimen	Blood agar Incubate aerobically Chocolate agar Incubate in CO_2 MacConkey's agar Incubate aerobically	Culture for *M. tuberculosis* Requires facilities of a Tuberculosis Reference Laboratory
5. Examine microscopically	Gram smear Look for pus cells and bacteria Ziehl-Neelsen Look for AFB	Wet preparation for crystals: When gout or pseudogout is suspected (joint fluid only) Cytology smear: When maligancy is suspected
Day 2 and onwards		
6. Examine and report cultures	Blood, chocolate, MacConkey's agar cultures Look particularly for: *S. aureus* *S. pyogenes* *S. pneumoniae* *H. influenzae* *Neisseria species* *Enterbacteria* *P. aeruginosa*	Antibiotic sensitivity test as required

Day 2 and Onwards

Examine and report the cultures

Chocolate agar, blood agar, and MacConkey agar cultures
Look especially for colonies that could be:
- *Staphylococcus aureus*
- *Streptococcus pyogenes*
- *Streptococcus pneumoniae*
- *Haemophilus influenzae*
- *Enterobacteria*, see subunit
- *Pseudomonas aeruginosa*
- *Neisseria* species.

Possible Pathogens

URETHRAL SWABS
Neisseria gonorrhoeae, Chlamydia trachomatis (serovars D-K), and occasionally *Ureaplasma, Mycoplasma,* and *Trichomonas vaginalis.*

CERVICAL SWABS
From non-puerperal women: *Neisseria gonorrhoeae, Chlamydia trachomatis* (serovars D-K), *Streptococcus pyogenes, herpes* simplex virus.

From Women with Puerperal Sepsis or Septic Abortion

Streptococcus pyogenes, other beta-haemolytic streptococci, *Enterococcus* species, anaerobic cocci, *Clostridium perfringens, Bacteroides, Proteus, Escherichia coli* and other coliforms, *Listeria monocytogens.*

VAGINAL SWABS
Vaginal discharge may be due to infection of the vagina or infection of the cervix or uterus. Pathogens causing vaginal infections include *Trichomonas vaginalis, Candida* species, and *Gardnerella vaginalis* with anaerobes.

GENITAL ULCER SPECIMENS
Treponema pallidum, Haemophilus ducreyi, Calymmatobacterium granulomatis, Chlamydia trachomatis (serovars L1, L2, L3), herpes simplex virus.

COLLECTION AND TRANSPORT OF UROGENITAL SPECIMENS

Urogenital specimens should be collected by a medical officer or an experienced nurse.

Amies medium is the most efficient medium for transporting urethral, cervical, and vaginal swabs. Specimens should be transported in a cool box.

Collection of Urethral Discharge from Male Patients

Cleanse around the urethral opening using a swab moistened with sterile physiological saline.

* Gently massage the urethra from above downwards. Using a swab, collect a sample of discharge. Make a smear of the discharge on a microscope slide by gently rolling the swab on the slide
* When culture is indicated collect a sample of pus on a sterile cotton-wool swab. If possible, before inserting the swab in a container of Amies transport medium, inoculate a plate of culture medium.

LABORATORY EXAMINATION OF UROGENITAL SPECIMENS

Culture the Specimen

Modified New York City (MNYC) or Thayer-Martin-medium

* Inoculate the specimen on MNYC medium or other selective enriched culture medium suitable for isolating *N. gonorrhoeae* from urogenital specimens such as Thayer-Martin medium.

ADDITIONAL

Blood agar (aerobic and anaerobic), MacConkey agar, and cooked meat medium when puerperal sepsis or septic abortion is suspected.

* Inoculate the specimen on two plates of blood agar and incubate one anaerobically and the other aerobically at 35 to 37°C overnight.
* Inoculate the specimen on MacConkey agar and incubate the plate aerobically at 35 to 37°C overnight.
* Inoculate the specimen in cooked meat medium and incubate at 35 to 37°C, subculturing as indicated at 24 hr, 48 hr and 72 hr.

Examine the Specimen Microscopically

Gram smear

Fix the smear with methanol and stain by the Gram technique . Using the 40X and 100X objectives, examine the smear for pus cells and bacteria.

Smear from a patient with suspected gonorrhoea. Look for pus cells containing gram-negative diplococci that could be *N. gonorrhoeae*.

Wet (saline) preparation to detect *T. vaginalis*

To detect motile *T. vaginalis* trophozoites, the preparation must be examined as soon as possible.

ADDITIONAL

Dark-field preparation to detect motile *T. pallidum.*

Day 2 and Onwards

Examine and report the cultures

MNYC and Thayer-Martin cultures

N. gonorrhoeae produces small raised, grey shiny colonies on MNYC medium and Thayer-Martin medium after overnight incubation.

- Perform an oxidase test. *Neisseriae* are strongly oxidase-positive.
- Gram stain the colonies. *N. gonorrhoeae* appears as a gram-negative coccus.
- Test the colonies for beta-lactamase production.

ADDITIONAL

Blood agar and MacConkey's agar cultures

Look for colonies that could be:

- *Streptococcus pyogenes* or other beta-haemolytic streptococci,
- *Staphylococcus aureus*
- *Clostridium perfringens*
- *Proteus species*
- *Enterococcus*
- *Escherichia coli*

Summary of Microbiological Examination of Urogenital Specimens		
Day 1		
1. Culture specimen	MNYC Medium or Thayer - Martin medium - Incubate in CO_2 (moist environment)	ADDITIONAL INVESTIGATIONS When puerperal or septic abortion is suspected Blood agar (2 plates) -Incubate aerobically -Incubate anaerobically -MacConkey's agar -Incubate aerobically -Cooked meat mediun -Incubate overnight -Subculture as indicated at 24 hr, 48 hr, 72 hr
2 Examine microscopically	Gram smear -Urethral: Intracellular Gram- negative diplococci -Vaginal: Yeast cells (candidiasis) Clue cells(bacterial vaginosis) -Vaginal/cervical ; Pus cells and bacteria associated with puerperal sepsis and septic abortion Wet preparation	Dark-field: When syphilis is suspected Giemsa smear When *C.granulomatis* infection (donovanosis) is suspected Cervical smear sent to histology/ cytology laboratory: When malignancy is suspected

Contd...

Contd...

Day 2 and onwards		
3. Examine and report cultures	MNYC plate or Thayer-Martin plate -Examine for *N. gonorrhoeae* Colonies resembling *N gonorrhoeae:* Oxidase test Gram stain colonies Beta-lactase test	Blood agar, MacConkey's agar plates -Look especially for: *S. pyogens* *S. aureus* *C. perfringes* *Proteus* *Enterococcus* *E. coli* *Bacteroides* Antimicrobial sensitivity tests as required

Examination of Faecal specimens
Possible pathogens

- *BACTERIA*

  Gram-positive

  Clostridium perfringens types a and C

  Clostridium difficile

  Bacillus cereus (toxin)

  Staphylococcus aureus toxin.

  Gram-negative

  Shigella species

  Salmonella spcies

  Campylobacter species

  Yersinia enterocolitica

  Escherichia coli (ETEC, EIEC, EPEC, VTEC)

  Vibrio chlorae 01, 0139

  Other *Vibrio* species

  Aeromonas species

  Also *Mycobacterium tuberclosis*

- *VIRUSES*

  Mainly rotaviruses and occasionally Norwalk agents, adenoviruses, calcivirus and coronavirus.

- *PARASITES*

  Entamoeba histolytica, Giardia lamblia, intestinal coccidian (*Isospora, Cryptosporidium, Cyclospora*) and other protozoan enteric pathogens.

COLLECTION AND TRANSPORT OF FAECES

Faeces for microbiological examination should be collected during the acute stage of diarrhoea.

Give the patient a clean, dry, disinfectant-free bedpan or suitable wide-necked container in which to pass a specimen.

Transfer a portion (about a spoonful) of the specimen, especially that which contains mucus, pus, or blood, into a clean, dry, leakproof container.

Worms and tapeworm segments: When the specimen contains worms or tapeworm segments, transfer these to a separate container and send them to the laboratory for identification.

Day 1

Describe the appearance of the specimen:
* Colour of the specimen.
* Whether it is formed, semiformed, unformed or fluid.
* Presence of blood, mucus or pus
* Presence of worms, e.g. *Enterobius vermicularis, Ascaris lumbricoides,* or tapeworm segments, e.g. *Taenia species.*

Appearance of Faecal Specimens in Some Diseases	
Appearance	*Possible cause*
Unformed, containing pus and mucus mixed with blood	• Shigellosis • E1EC dysentery • *Campylobacter* enteritis
Unformed with blood and mucus (acid pH)	• Amoebic dysentery
Unformed or semiformed, often with blood and mucus	• Schistosomiasis
Bloody diarrhoea:without pus cells.	• EHEC 0157 infection (haemorrhagic colitis)
Watery stools	• ETEC, EPEC diarrhoea • Cryptosporidiosis • Rotavirus enteritis
Rice water stools with mucous flakes	• Cholera
Unformed or watery and sometimes with blood, mucus, and pus	• *Salmonella* infection
Unformed, pale coloured, frothy, unpleasant smelling tools that float on water (high fat content)	• Giardiasis • Other conditions that cause malabsorption, e.g. post-infective tropical malabsorption
Fluid stools (containing lactose with pH below 6)	• Lactase deficiency
Unformed or semiformed black stools positive occult blood)	• Melaena (gastrointestinal bleeding) • Hookworm disease • Iron therapy

Normal faeces: Appear brown and formed or semi-formed. Infant are yellow-green and semiformed.

Examine the Specimen Microscopically

Saline and eosin preparations to detect *E. histolytica* and other parasites
Examine saline and eosin preparations to detect *E. histolytica* and other parasites

ADDITIONAL

Methylene blue preparation to detect faecal leucocytes when the specimen is unformed.

Motility test and Gram stained smear when cholera is suspected

Examine an alkaline peptone water culture (sample from the surface of the culture) for vibrios showing a rapid and darting motility. Examine also a Gram stained smear of the culture for Gram-negative vibrios.

Culture the Specimen

When the specimen is formed or semiformed, make a thick suspension of it in about 1 ml of sterile peptone water.

Xylose lysine deoxycholate (XLD) agar

- Inoculate a loopful of fresh emulsified faeces or a fluid specimen on XLD agar
- Incubate the XLD agar plate aerobically at 35 to 37°C overnight.

Alkaline peptone water and TCBS agar when cholera is suspected

- Inoculate several loopfuls of specimen in alkaline (pH 8.6) peptone water, and Incubate at 35 to 37°C for 5 to 8 hours.
  Note: Prior enrichment in alkaline peptone water is not necessary if the. specimen is likely to contain large numbers of vibrios (e.g. in acute cholera). Alkaline peptone water is a useful transport medium for *V. cholerae*
- Subculture several loopfuls of the peptone water culture (taken from the surface) on thiosulphate citrate bile-salt sucrose (TCBS) agar. Incubate aerobically at 35 to 37°C overnight.

Day 2 and Onwards

Examine and report the cultures

XLD agar culture

Look for colonies that could be species of *Shigella* or *Salmonella*. *Shigella* and H_2S negative strains of *Salmonella* produce 1 to 2 mm diameter red colonies on XLD agar. Red colonies with black centres are produced by H_2S positive salmonellae, e.g. strains of *S. typhimurium*.

Proteus, Providencia and Pseudomonas organisms may also produce red colonies on XLD agar. Some *Proteus* strains are also H_2S producing and form red colonies with black centres.

On MacConkey agar, shigellae, and salmonellae and other non-lactose fermenting organisms, produce colourless colonies. *E. coil* and other lactose-fermenting organisms produce pink colonies.

Identification of suspect *Salmonella* and *Shigella* isolates:

Perform a urease test using urea broth

A positive urease test within 2 to 4 hr indicates that the organism is probably *Proteus*.

When the urease test is negative at 4 hours, proceed as follows:

1. Perform indole and lysine decarboxylase (LDC)
2. Inoculate a tube of Kliger iron agar. Use a sterile straight wire, stab first the butt and then streak the slope. Close the tube with a loose-fitting cap and incubate at 35-37°C overnight.

Results

LDC   *Shigella* is LDC negative
          Samonella species are LDC positive except *S. paratyphi* A which is LDC negative

Indole *S. sonnei* is indole negative. Other shigellae give variable indole reactions
           Salmonella species are indole negative

KIA   *Salmonella* and *Shigella* species produces a pink-red slope and yellow butt. Many salmonellae also produce blackening due to hydrogen sulphide production and cracks in the medium due to gas production from glucose fermentation. *Salmonella* typhi produces only a small amount of blackening.

Serological Identification of Salmonellae and Shigellae

Test serologically any isolate giving reactions suggestive of *Salmonella* or *Shigella* species, using a slide technique.

TCBS Agar Culture

V. cholerae is sucrose fermenting and therefore produces yellow 2 to 3 mm in diameter shiny colonies on TCBS agar with yellow colour in the medium.

With prolonged incubation (48 hr or more) the colonies may become green.

Identification of a Suspect *V. cholerae* Isolate

1. Examine a Gram stained smear of the culture for gram-negative vibrios. The organisms may appear less curved after culture.

2. Subculture the organism on a slope of nutrient agar, and incubate for 4 to 6 hours.
3. Perform an oxidase test on the nutrient agar culture.

Antisera to identify V. cholerae 01 and 0139

With the rapid spread of *V cholerae* 01 antiserum and *V. cholerae* 0139 (Bengal) antiserum to identify isolates. There is no need to differentiate between classical and EI T or biovars.

Examination of Cerebrospinal fluid

In the case of suspected bacterial meningitis, cerebrospinal fluid is submitted to the laboratory for culture. Pathogenic bacteria commonly associated with meningitis are *Neisseria meningitides, Haemophilus influenzae, Streptococcus pneumoniae,* and *mycobacterium. Neisseria meningitides (Meningococcus)* is related to epidemic meningitis. *Streptococcus pneumoniae (pneumococcus)* causes meningitis in infants, alcoholics and in elderly persons. *Mycobacterium tuberculosis* causes tuberculous meningitis. The CSF is clear or opalescent, has high protein and low sugar concentrations, has plenty of lymphocytes some PMNs present), and may have acid-fast positive rod shaped organisms present. Cryptococcus causes mycotic meningitis. While the causal organism of sleeping sickness (*Typanosoma*) is a protozoan. Since it is virtually impossible to differentiate various types of meningitis by clinical symptoms, and the condition of the patient can rapidly deteriorate, fast and accurate means of identifying the etiologic agent must be available in the laboratory. CSF is a valuable specimen and should never be refrigerated. The pathogenic organism will die, cells will lyse, and the glucose concentration will fall in standing CSF specimens. Thus the results will not be reliable. The amount of a CSF specimen is also small and hence must be economical in handling the CSF specimen. A norman CSF specimen is sterile, and does not have any commensal organisms. Hence, if a large number of organisms are seen in the direct examination of the specimen, set up a preliminary sensitivity test on chocolate agar with a suspension of the CSF sediment, in order to recommend to the physician the most effective antibiotics.

A complete work up of spinal fluid includes the following :

a. Physical examination
b. Microscopic examination—wet mount (e.g. India ink preparation), and study of a stained smear using hamatological stain and gram stain,
c. Chemical examination and
d. Culture of the spinal fluid sediment. Seradiagnosis of syphilis is also done with CSF.

Specimen Collection

CSF is always collected by a physician or a trained nurse through lumbar puncture. Usually 2 to 3 tubes are taken for specimen collection. The second or the third tube is taken for microbiological study. The collection tube used for microbiological study must be sterile.

Laboratory Investigation

The spinal fluid collected in sterile collection tubes is first used for an India ink preparation without centrifugation. If yeast-like organisms are seen, part of the specimen is taken for the culture of *Cryptococcus*. After that, the rest of the specimen is centrifugal (3000 rpm for 15 minutes) and the sediment is used for study of a stained smear (Romanowsky stain, Gram's stain and acid-fast stain), and for culture of the infectious agent.

A CSF specimen from a patient with acute bacterial meningitis is turbid with higher total protein and lower sugar than normal. Microscopic examination reveals an increased while cell count (predominantly polymor-phonuclear cells). In purulent meningitis, caused by bacterial infections, a large clot may form. The clot may be used to prepare a smear. Gram staining is done by the routine procedure and the gram reaction reported immediately, when available. In reporting the gram reaction, mention the number of bacteria seen per oil-immersion field. Observation of the presence of gram-negative diplococci which may be found inside the neurophils, or extracellular, suggests the possibility of *Neisseria meningitides* infection. Further confirmation will be required for a positive diagnosis. Presumptive identification for *Streptococcus pneumoniae* comes from the presence of gram-positive diplococci, and occurrence of pleomorphic gram negative cocco-bacilli suggests a possible *Haemophilus influenzae* infection.

A CSF specimen should also be subjected to acid-fast staining for the possible diagnosis of tuberculous meningitis (*Mycobacterium tuberculosis*). For positive identification of the pathogen, laboratory culture is necessary, although the physician is given the results of presumptive identification, based on direct examination. A CSF is routinely inoculated on the following media; blood agar, chocolate agar, MacConkey's agar (or EMB) and thioglycollate broth. All culture media are kept under aerobic conditions except chocolate agar which is kept in a candle jar. Following 18 to 24 hours of incubation at 37°C, colonies are examined. If there is no growth, continue to incubate for 48 hours.

The blood agar medium serves as a general purpose medium and growth of *S. pneumoniae* can be recognised on the primary plate. MacConkey plate will allow gram-negative rods to grow, and further biochemical reactions will be able to identify them. Growth in thioglycollate broth, with the presence of gram-negative rods, suggest the possibility of *bacteroides* infection. If there is no growth, observe the thioglycollate broth for 7 days.

Examination of Blood

Blood is probably the most important kind of specimen submitted to the bacteriology laboratory for direct examination and culture. Bacteria found in the circulating blood in septicaemia are associated with various diseases. Blood is collected by venipuncture during the acute phase of the disease and before any antibiotic therapy.

Blood Collection Procedure

The following procedure is applicable specifically for blood culture.

Choose the vein to be drawn by touching the skin before it has been disinfected.

Cleanse the skin over the venipuncture site in a circle approximately 5 cm in diameter with 70 per cent alcohol, rubbing vigorously.

Starting in the centre of the circle, apply 2 per cent iodine solution in ever widening circles until the entire circle has been saturated with iodine. Allow the iodine to remain on the skin for at least one minute. The timing critical; a watch or a timer should be used.

If the site need to be touched for palpation, the phlebotomist must disinfect the fingers in identical fashion described above.

Insert the needle into the vein and withdraw blood.

After the needle has been removed, the site should be cleansed with 70 per cent alcohol again, as many patients are sensitive to iodine.

Promptly introduce the blood specimen directly into the special blood collection bottle through the hole in the cap (lift the adhesive tape and then introduce the needle; make sure that the needle does not touch any expose surface). If the bottle has a rubber lining, first disinfect the surface before piercing with the needle.

Use of an anticoagulant like sodium polyanethol sulphonate (SPS) is widely used in advanced laboratories. The amount of blood collected (10 ml, for adults and 1 to 5 ml for paediatric patients) is diluted into at least 10 times its volume with liquid medium like nutrient broth. Thus the bottle must have at least 50 ml of broth for in oculating 5 ml of the specimen. Do not use cold medium, wark the bottle by placing at room temperature or at 37°C for 30 minutes before inoculation. Three or four sets of blood cultures may be obtained in 24 to 48 hours in order to increase bacterial yield.

Blood collection bottles can be monophasic with one type of medium or can be biphasic with two types of media in the same bottle-liquid an solid. The biphasic medium is prepared by first putting 10 ml of nutrient agar or trypticase soy agar (at 50°C) into a "flat dispensary bottle (120 ml), followed by autoclaving. Cool the bottlle by lying flat. Later introduce 50 ml of strerilised nutrient broth (trypticase soy broth). In the case of the monophasic liquid medium only the nutrient broth is used (50 ml) without the solid phase. The diphasic medium is flushed everyday over the solid phase agar and incubated in an upright position. Daily observation of colonies on the solid phase is needed. Many laboratories used two biphasic media, one with nutrient agar and the other with MacConkey's agar as the solid phase. This facilitates a ready recovery of gram-positive cocci and members of Enterobactriaceae respectively in the primary culture.

A portion of the blood specimen should also be put into a thioglycollate broth tube for the cultivation of any anaerobes present. A blind subculture can be made between 6 to 14 hours after the blood is obtained and after 48 hours.

Incubate the blood broth routinely for 7 days at 37°C. After 24 hours, if there is any turbidity, discolouration or clotting, perform Gram staining and report the result of gram staining and preliminary observations to the physician without delay. Subculture the blood broth with suspected growth of organisms for the follow-up work. This is done by transferring, under aseptic conditions, a few drops of the blood-broth onto a chocolate agar plate (for *Neisseria* and *Haemophilus*) and on MacConkey, if this is not used on the biphasic medium. Incubate the chocolate agar plate in a candle jar. The subculture helps in the recovery, isolation and presumptive identification of the infectious agent. Examine the primary blood culture for 48 hours, after daily agitation and then leave it for 7 days without agitation, unless growth occurs earlier. If no change is noted after 7 days, report as: "no growth, aerobic or anaerobic, after 7 days of incubation." If *Brucella* infection is suspected, incubate the culture bottles in a carbon dioxide atmosphere and keep for 30 days; shake the bottle every third day.

Laboratory Observation of Blood-broth Culture	
Observation	**Presumptive diagnosis**
Turbidity, haemolysis, discolouration or clotting	Positive culture
Union turbidity, gas bubble formation, growth on MacConkey's agar (biphasic)	Enteric organism (gram –, bacilli)
Blood culture with greenish tinge	*Streptococcus pneumoniae* (gram+, Cocci)
Haemolysis with "cotton Ball colonies"	*Streptococcus pyogenes* (gram +, Cocci)
Jelly-like coagulum throughout the broth; discrete colonies on the surface of red cell layer and in the broth	*Streptococcus aureus* (gram +, Cocci)
Unpleasant odour and gas formation	*Bacteriodes* (gram –, rods)
	Clostridium (gram+. spore forming rods)

Index